Contents

About the Editor vii

Mojmír Šob
Editorial for the Special Issue on Computational Quantum Physics and Chemistry of Nanomaterials
Reprinted from: *Nanomaterials*, 10, 2395, doi:10.3390/nano10122395 1

Qi Liang, Xi Nie, Wenzheng Du, Pengju Zhang, Lin Wan, Rajeev Ahuja, Jing Ping and Zhao Qian
First-Principles Exploration of Hazardous Gas Molecule Adsorption on Pure and Modified $Al_{60}N_{60}$ Nanoclusters
Reprinted from: *Nanomaterials* 2020, 10, 2156, doi:10.3390/nano10112156 3

Jakub Šebesta, Karel Carva, Dominik Kriegner and Jan Honolka
Twin Domain Structure in Magnetically Doped Bi_2Se_3 Topological Insulator
Reprinted from: *Nanomaterials* 2020, 10, 2059, doi:10.3390/nano10102059 17

Md Al Mamunur Rashid, Dini Hayati, Kyungwon Kwak and Jongin Hong
Theoretical Investigation of Azobenzene-Based Photochromic Dyes for Dye-Sensitized Solar Cells
Reprinted from: *Nanomaterials* 2020, 10, 914, doi:10.3390/nano10050914 35

Svatava Polsterová, Martin Friák, Monika Všianská and Mojmír Šob
Quantum-Mechanical Assessment of the Energetics of Silver Decahedron Nanoparticles
Reprinted from: *Nanomaterials* 2020, 10, 767, doi:10.3390/nano10040767 59

David Holec, Phillip Dumitraschkewitz, Dieter Vollath and Franz Dieter Fischer
Surface Energy of Au Nanoparticles Depending on Their Size and Shape
Reprinted from: *Nanomaterials* 2020, 10, 484, doi:10.3390/nano10030484 75

Takahiro Shimada, Koichiro Minaguro, Tao Xu, Jie Wang and Takayuki Kitamura
Ab Initio Study of Ferroelectric Critical Size of SnTe Low-Dimensional Nanostructures
Reprinted from: *Nanomaterials* 2020, 10, 732, doi:10.3390/nano10040732 91

Martina Mazalová, Monika Všianská, Jana Pavlů and Mojmír Šob
The Effect of Vacancies on Grain Boundary Segregation in Ferromagnetic *fcc* Ni
Reprinted from: *Nanomaterials* 2020, 10, 691, doi:10.3390/nano10040691 105

Xun Sun, Hualei Zhang, Wei Li, Xiangdong Ding, Yunzhi Wang and Levente Vitos
Generalized Stacking Fault Energy of Al-Doped CrMnFeCoNi High-Entropy Alloy
Reprinted from: *Nanomaterials* 2020, 10, 59, doi:10.3390/nano10010059 145

Rui Yun, Li Luo, Jingqi He, Jiaxi Wang, Xiaofen Li, Weiren Zhao, Zhaogang Nie and Zhiping Lin
Mixed-Solvent Polarity-Assisted Phase Transition of Cesium Lead Halide Perovskite Nanocrystals with Improved Stability at Room Temperature
Reprinted from: *Nanomaterials* 2019, 9, 1537, doi:10.3390/nano9111537 153

Xi Nie, Zhao Qian, Wenzheng Du, Zhansheng Lu, Hu Li, Rajeev Ahuja and Xiangfa Liu
Structural Evolution of AlN Nanoclusters and the Elemental Chemisorption Characteristics: Atomistic Insight
Reprinted from: *Nanomaterials* 2019, 9, 1420, doi:10.3390/nano9101420 165

Lei Zhang, Yi Yu and Meizhen Xiang
A Study of the Shock Sensitivity of Energetic Single Crystals by Large-Scale Ab Initio Molecular Dynamics Simulations
Reprinted from: *Nanomaterials* **2019**, *9*, 1251, doi:10.3390/nano9091251 . . . . **175**

About the Editor

Mojmír Šob is a Professor at the Department of Chemistry at the Masaryk University in Brno, at the Institute of Physics of Materials of the Czech Academy of Sciences, Brno and at the Central European Institute of Technology at the Masaryk University in Brno. His research interests include applications of electronic structure calculations in solid-state physics, chemistry and materials science as well as problems regarding positron annihilation in solids. He is known as a pioneer in application of quantum-mechanical approaches to understanding the properties of new prospective materials, e.g. their ultimate strength, magnetic behavior, phase diagrams and role of extended defects.

Editorial

Editorial for the Special Issue on Computational Quantum Physics and Chemistry of Nanomaterials

Mojmír Šob [1,2,3]

1 Department of Chemistry, Faculty of Science, Masaryk University, Kotlářská 2, CZ-611 37 Brno, Czech Republic; sob@chemi.muni.cz or mojmir@ipm.cz or mojmir.sob@ceitec.muni.cz
2 Institute of Physics of Materials, v.v.i., Czech Academy of Sciences, Žižkova 22, CZ-616 62 Brno, Czech Republic
3 Central European Institute of Technology, CEITEC MU, Masaryk University, Kamenice 5, CZ-625 00 Brno, Czech Republic

Received: 4 November 2020; Accepted: 27 November 2020; Published: 30 November 2020

Nanomaterials have become increasingly important both in basic research and in applications. Some properties may be understood only at the level of the quantum mechanical study of these materials. The purpose of this Special Issue is to advance our fundamental understanding of the structure and technologically important properties of nanomaterials with the help of computational quantum solid-state physics and chemistry. There is no doubt that quantum mechanical approaches are indispensable in comprehensive studies of nanomaterials and will be increasingly crucial in the future. Of course, this field is too extensive and too diverse to be described in a single volume. Nevertheless, this Special Issue provides at least a partial snapshot of the state of the art of computational quantum mechanical studies of nanomaterials and covers some recent advances and problems.

The scope of the articles included in this Special Issue is quite diverse, including adsorption of gas molecules on nanoclusters [1], domain structure of magnetically doped topological insulators [2], properties of dye-sensitized solar cells [3], energetics of silver decahedron nanoparticles [4], the effect of the size and shape on the surface energy of Au nanoparticles [5], critical size of ferroelectric SnTe low-dimensional nanostructures [6], the effect of vacancies on grain boundary segregation in ferromagnetic nickel [7], generalized stacking-fault energy in selected high-entropy alloys [8], phase transition of cesium lead halide perovskite nanocrystals [9], structural evolution of AlN nanoclusters [10] and the shock sensitivity of selected energetic materials [11].

In all these cases, application of the quantum methods was indispensable to determine how various features of atomic configuration of these materials are reflected in their properties and experimentally ascertained quantities.

In summary, this Special Issue of *Nanomaterials* collects a series of original research articles providing new insight into the application of computational quantum physics and chemistry in research on nanomaterials. It illustrates the extension and diversity of the field and indicates some future directions. I am confident that this Special Issue will provide the reader with an overall view of the latest prospects in this fast evolving and cross-disciplinary field.

Funding: M. Š. acknowledges the financial support from the Ministry of Education, Youth and Sports of the Czech Republic in the range of the Project CEITEC 2020 (Project No. LQ1601) and from the Institute of Physics of Materials of the Czech Academy of Sciences in Brno, Czech Republic.

Acknowledgments: The Guest Editor thanks all the authors for submitting their work to this Special Issue and for contributing to its successful completion. A special thank you belongs to all the reviewers participating in the peer-review process of the submitted manuscripts for enhancing their quality and impact. I am also grateful to Ms. Tracy Jin and the editorial assistants who made the entire Special Issue creation a smooth and efficient process.

Conflicts of Interest: The author declares no conflict of interest.

References

1. Liang, Q.; Nie, X.; Du, W.; Zhang, P.; Wan, L.; Ahuja, R.; Ping, J.; Qian, Z. First-Principles Exploration of Hazardous Gas Molecule Adsorption on Pure and Modified $Al_{60}N_{60}$ Nanoclusters. *Nanomaterials* **2020**, *10*, 2156. [CrossRef] [PubMed]
2. Šebesta, J.; Carva, K.; Kriegner, D.; Honolka, J. Twin Domain Structure in Magnetically Doped Bi_2Se_3 Topological Insulator. *Nanomaterials* **2020**, *10*, 2059. [CrossRef] [PubMed]
3. Rashid, M.A.M.; Hayati, D.; Kwak, K.; Hong, J. Theoretical Investigation of Azobenzene-Based Photochromic Dyes for Dye-Sensitized Solar Cells. *Nanomaterials* **2020**, *10*, 914. [CrossRef]
4. Polsterová, S.; Friák, M.; Všianská, M.; Šob, M. Quantum-Mechanical Assessment of the Energetics of Silver Decahedron Nanoparticles. *Nanomaterials* **2020**, *10*, 767. [CrossRef] [PubMed]
5. Holec, D.; Dumitraschkewitz, P.; Vollath, D.; Fischer, F.D. Surface Energy of Au Nanoparticles Depending on Their Size and Shape. *Nanomaterials* **2020**, *10*, 484. [CrossRef] [PubMed]
6. Shimada, T.; Minaguro, K.; Xu, T.; Wang, J.; Kitamura, T. Ab Initio Study of Ferroelectric Critical Size of SnTe Low-Dimensional Nanostructures. *Nanomaterials* **2020**, *10*, 732. [CrossRef]
7. Mazalová, M.; Všianská, M.; Pavlů, J.; Šob, M. The Effect of Vacancies on Grain Boundary Segregation in Ferromagnetic *fcc* Ni. *Nanomaterials* **2020**, *10*, 691. [CrossRef]
8. Sun, X.; Zhang, H.; Li, W.; Ding, X.; Wang, Y.; Vitos, L. Generalized Stacking Fault Energy of Al-Doped CrMnFeCoNi High-Entropy Alloy. *Nanomaterials* **2020**, *10*, 59. [CrossRef]
9. Yun, R.; Luo, L.; He, J.; Wang, J.; Li, X.; Zhao, W.; Nie, Z.; Lin, Z. Mixed-Solvent Polarity-Assisted Phase Transition of Cesium Lead Halide Perovskite Nanocrystals with Improved Stability at Room Temperature. *Nanomaterials* **2019**, *9*, 1537. [CrossRef]
10. Nie, X.; Qian, Z.; Du, W.; Lu, Z.; Li, H.; Ahuja, R.; Liu, X. Structural Evolution of AlN Nanoclusters and the Elemental Chemisorption Characteristics: Atomistic Insight. *Nanomaterials* **2019**, *9*, 1420. [CrossRef] [PubMed]
11. Zhang, L.; Yu, Y.; Xiang, M. A Study of the Shock Sensitivity of Energetic Single Crystals by Large-Scale Ab Initio Molecular Dynamics Simulations. *Nanomaterials* **2019**, *9*, 1251. [CrossRef]

Publisher's Note: MDPI stays neutral with regard to jurisdictional claims in published maps and institutional affiliations.

© 2020 by the author. Licensee MDPI, Basel, Switzerland. This article is an open access article distributed under the terms and conditions of the Creative Commons Attribution (CC BY) license (http://creativecommons.org/licenses/by/4.0/).

Article

First-Principles Exploration of Hazardous Gas Molecule Adsorption on Pure and Modified Al$_{60}$N$_{60}$ Nanoclusters

Qi Liang [1], Xi Nie [1], Wenzheng Du [1], Pengju Zhang [1], Lin Wan [1], Rajeev Ahuja [2,3], Jing Ping [4] and Zhao Qian [1,*]

[1] Key Laboratory of Liquid-Solid Structural Evolution and Processing of Materials (Ministry of Education) & School of Software, Shandong University, Jinan 250061, China; 18536228352@139.com (Q.L.); 17861412028@139.com (X.N.); 201813740@mail.sdu.edu.cn (W.D.); zpj201813800@mail.sdu.edu.cn (P.Z.); wanlin@sdu.edu.cn (L.W.)
[2] Condensed Matter Theory, Department of Physics and Astronomy, Ångström Laboratory, Uppsala University, 75120 Uppsala, Sweden; rajeev.ahuja@physics.uu.se
[3] Applied Materials Physics, Department of Materials Science and Engineering, KTH Royal Institute of Technology, 10044 Stockholm, Sweden
[4] College of Traditional Chinese Medicine, Shandong University of Traditional Chinese Medicine, Jinan 250355, China; pingjing@sdutcm.edu.cn
* Correspondence: qianzhao@sdu.edu.cn

Received: 17 September 2020; Accepted: 4 October 2020; Published: 29 October 2020

Abstract: In this work, we use the first-principles method to study in details the characteristics of the adsorption of hazardous NO_2, NO, CO_2, CO and SO_2 gas molecules by pure and heteroatom (Ti, Si, Mn) modified Al$_{60}$N$_{60}$ nanoclusters. It is found that the pure Al$_{60}$N$_{60}$ cluster is not sensitive to CO. When NO_2, NO, CO_2, CO and SO_2 are adsorbed on Al$_{60}$N$_{60}$ cluster'stop.b, edge.a$_p$, edge.a$_h$, edge.a$_p$ andedge.a$_h$ sites respectively, the obtained configuration is the most stable for each gas. Ti, Si and Mn atoms prefer to stay on the top sites of Al$_{60}$N$_{60}$ cluster when these heteroatoms are used to modify the pure clusters. The adsorption characteristics of above hazardous gas molecules on these hetero-atom modified nanoclusters are also revealed. It is found that when Ti-Al$_{60}$N$_{60}$ cluster adsorbs CO and SO_2, the energy gap decreases sharply and the change rate of gap is 62% and 50%, respectively. The Ti-modified Al$_{60}$N$_{60}$ improves the adsorption sensitivity of the cluster to CO and SO_2. This theoretical work is proposed to predict and understand the basic adsorption characteristics of AlN-based nanoclusters for hazardous gases, which will help and guide researchers to design better nanomaterials for gas adsorption or detection.

Keywords: environment and health; first-principles physics; DFT; electronic structure; hazardous gas

1. Introduction

In recent years, industrial and fossil-fuel motor exhaust gases and the flue gas from the burning of garbage and straw crops in some areas have come to harm humans and the environment. These gases accumulate in the air and undergo a series of complex chemical reactions with the dust, small particles, bacteria, etc. in the air to form small agglomerated particles. When their concentration reaches a certain level and is further affected by weather, smog could appear. These hazardous gases include nitrogen dioxide (NO_2), carbon monoxide (CO), nitric oxide (NO), carbon dioxide (CO_2) and sulfur dioxide (SO_2), etc. Some of these gases are irritating and toxic gases which are the main source of acid rain. Some have strong oxidizing properties and are strong combustion aids and some gases can cause greenhouse effects. They harm people's health and pollute the environment, thus the adsorption and detection of these gases are of great significance to environmental protection and human health.

From the perspective of materials, it is very important to find material substrates with high sensitivity to these gases. At present, nanoclusters, nanosheets, nanotubes and other low-dimensional nanomaterials are regarded to be potential candidates for these applications [1–3].

AlN is a kind of semiconducting material with excellent physical and chemical properties and its unique nanostructures have attracted the attentions of gas sensing researchers [4–11]. The electronic properties of nanomaterials can be improved by atomic modification or doping [12–15]. For example, Rezaei-Sameti et al. studied the adsorption characteristics of pure, B-, As-doped AlN nanotubes for CO adsorption [16]. The results showed that doping with B and As atoms was conducive to CO gas adsorption and effectively improved the sensitivity of AlN nanotubes to CO gas. Saedi et al. [17] considered the adsorption of H_2S, COS, CS_2 and SO_2 gases by $Al_{12}N_{12}$ nanoclusters using density functional theory and it had been revealed that the $Al_{12}N_{12}$ nanocluster was a promising SO_2 gas and electronic sensor and a CS_2 gas electronic sensor, while they had different effects on conductivity. The theoretical investigations of AlN-based low-dimensional nanomaterials in this field are not only interesting to unveil some physics in the area but also of significance to move forward the field.

Based on previous research on AlN nanoclusters [18], in this work we have investigated the adsorption behaviors of NO_2, NO, CO_2, CO, and SO_2 molecules on the large $Al_{60}N_{60}$ cluster. We select more stable adsorption sites to study their adsorption characteristics and electronic structures. Beside the pure cluster, some heteroatoms such as Ti, Si and Mn are also used to modify the $Al_{60}N_{60}$ nanocluster as substrates, for each of which the most stable sites are respectively considered to adsorb the above five gases and the sensitivity of each modified cluster to the gas molecules are also studied systematically.

2. Theoretical Methods

In this work, the first-principles method based on Density Functional Theory (DFT) [19,20] is used. The Vienna ab-initio simulation package [21–23] is employed to optimize the geometry structures and calculate the energetics and electronic structures of various gas adsorption on pure and hetero-atom modified nanoclusters. The projector augmented wave (PAW) [21] pseudopotentials are used to describe the ion-electron interactions. The exchange correlation potentials are treated by the Perdew-Burke-Ernzerhof (PBE) functional within the framework of the generalized gradient approximation [24]. Due to the van der Waals interaction between the atomic cluster and the gas molecule, we have used DFT-D2 method of Grimme [25] to make the correction. The plan wave basis set with an energy cut-off of 520 eV has been used. The Brillouin zone is sampled using the Monkhorst-Pack method and the supercell approach is used for structural optimizations and electronic structure calculations. The conjugate gradient algorithm is used to optimize the geometry structures and the convergence criterion of the atomic force is less than 0.02 eV/Å. The charge transfer between $Al_{60}N_{60}$ nanocluster and hazardous gas molecules were analyzed by the Bader charge analysis using the previous method [26–28]. In order to avoid interaction between clusters in x, y and z directions, all the models are separated by the vacuum space of 20 Å. The following formula is utilized to calculate the adsorption energies:

$$E_{ad} = E_{cluster+gas} - E_{cluster} - E_{gas} \quad (1)$$

where $E_{cluster+gas}$ represents the total energy of the cluster (pure or hetero-atom modified) after gas molecule adsorption, $E_{cluster}$ and E_{gas} represent the energies of the bare cluster and gas molecule respectively.

3. Results and Discussion

3.1. Adsorption of NO_2, NO, CO_2, CO and SO_2 by Pure $Al_{60}N_{60}$ Nanocluster

Firstly, we have systematically considered the adsorption of NO_2, NO, CO_2, CO and SO_2 on the surface of pure $Al_{60}N_{60}$ nanoclusters, the structure of which had been revealed by us in an earlier study. There are different adsorption positions (we designate them as the edge, side and top position in this work) on the surface of clusters, and these three positions have four, two and two different adsorption sites, respectively. In Figure 1, Figure 1a shows the four adsorption sites at the edge position: edge.a (N-site) and edge.b (Al-site) have four coordination, edge.c (N-site) and edge.d (Al-site) have three coordination. Figure 1b shows the adsorption sites of the side position: side.a and side.b are the sites at the concave N atom and the Al-N bond, respectively. Figure 1c illustrates the two sites at the top position: top.a and top.b are the sites at the center and the Al-N bond, respectively. The NO_2, NO, CO_2, CO and SO_2 molecules are horizontally or vertically placed in front of the edge and side sites and are horizontally placed directly above the top two sites for adsorption. Here, the readers may ask why we don't adopt the cluster model with passivated heteroatoms or functional group such as H, O, OH, etc. in this study, since the cluster may react in air or solution. There are mainly two reasons: one is that the real cluster structure after passivation in air or solution would be very complicated, i.e., it is hard to define what kinds of heteroatoms or functional group (H, O, OH, etc.) passivate which specific sites of the cluster. It may be argued that this can be run through in every possible "combination", but the computational time and cost would be huge considering that only for one model there will be 14 different adsorption configurations for each kind of gas molecule (more details will be in the following paragraph) and we have five different gas molecules to investigate in total. The other is that in our intrinsic cluster model, most surface Al/N atoms have three or four coordinations and the Al-N bonds are almost saturated, similar case also exists in other research [17]. Thus in this fundamental study we employ the current model to explore the intrinsic adsorption properties of $Al_{60}N_{60}$ nanocluster.

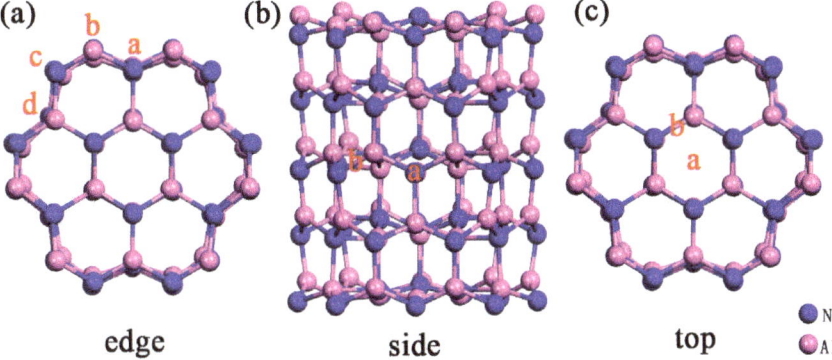

Figure 1. Various adsorption sites (shown in red letters) at each position (a) edge, (b) side or (c) top of pure $Al_{60}N_{60}$ cluster.

Based on the above model, there are 14 different adsorption configurations for each kind of gas molecule. We have systematically studied the adsorption characteristics of each gas at various sites, as shown in Table 1. The subscripts h and p represent the horizontal and perpendicular adsorption of gas molecules at each site. From the data of adsorption energy, it can be seen that compared with other four gases, the adsorption energy absolute values of pure $Al_{60}N_{60}$ cluster towards CO are generally smaller. The adsorption energies of NO_2, NO, CO_2 and CO are relatively high at edge.a site. In addition, it is also found that when these five gas molecules are adsorbed at the top position, the adsorption energy value obtained at the Al-N bond site is high. When NO_2 and CO_2 are adsorbed at the side sites and SO_2 is adsorbed at the edge.c_h and side.a_p sites, the adsorption energy values are more than 10 eV. Therefore, we have studied the adsorption details of these five gases at these sites. The optimized configurations corresponding to the higher energy values are shown in Figure 2. It can be seen from the Figure 2 that when the adsorption energy values are large, the corresponding optimized configurations have obvious shrinkage deformation such as NO_2 adsorption at the side.a_h and side.b_h sites, CO_2 adsorption at the side.a_p and side.b_h sites, and SO_2 adsorption at the edge.c_h and side.a_p. When NO_2 is adsorbed at the edge.a_h site, the structure of the edge position is obviously deformed. From the optimized model structure, it can be seen that when the gas molecules are located at the edge and side sites, the structure is prone to deform.

Table 1. The adsorption energies of various gas molecules adsorbed on different sites of pure $Al_{60}N_{60}$ nanocluster.

E_{ad} (eV)	Edge								Side				Top	
	a_h	a_p	b_h	b_p	c_h	c_p	d_h	d_p	a_h	a_p	b_h	b_p	a	b
E_{ad+NO_2}	−4.77	−4.41	−4.41	−3.39	−2.98	−2.88	−4.22	−3.17	−18.21	−0.22	−10.77	−2.79	−0.31	−3.32
E_{ad+NO}	−2.02	−2.60	−2.01	−1.99	−2.01	−1.97	−1.58	−0.17	−0.81	−1.38	−0.83	−0.83	−0.36	−0.38
E_{ad+CO_2}	−2.36	−0.12	−0.11	−0.09	−0.17	−0.28	−0.90	−0.37	−0.35	−10.04	−10.04	−0.25	−0.29	−0.30
E_{ad+CO}	−0.99	−1.47	−0.97	−0.97	−0.09	−0.75	−0.08	−0.26	−0.74	−0.16	−0.75	−0.12	−0.13	−0.23
E_{ad+SO_2}	−4.49	−3.95	−2.36	−1.00	−13.90	−1.78	−2.18	−1.79	−3.76	−10.82	−1.61	−3.23	−0.44	−3.43

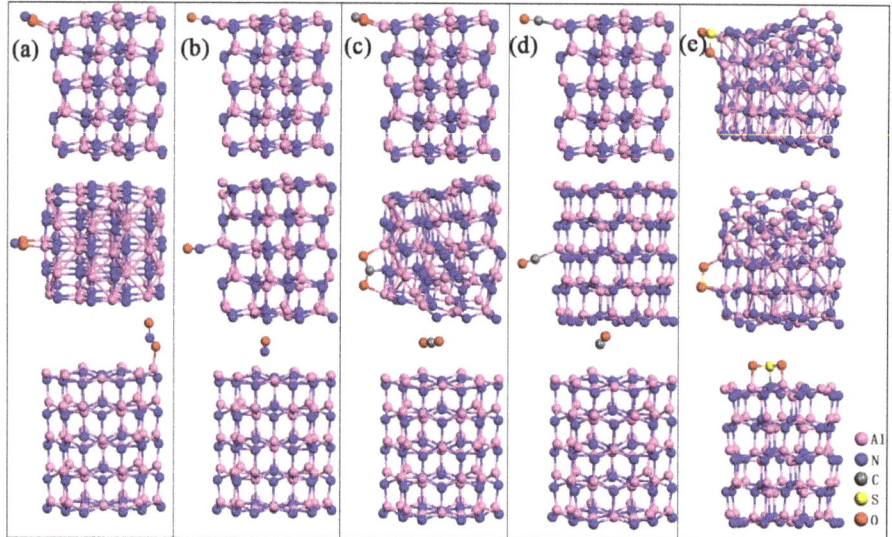

Figure 2. The configurations of pure $Al_{60}N_{60}$ clusters adsorbing hazardous gas molecules at three positions with higher adsorption energy values: (a) NO_2, (b) NO, (c) CO_2, (d) CO, (e) SO_2. Top down in each subbox: the edge, side and top adsorption positions of clusters respectively.

In order to further study the electronic structures of the pure $Al_{60}N_{60}$ cluster adsorbing these five gases, we have selected the relatively stable structures when the $Al_{60}N_{60}$ cluster adsorbs those gas molecules. After screening, five configurations were obtained: NO_2, NO, CO_2, CO and SO_2 adsorbed at the cluster top.b, edge.a_p, edge.a_h, edge.a_p and edge.a_h sites, respectively. Compared with other sites of the same gas, these structures are more stable and the adsorption energy values are higher. Based on this, the electronic properties are further studied.

In Figure 3, the differential charge density and charge transfer of the five gas molecules adsorbed at the relatively stable sites are illustrated. The O atoms of NO_2, CO_2 and SO_2 are bonded with the Al atom at the edge of cluster, and the charge density of O atom after bonding is spindle-shaped. When the pure $Al_{60}N_{60}$ cluster adsorbs NO_2, NO, CO_2, CO and SO_2, charge is transferred from the nanocluster to the gas molecules. The charge transfer of these five systems is not small. We also calculated the electronic structures of these five systems, as shown in Figure 4. After the pure $Al_{60}N_{60}$ cluster adsorbs NO_2, the electronic structure changes significantly. From the density of states curve, it can be seen that after the adsorption of NO_2, a new energy level appears between the Fermi level and the conduction band bottom. Before adsorption, the energy gap of pure $Al_{60}N_{60}$ cluster is 1.297 eV. After adsorption, the energy gap of the system becomes 0.721 eV and the change rate of gap is 44.40%. The pure $Al_{60}N_{60}$ cluster is relatively sensitive to NO_2. When the pure $Al_{60}N_{60}$ cluster adsorbs CO_2, NO and SO_2, the energy gap change by 0.035 eV, 0.08 eV and 0.035 eV, respectively. Before and after the adsorption of CO by pure $Al_{60}N_{60}$, the energy gap changes by 0.945 eV and the change rate is 27.14%.

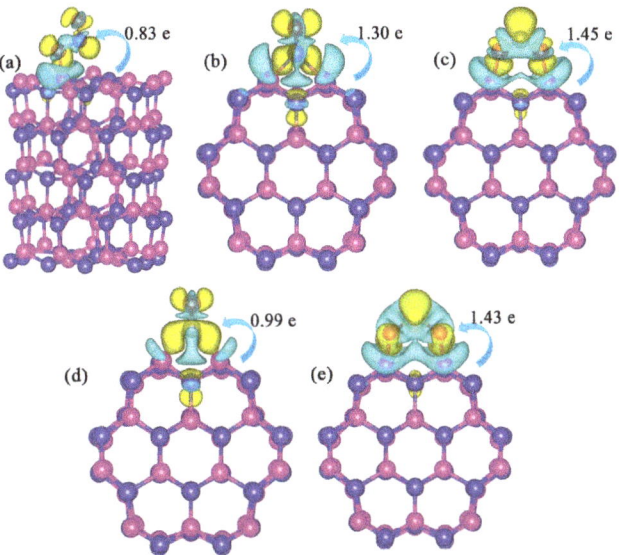

Figure 3. Differential charge density and charge transfer of (**a**) NO_2 adsorbed at top.b site, (**b**) NO adsorbed at edge.a_p site, (**c**) CO_2 adsorbed at edge.a_h site, (**d**) CO adsorbed at edge.a_p site and (**e**) SO_2 adsorbed at edge.a_h site. Yellow and blue represent the charge accumulation and depletion respectively. The isosurface value is 0.0025 e/Å^3. The arrow and charge number in the figure indicate the direction and value of the charge transfer between the nanocluster and the gas molecule.

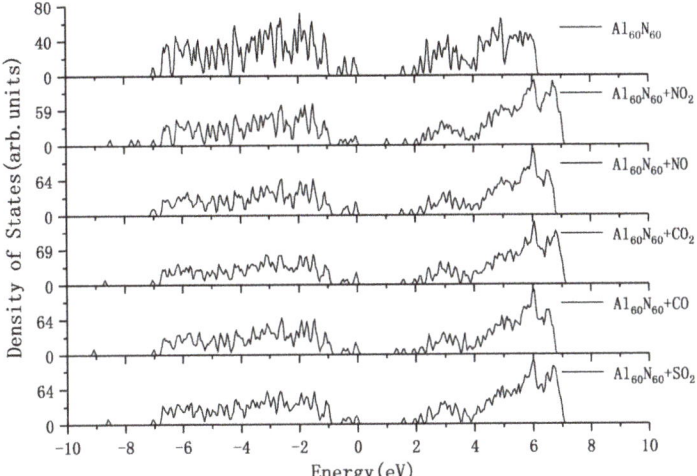

Figure 4. The electronic density of states of pure $Al_{60}N_{60}$ cluster before and after adsorption of various gas molecules. The Fermi level is set at zero.

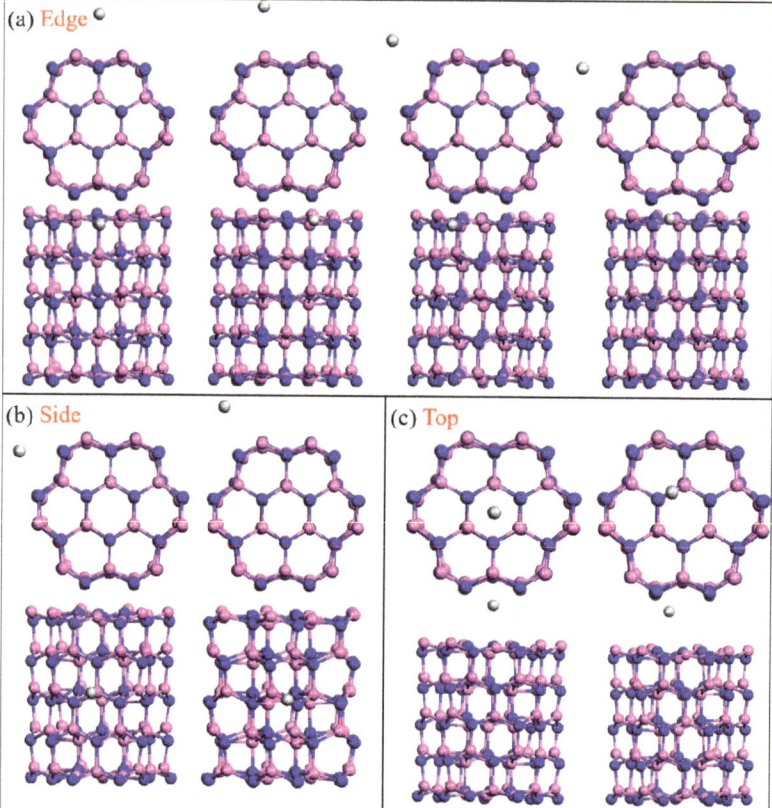

Figure 5. Top view and side view of the edge, side and top positions of $Al_{60}N_{60}$ cluster adsorbing single M atom (M atom is Ti, Si and Mn respectively): (**a**) Edge: From left to right, the M atom is located in front of the four-coordinated N atom, two-coordinated Al atom, three-coordinated N atom and three-coordinated Al atom at the edge position, respectively. (**b**) Side: From left to right, M atom is located in front of the Al atom recessed and Al-N bond at the side position, respectively. (**c**) Top: From left to right, M atom is located above the center of the top six-member ring and the top Al-N bond, respectively. The pink, blue and silver-white spheres represent Al, N, and M atoms, respectively.

3.2. Atomic Modification of $Al_{60}N_{60}$ Cluster by Ti, Si and Mn

In order to further improve the adsorption and sensitivity of the cluster to those five gases, in this work we have tried to modify the $Al_{60}N_{60}$ cluster with Ti, Si and Mn heteroatoms and systematically calculated the adsorption energy, adsorption distance, bond length, bond angle and electronic structure of each system. Since the surface of the $Al_{60}N_{60}$ cluster has many different sites, first of all, we have to determine at which site each heteroatom is prone to occupy and which structure is relatively stable. These two aspects are discussed in the following research. When Ti, Si or Mn heteroatoms are adsorbed at the edge, side and top positions of $Al_{60}N_{60}$ cluster surface, there are four, two, and two adsorption sites, respectively, as shown in Figure 5. We have studied these eight adsorption structures and calculated the adsorption energy of each structure as shown in Table 2. It is found that when a Ti, Si or Mn atom is adsorbed at the side.a site the respective adsorption energies are −14.17 eV, −12.35 eV and −11.38 eV, which shows that the adsorption energy values of these three adsorption systems are generally higher. When these heteroatoms are adsorbed at some sites of the cluster, the adsorption energy is about 10 eV higher than other sites. To know the reason why the adsorption energy changes so much, we have studied the structures after adsorption and selected the geometrically optimized configurations with stable adsorption of the three hetero-atoms at the edge, side and top positions of the cluster as shown in Figure 6. When the cluster adsorbs Ti, Si or Mn atom at the side position, the structures shrink greatly. After the cluster edge.d site adsorbs a Ti atom, the structure also deforms obviously. The system with obvious deformation of structure after optimization has a relatively high adsorption energy value and unstable structure. Next, we will further study them from the aspect of electronic structures.

Table 2. Adsorption energies of Ti, Si or Mn hetero-atom at different sites of $Al_{60}N_{60}$ cluster.

E_{ad} (eV)	Edge				Side		Top	
	a	b	c	d	a	b	a	b
E_{ad+Ti}	−3.21	−3.20	−3.36	−14.67	−14.17	−5.53	−4.80	−4.82
E_{ad+Si}	−3.64	−3.63	−2.55	−2.82	−12.35	−6.63	−4.33	−3.40
E_{ad+Mn}	−2.58	−2.59	−1.61	−1.94	−11.38	−3.02	−2.95	−2.67

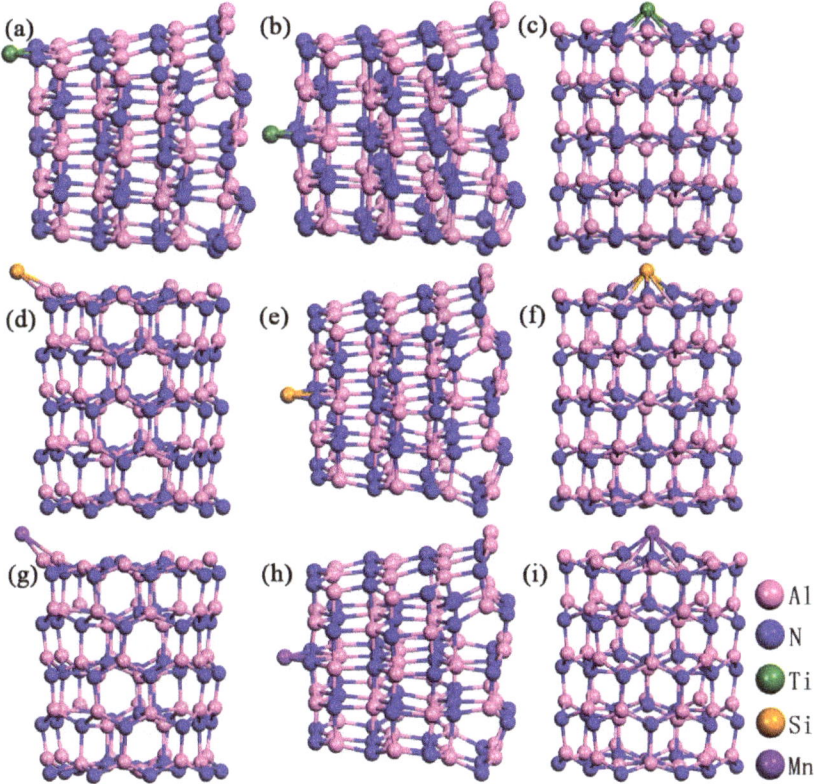

Figure 6. The optimized configurations of $Al_{60}N_{60}$ cluster adsorbing M atom at three positions with higher adsorption energy values: Ti atom is located at the (a) edge, (b) side, (c) top position of $Al_{60}N_{60}$ cluster. Si atom is located at the (d) edge, (e) side, (f) top of $Al_{60}N_{60}$ cluster. Mn atom is located at the (g) edge, (h) side, (i) top of $Al_{60}N_{60}$ cluster.

Figure 7 is the differential charge density diagram corresponding to the higher adsorption energy values when Ti, Si or Mn atom is adsorbed on an $Al_{60}N_{60}$ cluster. When three kinds of atoms are adsorbed at the side position of the cluster and Ti atom is adsorbed at the edge, a small amount of charge accumulation will appear at the No.60 and No.57 Al atoms at the far end, and charge depletion will occur at the No.12, No.19 and No.23 N atoms, and the charge distribution will be spindle-like. When a Ti, Si or Mn atom is adsorbed at the top of the cluster, the charge distribution is more uniform. There is a charge aggregation phenomenon between the three kinds of atoms and the $Al_{60}N_{60}$ cluster, and the Ti, Si and Mn atoms are bonded with the N atom at the top position of the cluster. Thus, when Ti, Si or Mn atom is adsorbed at the top of the $Al_{60}N_{60}$ cluster, they can have a relatively stable structure and the charge distribution is uniform. In the next section, we will select these configurations of hetero-atom modified at the top position of the $Al_{60}N_{60}$ cluster to carry out their adsorption properties towards hazardous gas molecules.

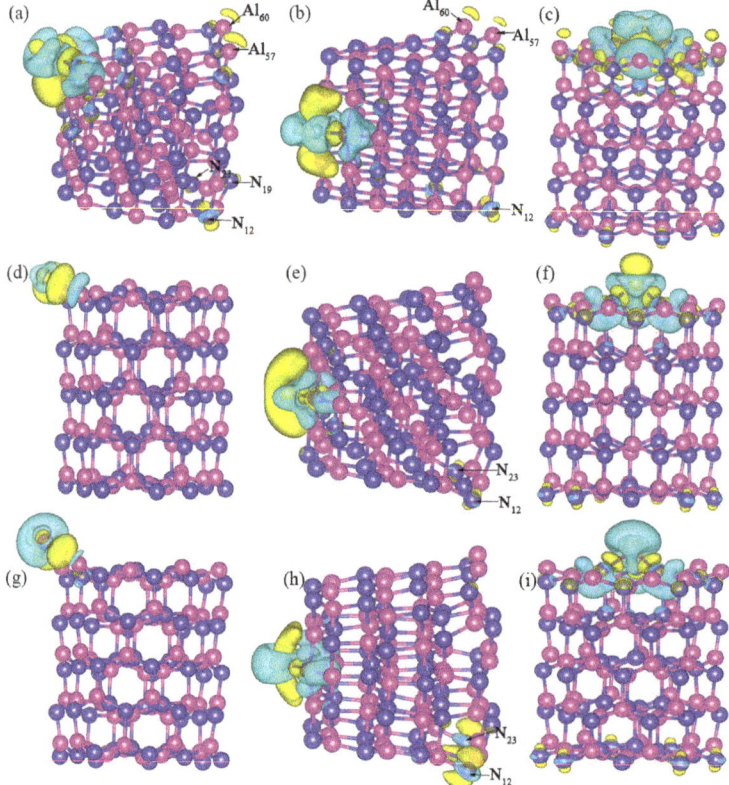

Figure 7. Differential charge density of $Al_{60}N_{60}$ cluster with higher adsorption energy values for single Ti, Si or Mn atom at three positions: (**a**) edge, (**b**) side, (**c**) top adsorption of Ti atom. (**d**) edge, (**e**) side, (**f**) top adsorption of Si atom. (**g**) edge, (**h**) side, (**i**) top adsorption Mn atom. Yellow and blue represent the charge accumulation and depletion (isosurface = 0.0025 e/Å^3).

3.3. Investigation on the Adsorption of Hazardous Gas Molecules by Ti, Si or Mn-Modified $Al_{60}N_{60}$ Cluster

NO_2, NO, CO_2, CO, and SO_2 gas molecules are adsorbed at the stable sites of $Al_{60}N_{60}$ clusters modified by Ti, Si or Mn atom. After structural relaxations, the optimized configurations are shown in Figure 8. After relaxations, the heteroatom-modified $Al_{60}N_{60}$ clusters are not sensitive to CO_2 and do not easily adsorb CO_2; the Si-$Al_{60}N_{60}$ cluster does not easily adsorb CO and SO_2, which can be further proved from Table 3. The M-$Al_{60}N_{60}$ cluster adsorbs CO_2 with a smaller adsorption energy value, which is physical adsorption. Moreover, CO_2 has a longer adsorption distance from the substrate.

Compared with the structure before adsorption, CO_2 is repelled by the substrate. The adsorption energy value of Si-$Al_{60}N_{60}$ cluster towards CO or SO_2 is relatively small, and the adsorption distance is relatively long, which is weak adsorption. When the M-$Al_{60}N_{60}$ cluster adsorbs these five gas molecules, the bond length of the gas is elongated. Except the linear NO and CO, the bond angles of other molecules shrink. As can be seen from the table, the Ti-modified $Al_{60}N_{60}$ cluster has higher adsorption energy values when adsorbing NO_2, NO, CO and SO_2.

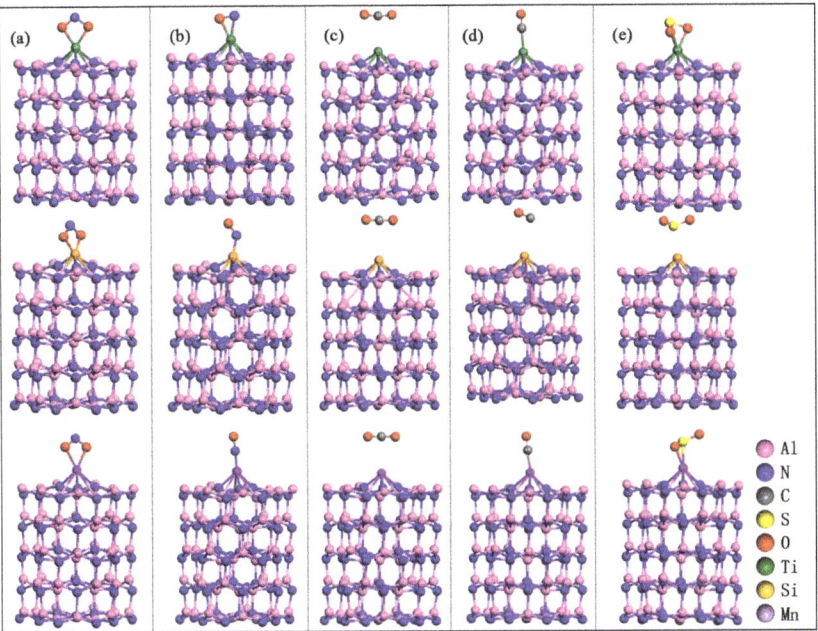

Figure 8. The optimal configurations of (a) NO_2, (b) NO, (c) CO_2, (d) CO and (e) SO_2 adsorption on $Al_{60}N_{60}$ cluster modified by Ti, Si and Mn atoms. Top down: Ti, Si and Mn adsorption systems respectively.

Table 3. The adsorption energy (E_{ad}), adsorption distance (d), bond length ($d_{(X-O)}$), and bond angle ($\angle_{(O-X-O)}$), (X represents N, C or S atom). Adsorption distance is defined as the shortest distance from an atom of gas molecule to an atom in the modified matrix.

Systems	E_{ad} (eV)	d (Å)	$d_{(x-o)}$ (Å)	$\angle_{(O-X-O)}$ (°)
Ti-$Al_{60}N_{60}$+NO_2	−3.99	1.98	1.35	105.95
Si-$Al_{60}N_{60}$+NO_2	−2.25	1.80	1.36	101.44
Mn-$Al_{60}N_{60}$+NO_2	−2.89	2.17	1.28	111.62
Ti-$Al_{60}N_{60}$+NO	−3.37	1.88	1.37	180
Si-$Al_{60}N_{60}$+NO	−0.93	1.72	1.27	180
Mn-$Al_{60}N_{60}$+NO	−2.97	1.70	1.21	180
Ti-$Al_{60}N_{60}$+CO_2	−0.12	2.98	1.19	172.22
Si-$Al_{60}N_{60}$+CO_2	−0.15	3.35	1.18	177.44
Mn-$Al_{60}N_{60}$+CO_2	−0.08	3.04	1.18	178.19
Ti-$Al_{60}N_{60}$+CO	−1.74	2.04	1.18	180
Si-$Al_{60}N_{60}$+CO	−0.07	3.35	1.15	180
Mn-$Al_{60}N_{60}$+CO	−1.44	1.90	1.17	180
Ti-$Al_{60}N_{60}$+SO_2	−3.59	2.02	1.62	98.31
Si-$Al_{60}N_{60}$+SO_2	−0.43	2.74	1.47	116.86
Mn-$Al_{60}N_{60}$+SO_2	−2.06	1.93	1.54	113.89

In order to further study the mechanism of M-Al$_{60}$N$_{60}$ clusters adsorption of those gas molecules, we have analyzed the differential charge density and charge transfer of M-Al$_{60}$N$_{60}$ clusters adsorbing NO$_2$, NO, CO$_2$, CO and SO$_2$, respectively. As can be seen from Figure 9, the charge in all systems is transferred from the modified Al$_{60}$N$_{60}$ cluster to the gas molecule. The system of M-Al$_{60}$N$_{60}$ cluster adsorbing CO$_2$ and Si-Al$_{60}$N$_{60}$ cluster adsorbing CO have smaller charge transfer and weaker interactions between gas and matrix. The system of Si-Al$_{60}$N$_{60}$ cluster adsorbing CO$_2$ and CO hardly shows charge aggregation and loss. The Al$_{60}$N$_{60}$ clusters modified with Ti, Si or Mn atom do not easily adsorb CO$_2$. When M-Al$_{60}$N$_{60}$ clusters adsorb NO$_2$, NO and SO$_2$, the charge transfers are large and there are strong interactions between gas and matrix. When Ti-Al$_{60}$N$_{60}$ and Mn-Al$_{60}$N$_{60}$ cluster adsorb those molecules, charge aggregation occurs between the gas and the matrix. When the Si-Al$_{60}$N$_{60}$ cluster adsorbs NO$_2$ and NO, there is charge depletion between gas and Si-Al$_{60}$N$_{60}$ cluster. When M-Al$_{60}$N$_{60}$ clusters adsorb NO$_2$ and SO$_2$, the O atoms can form bonds with the heteroatom M. When the M-Al$_{60}$N$_{60}$ clusters adsorb NO, the N atoms easily form bonds with the M atom and the N atoms gain electrons.

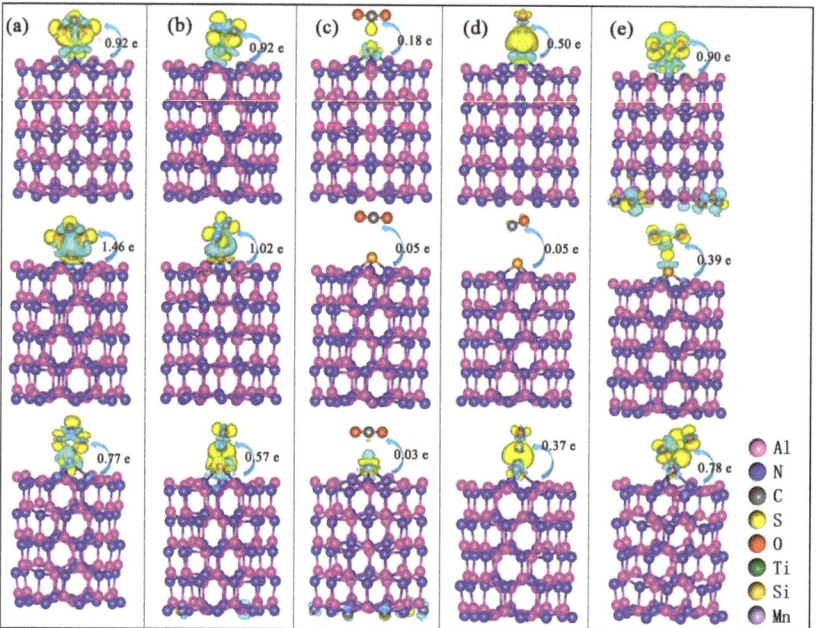

Figure 9. Differential charge density and charge transfer of (a) NO$_2$, (b) NO, (c) CO$_2$, (d) CO and (e) SO$_2$ adsorption on the Al$_{60}$N$_{60}$ cluster modified by Ti, Si or Mn atom. Top down in each subbox: Ti, Si and Mn adsorption systems respectively. Yellow and blue represent the charge accumulation and depletion (isosurface = 0.0025 e/Å^3). Arrows and values represent the direction and amount of charge transfer.

To further figure out the electronic structure changes of the Ti, Si or Mn-modified Al$_{60}$N$_{60}$ clusters adsorbing those five hazardous gas molecules, we have calculated their corresponding density of states as shown in Figure 10. It can be seen from the figure that for the M-Al$_{60}$N$_{60}$ clusters adsorbing NO$_2$, NO, CO$_2$, CO and SO$_2$, the energy gap of the system varies and the effects of three modifying atoms on the electronic structures are also different. In the figure, ΔE_g represents the energy gap change of the system before and after hetero-atom modification and the formula is as follows:

$$\Delta E_g = E_{g(M-Al_{60}N_{60}+gas)} - E_{g(Al_{60}N_{60}+gas)} \tag{2}$$

where $E_{g(M-Al_{60}N_{60}+gas)}$ represents the energy gap of the gas adsorption system by modified $Al_{60}N_{60}$ cluster with M heteroatoms, $E_{g(Al_{60}N_{60}+gas)}$ represents the gap of the system by pure $Al_{60}N_{60}$ cluster. It is found that for the Ti-$Al_{60}N_{60}$ cluster adsorbing CO_2, CO, and SO_2 and the Mn-$Al_{60}N_{60}$ cluster adsorbing CO_2, the energy gap of these four systems are sharply reduced and the energy gap change rate are 75%, 62%, 50% and 57% respectively. The change rate of energy gap of Ti-$Al_{60}N_{60}$ cluster adsorbing NO_2 was 25%. While, the energy gap of Si-modified $Al_{60}N_{60}$ cluster after adsorbing NO is unchanged. The energy gaps of the Ti-$Al_{60}N_{60}$ cluster adsorbing CO_2, CO and SO_2 and the Mn-$Al_{60}N_{60}$ cluster adsorbing CO_2 are decreased, but the $Al_{60}N_{60}$ clusters modified by Ti or Mn atom do not easily adsorb CO_2. Therefore, it is regarded that the Ti-atom modification can improve the sensitivity and sensing of $Al_{60}N_{60}$ cluster towards CO and SO_2 and have potential to be used as a choice of related gas-sensing materials design.

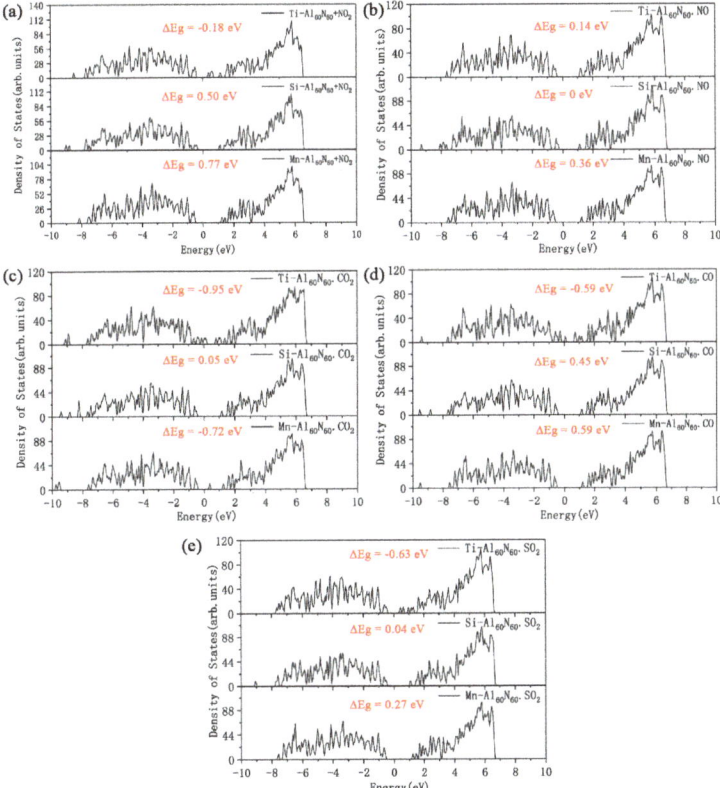

Figure 10. Density of states of various M-$Al_{60}N_{60}$ clusters adsorbing five gas molecules respectively: (**a**) NO_2; (**b**) NO; (**c**) CO_2; (**d**) CO; (**e**) SO_2. The Fermi level is set at zero. ΔEg represents the energy gap change of the system before and after hetero-atom (M) modification.

4. Summary and Outlook

We have systematically investigated the atomic scale configurations and electronic structures of various hazardous gas molecules adsorbed by pure and heteroatom-modified $Al_{60}N_{60}$ clusters with Ti, Si and Mn using Density Functional Theory. It is found that for the pure $Al_{60}N_{60}$ cluster adsorbing those gas molecules, when NO_2, NO, CO_2, CO and SO_2 are adsorbed on the top.b, edge.a_p, edge.a_h, edge.a_p and edge.a_h sites of the nanocluster, respectively, the corresponding structures are relatively stable and the charge transfer is relatively large. The chemisorption between $Al_{60}N_{60}$ cluster and gas molecule in these five systems is strong. The energy gap of pure $Al_{60}N_{60}$ cluster is reduced by 44% when adsorbing NO_2 at its stable site, which is expected to be a candidate gas sensing material for NO_2. When Ti, Si and Mn atoms are respectively adsorbed on the top sites of $Al_{60}N_{60}$ cluster, the corresponding structures are relatively stable and the charge distribution is uniform. When Ti-modified $Al_{60}N_{60}$ cluster adsorbs CO and SO_2 gas molecules, the energy gap decreases sharply. Compared with the pure cluster adsorption system, the energy gap changes by 62% and 50% respectively, which greatly improves the sensitivity of $Al_{60}N_{60}$ nanocluster to CO and SO_2. This theoretical work is proposed to supply fundamental clues and guide for experimentalists to develop appropriate nanoclusters for environmental and health applications.

Author Contributions: Formal analysis, X.N., Z.Q.; Project administration, Z.Q., W.D., P.Z., L.W., R.A. and J.P.; Supervision, Z.Q.; Writing—original draft, X.N., Q.L. and Z.Q. All authors have read and agreed to the published version of the manuscript.

Funding: We thank the support from the Natural Science Foundation of China (51801113), the Natural Science Foundation of Shandong Province (ZR2018MEM001), the Young Scholars Program of Shandong University (YSPSDU) and the Education Program of Shandong University (2020Y299).

Acknowledgments: The National Supercomputer Centre (NSC) and the HPC Cloud Platform of Shandong University are acknowledged. R.A. thanks to SNIC & HPC2N, Sweden for providing the computing facilities.

Conflicts of Interest: The authors declare no conflict of interest.

References

1. Soltani, A.; Peyghan, A.A.; Bagheri, Z. H_2O_2 Adsorption on the BN and SiC Nanotubes: A DFT Study. *Phys. E Low-Dimens. Syst. Nanostruct.* **2013**, *48*, 176–180. [CrossRef]
2. Salih, E.; Mekawy, M.; Hassan, R.Y.A.; El-Sherbiny, I.M. Synthesis, Characterization and Electrochemical-sensor Applications of Zinc Oxide/graphene Oxide Nanocomposite. *J. Nanostruct. Chem.* **2016**, *6*, 137–144. [CrossRef]
3. Beheshtian, J.; Peyghan, A.A.; Bagheri, Z.; Kamfiroozi, M. Interaction of Small Molecules (NO, H_2, N_2, and CH_4) with BN Nanocluster Surface. *Struct. Chem.* **2012**, *23*, 1567–1572. [CrossRef]
4. Ruiz, E.; Alvarez, S.; Alemany, P. Electronic Structure and Properties of AlN. *Phys. Rev. B* **1994**, *49*, 7115–7123. [CrossRef] [PubMed]
5. Tang, Y.B.; Liu, Y.Q.; Sun, C.H.; Cong, H.T. AlN Nanowires for Al-based Composites with High Strength and Low Thermal Expansion. *J. Mater. Res.* **2007**, *22*, 2711–2718. [CrossRef]
6. Rubio, A.; Corkill, J.L.; Cohen, M.L.; Shirley, E.L.; Louie, S.G. Quasi particle Band Structureof AlN and GaN. *Phys. Rev. B* **1993**, *48*, 11810–11816. [CrossRef]
7. Taylor, K.M.; Lenie, C. Some Properties of Aluminum Nitride. *J. Electrochem. Soc.* **1960**, *107*, 308–314. [CrossRef]
8. Ouyang, T.; Qian, Z.; Hao, X.; Ahuja, R.; Liu, X. Effect of Defects on Adsorption Characteristics of AlN Monolayer towards SO_2 and NO_2: Ab initio Exposure. *Appl. Surf. Sci.* **2018**, *462*, 615–622. [CrossRef]
9. Samadizadeh, M.; Rastegar, S.F.; Peyghan, A.A. F^-, Cl^-, Li^+ and Na^+ Adsorption on AlN Nanotube Surface: A DFT Study. *Phys. E Low-Dimens. Syst. Nanostruct.* **2015**, *69*, 75–80. [CrossRef]
10. Wang, Y.; Song, N.; Song, X.; Zhang, T.; Yang, D. A First-principles Study of Gas Adsorption on Monolayer AlN Sheet. *Vacuum* **2018**, *147*, 18–23. [CrossRef]
11. Ouyang, T.; Qian, Z.; Ahuja, R.; Liu, X. First-principles Investigation of CO Adsorption on Pristine, C-doped and N-vacancy Defected Hexagonal AlN Nanosheets. *Appl. Surf. Sci.* **2018**, *439*, 196–201. [CrossRef]
12. Beheshtian, J.; Baei, M.T.; Peyghan, A.A.; Bagheri, Z. Nitrous Oxide Adsorption on Pristine and Si-doped AlN Nanotubes. *J. Mol. Modeling* **2013**, *19*, 943–949. [CrossRef] [PubMed]
13. Rastegar, S.F.; Peyghan, A.A.; Hadipour, N.L. Response of Si- and Al-doped Graphenes toward HCN: A Computational Study. *Appl. Surf. Sci.* **2013**, *265*, 412–417. [CrossRef]
14. Hadipour, N.L.; Peyghan, A.A.; Soleymanabadi, H. Theoretical Study on the Al-doped ZnO Nanoclusters for CO Chemical Sensors. *J. Phys. Chem. C* **2015**, *119*, 6398–6404. [CrossRef]
15. Aslanzadeh, S. Transition Metal Doped ZnO Nanoclusters for Carbon Monoxide Detection: DFT Studies. *J. Mol. Modeling* **2016**, *22*, 160. [CrossRef]

16. Sameti, M.R.; Jamil, E.S. The Adsorption of CO Molecule on Pristine, As, B, BAs Doped (4, 4) Armchair AlNNTs: A Computational Study. *J. Nanostruct. Chem.* **2016**, *6*, 197–205. [CrossRef]
17. Saedi, L.; Javanshir, Z.; Khanahmadzadeh, S.; Maskanati, M.; Nouraliei, M. Determination of H_2S, COS, CS_2 and SO_2 by An Aluminium Nitride Nanocluster: DFT Studies. *Mol. Phys.* **2020**, *7*, 118. [CrossRef]
18. Nie, X.; Qian, Z.; Du, W.; Lu, Z.; Li, H.; Ahuja, R.; Liu, X. Structural Evolution of AlN Nanoclusters and the Elemental Chemisorption Characteristics: Atomistic Insight. *Nanomaterials* **2019**, *9*, 1420. [CrossRef]
19. Kohn, W.; Sham, L.J. Self-Consistent Equations Including Exchange and Correlation Effects. *Phys. Rev.* **1965**, *140*, A1133–A1138. [CrossRef]
20. Hohenberg, P.; Kohn, W. Inhomogeneous Electron Gas. *Phys. Rev.* **1964**, *136*, B864–B871. [CrossRef]
21. Kresse, G.; Joubert, D. From Ultrasoft Pseudopotentials to the Projector Augmented Wave Method. *Phys. Rev. B* **1999**, *59*, 1758–1775. [CrossRef]
22. Kresse, G.; Furthmüller, J. Efficiency of Ab-Initio Total Energy Calculations for Metalsand Semiconductors Using A Plane-Wave Basis Set. *Comput. Mater. Sci.* **1996**, *6*, 15–50. [CrossRef]
23. Kresse, G.; Furthmüller, J. Efficient Iterative Schemes for ab Initio Total-Energy Calculations Using A Plane-Wave Basis Set. *Phys. Rev. B* **1996**, *54*, 169–186. [CrossRef]
24. Perdew, J.P.; Burke, K.; Ernzerhof, M. Generalized Gradient Approximation Made Simple. *Phys. Rev. Lett.* **1996**, *77*, 3865–3868. [CrossRef]
25. Grimme, S. Semiempirical GGA-type Density Functional Constructed with A Longrange Dispersion Correction. *J. Comput. Chem.* **2006**, *27*, 1787–1799. [CrossRef]
26. Cioslowski, J. Atoms in Molecules-a Quantum Theory-bader, RFW. *Science* **1991**, *252*, 1566–1567. [CrossRef]
27. Tang, W.; Sanville, E.; Henkelman, G. A Grid-based Bader Analysis Algorithm without Lattice Bias. *J. Phys. Condens. Matter* **2009**, *21*, 084204. [CrossRef]
28. Nie, X.; Qian, Z.; Li, H.; Ahuja, R.; Liu, X. Theoretical Prediction of A Novel Aluminum Nitride Nanostructure: Atomistic Exposure. *Ceram. Int.* **2019**, *45*, 23690–23693. [CrossRef]

Publisher's Note: MDPI stays neutral with regard to jurisdictional claims in published maps and institutional affiliations.

© 2020 by the authors. Licensee MDPI, Basel, Switzerland. This article is an open access article distributed under the terms and conditions of the Creative Commons Attribution (CC BY) license (http://creativecommons.org/licenses/by/4.0/).

Article

Twin Domain Structure in Magnetically Doped Bi$_2$Se$_3$ Topological Insulator

Jakub Šebesta [1,*], Karel Carva [1,*], Dominik Kriegner [2,3] and Jan Honolka [2]

[1] Department of Condensed Matter Physics, Faculty of Mathematics and Physics, Charles University, Ke Karlovu 5, 121 16 Praha 2, Czech Republic
[2] Institute of Physics, Academy of Science of the Czech Republic, Na Slovance 2, 182 21 Praha 8, Czech Republic; dominik.kriegner@tu-dresden.de (D.K.); honolka@fzu.cz (J.H.)
[3] Institute of Solid State and Materials Physics, Technical University of Dresden, 01062 Dresden, Germany
* Correspondence: jakub.sebesta@mff.cuni.cz (J.Š.); carva@karlov.mff.cuni.cz (K.C.)

Received: 18 September 2020; Accepted: 12 October 2020; Published: 19 October 2020

Abstract: Twin domains are naturally present in the topological insulator Bi$_2$Se$_3$ and strongly affect its properties. While studies of their behavior in an otherwise ideal Bi$_2$Se$_3$ structure exist, little is known about their possible interaction with other defects. Extra information is needed, especially for the case of an artificial perturbation of topological insulator states by magnetic doping, which has attracted a lot of attention recently. Employing ab initio calculations based on a layered Green's function formalism, we study the interaction between twin planes in Bi$_2$Se$_3$. We show the influence of various magnetic and nonmagnetic chemical defects on the twin plane formation energy and discuss the related modification of their distribution. Furthermore, we examine the change of the dopants' magnetic properties at sites in the vicinity of a twin plane, and the dopants' preference to occupy such sites. Our results suggest that twin planes repel each other at least over a vertical distance of 3–4 nm. However, in the presence of magnetic Mn or Fe defects, a close twin plane placement is preferred. Furthermore, calculated twin plane formation energies indicate that in this situation their formation becomes suppressed. Finally, we discuss the influence of twin planes on the surface band gap.

Keywords: topological insulators; magnetic doping; defects; ab initio

1. Introduction

Some of the most characteristic representatives of topological insulators (TI) are three-dimensional compounds with non-trivial topology protected by time reversal symmetry (TRS) [1–4]. Although their bulk band structure contains a band gap, the surface of such materials hosts unique conductive states, which intersect in the so-called Dirac point possessing a linear dispersion [1,4–7]. The formation of metallic surface electron states originates from the occurrence of band inversion driven by the strong spin orbit coupling (SOC) [8,9], which leads to non-trivial band topology protected by the TRS. It ensures band crossing at high symmetry points of the Brillouin zone [10,11], while no extra crystal symmetry is required (compare e.g., topological crystal insulators [1,5]). A combination of strong SOC, brought about by the occurrence of heavy elements as e.g., Bi, Se or Te, and TRS leads to the spin polarization of surface bands [3,5,10]. Electrons occupying states in the proximity of the Dirac cone with an opposite momentum also possess opposite spins (so-called spin-momentum locking). It ensures e.g., the suppression of back-scattering and related outstanding surface transport properties [11,12].

In real applications, the influence of defects could be important, since they might significantly alter properties of the ideal matter. They could be varied intentionally by chemical doping. In the case of TIs, it includes particularly magnetic doping. It can lead to the breaking of TRS, and therefore it opens a surface gap [13–15]. Then, not only does a possible control of the surface conductivity appear due to increased surface scattering, but also it could bring about the occurrence of new unique phenomena e.g., quantum anomalous Hall effect (QAHE) [16]. Besides, native defects occur there, naturally. Their presence is hardly controllable and they might have a significant impact on physical properties [17,18] as well. There could exist several kinds of native defects depending on the actual compound and the growth process. These include twin planes (TP), which are the focus of this article.

An important group of 3D topological insulators are bismuth chalcogenides, such as Bi_2Se_3 [8,19], which have been shown to be prone to the formation of twin domains [20,21]. This compound possesses a relatively simple band structure, convenient for experimental and theoretical studies, with a Dirac cone appearing at the Γ point. The crystal structure of Bi_2Se_3, belonging to the $R\bar{3}m$ space group, consists of Bi and Se hexagonal layers. These form quintuple layers (QLs) with alternating Bi and Se layers (Figure 1). Due to the coupling of QLs only by van der Waals (vdW) forces, there appears to be a gap between QLs, the so-called 'van der Waals' gap [19] (Figure 2). The gap consists of unoccupied tetrahedral and octahedral positions and its thickness is about 2.25 Å for Bi_2Se_3 [19]. The vdW gap allows cleaving the structure without breaking covalent bonds, which is important for the observation of topological surface states, as it does not yield any extra ingap states [22]. The presented crystal structure offers several sites, which could be occupied by magnetic atoms. Based on the theoretical and experimental studies, the most probable site for a magnetic dopant (Cr, Fe, or Mn) is the substitutional position, where magnetic atoms replace Bi ones [23–27]. Recently, a formation of septuple layers induced by magnetic defects was described. However, it was shown that it is negligible in Bi_2Se_3 for small concentrations of magnetic dopants [28]. Besides, there are studies which suggest an occupation of the interstitial positions within vdW gap concerning the related Bi_2Te_3 compound [29]. In our calculation, we especially employ Mn^{Bi} [23,27,30,31] as well as Fe^{Bi} [27,31,32] magnetic dopants. In addition in our calculations we also assume native defects like Bi or Se antisites (Bi^{Se} resp. Se^{Bi}), where Bi atoms replace Se ones and vice versa. This non-stoichiometry arises due to difficulties in controlling growth conditions, which result in Bi- or Se- rich samples [23,33–36]. It has been shown that not only the magnetic defects, but also nonmagnetic disorder can substantially modify the dispersion of the Dirac surface states. [37]. Furthermore, regarding magnetically doped Bi_2Se_3, there exist suggestions that the surface band gap is not caused by magnetic ordering [38].

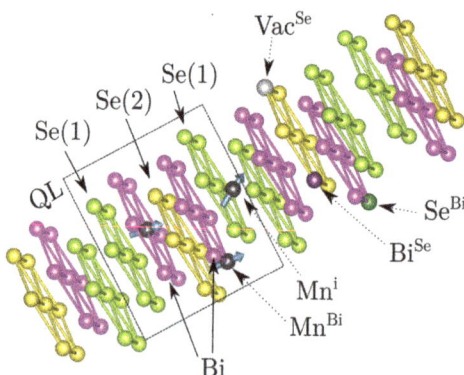

Figure 1. Layered crystal structure of Bi_2Se_3. Se and Bi layers gathered into QLs are depicted. Examples of (non)-magnetic defects are shown: (Mn^{Bi}) substitutional Mn atoms, (Mn^i) interstitial Mn atoms, (Bi^{Se}) resp. (Se^{Bi}) Bi and Se antisites, (Vac^{Se}) Se vacancies.

The above mentioned TPs represent a stacking fault of the layered structure of bismuth chalcogenides [20,21,39]. From symmetry arguments there are three three possible stacking positions of hexagonal layers alternating similarly to the fcc stacking sequence. During the formation of the crystal, there exist two energetically almost equivalent sites, which the atoms in the new layer can choose to occupy. Therefore, mirrored stacking might arise, which could be represented by a 60° rotation of new layers in relation to the ideal ones [21]. It results in the inverse order of the *abc*-like stacking beyond the TP (Figure 2). Generally, there exist a few positions where TP could occur, but the most probable ones lie at outer chalcogenides of quintuples [40]. This means that stacking order inside each separate QL contains no defect; the perturbation occurs in the vdW gap between them. QLs after the TP are then constructed with a mirrored stacking order (Figure 2). The reported experiments show that the presence of TPs strongly depends on the used substrate [41,42].

Similarly to point defects, TPs might have a significant influence on the physical properties [43]. Therefore, in this paper, we focus on the influence of TPs on 3D TI Bi_2Se_3 behavior and their interplay with chemical disorder, especially the magnetic one. First, we describe a distribution of TPs in a nanoscopically thin Bi_2Se_3 slab. Then, we discuss their behavior under the presence of magnetic and nonmagnetic defects. Finally, the influence of TPs on the surface states is shown.

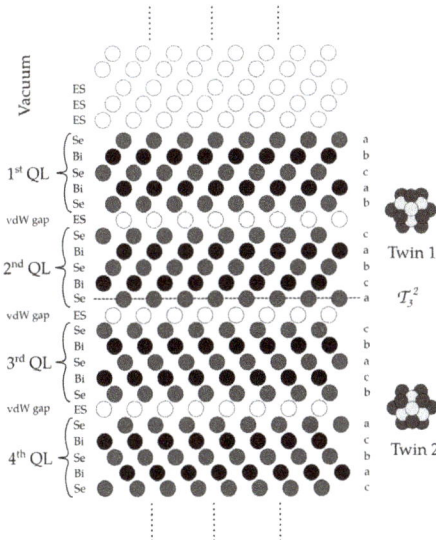

Figure 2. Layout of the simulated multilayer Bi_2Se_3 structure including twin planes. Proportions of atoms are not realistic within this schematic figure. QL—quintuple layer, ES—empty sphere, T_3^2 twin plane between the second QL and the third QL.

2. Methods

The study employs ab initio calculations done in the framework of the tight-binding linear muffin-tin orbital method within the atomic sphere approximation (TB-LMTO-ASA) formulated in terms of Green's functions [44,45]. It involves the local spin density approximation with the Vosko–Wilk–Nursain exchange-correlation potential [46] and the use of a *s,p,d* atomic model. Calculations were treated in the scalar relativistic approximation, where on-site spin-orbit coupling was involved into the scalar relativistic Hamiltonian as a perturbation. An inclusion of spin-orbit coupling is needed to achieve a proper description of the electronic structure [5,8,47]. A basic screened impurity model was included to improve treating electrostatics of disordered systems [48]. Thanks to the use of the Green's function formalism, chemical disorder could be included by the coherent potential approximation (CPA) [49]. It allows one to avoid using large statistical ensembles and it is suitable for small perturbation in the system. To simulate a layered structure with TPs, layered Green's functions reflecting translation symmetry only within an atomic layer were employed [45,50–52]. In calculations, a multilayer system is attached to the semi-infinite leads, which have to satisfy self-consistent conditions.

Due to the coupling of the multilayer to the attached leads, it is possible to obtain a self-consistent solution also for the inner layers. Based on the down-folding method, one is able to construct recursively embedding potentials acting from both sides on the particular layer, which are related to the interlayer coupling. For a detailed description, we refer the reader to Ref. [45,50].

The crystal structure is based on experimental Bi_2Se_3 lattice parameters (unit cell a = 4.138 Å and c = 28.64 Å [19]), which were used to build Bi_2Se_3 multilayer structures. The vdW gap between QLs is included within ASA by placing appropriate empty spheres (ES). To avoid effects of the substrate (or leads) and to concentrate only on the behavior of proper Bi_2Se_3 layers, we surround it by vacuum, which is treated in a similar sense to the vdW gap. It is formed from the fcc-like stacked empty sphere layers keeping the three-fold symmetry of Bi_2Se_3 layers. Furthermore, because leads should fit to the simulated structure, slightly modified scandium is selected. Its hcp crystal structure suits fcc-like stacking within QLs and it possesses lattice parameters that are not too distinct [53]. However, leads are much less unimportant thanks to used vacuum spacers.

Finally, one is able to construct a layer structure, which consists of intermediate Sc layers at borders, coupled to semi-infinite leads, and several Bi_2Se_3 QLs enclosed by the vacuum spacer. In our calculation, we employed ten or twenty QLs wide Bi_2Se_3 structures and the vacuum spacers about ten ES layers width. These dimensions are sufficient to simulate the vacuum and to obtain surface gapless states. Native defects (Se^{Bi}), as well as magnetic doping by either Mn^{Bi} or Fe^{Bi}, are included. In general, we assumed a homogeneous disorder, where mentioned defects occupy the appropriate sites with the same probability, unless otherwise stated. This assumption is supported by synchrotron experiments which show that Mn is not metallic in Bi_2Se_3 and thus does not segregate there [54]. The influence of the magnetic defects on the crystal structure is reflected by local lattice relaxation similar to the previous bulk calculation of Bi_2Te_3 and Bi_2Se_3 [17,30]. According to supercell calculations [30], we have modified and used the Wigner–Seitz radii locally while the total volume of Bi-sublattice is retained. Concerning details, we refer the reader to Ref. [30] and the Supplementary of Ref. [17]. Small changes which stem from the presence of the intrinsic Se^{Bi} defects are neglected in correspondence to our previous work [30]. Besides, the relaxation corresponding to the presence of surfaces is not included there. In our calculation, we simulate TPs in the vdW gaps with respect to the required 2D periodicity in the layer. Hence no structure boundaries within a layer, which are related to the presence of TP [39], are involved.

The formation energy $E_{form}^{def(x)}$ of the extra stacking defect at x with respect to the unperturbed system is given as $E_{form}^{def(x)} = E_{total}^{def(x)} - E_{total}$, where the energies on the right hand side correspond to total energies of the system with and without the defect. The selected approach unfortunately introduces numerical artifacts within the employed TB-LMTO-ASA framework, which renders calculations of the formation energy E_{form} not reliable. Therefore, we show only relative formation energies ΔE_{form} considering the same number and similar type of stacking faults, while we focus on various composition and TPs distribution. These are defined so that zero energy level corresponds to a selected defect placement x_0, often with the lowest energy. Then

$$\Delta E_{form}^{def(x)} = E_{form}^{def(x)} - E_{form}^{def(x_0)} = E_{total}^{def(x)} - E_{total} - (E_{total}^{def(x_0)} - E_{total}) = E_{total}^{def(x)} - E_{total}^{def(x_0)}. \quad (1)$$

The energy differences w.r.t. the unperturbed system are thus canceled. Furthermore, the systematic error depends on the distance of the twin plane from the surface, as discussed in the Appendix A. The dependence obtained there is thus subtracted from data presented in Results where applicable. Since we deal with various positions of TPs within the multilayer sample, we describe their location by sub- and superscript, denoting adjacent QLs for clarity. The notation T_{x+1}^{x} is used for simplicity, where QLs are enumerated from the top interface.

3. Results and Discussion

Stacking fault energies related to TP formation in ideal Bi_2Se_3 have already been studied elsewhere [39,40]. Here, we focus on their mutual interaction within Bi_2Se_3 slabs and subsequently on their interplay with chemical disorder.

3.1. Inter Twin Plane Interactions

3.1.1. Structure without Disorder

One of the simplest ways to study interactions between TPs is the introduction of two TPs in the pristine Bi_2Se_3 multilayer structure. Keeping the position of one TP fixed while another one being independent allows us to determine corresponding relative formation energies over all possible mutual positions of TPs (Figure 3).

When studying multiple TPs, we have to consider that TPs can occupy the same or distinct sides of particular QLs (Figure 4). Different possible cases are compared in Section 3.4. In the remaining text, we show, for clarity, the simplest case, with identical orientations of TPs (denoted as AA according the Figure 4, resp. AAA in the case of two extra TPs). This situation represents qualitatively the most probable behavior, as it also resembles the lowest energy case of systems with non-identical TPs (Section 3.4).

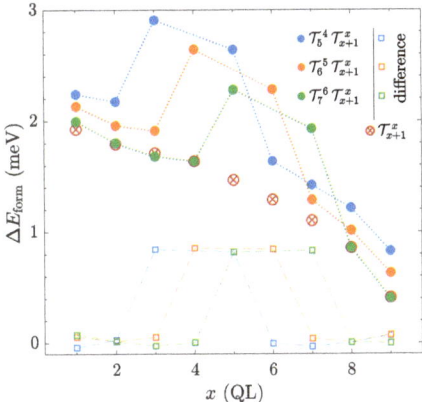

Figure 3. (●) Relative formation energy of two TPs as a function of their position within the structure, which consists of 10 Bi_2Se_3 QLs. (⊗) Relative formation energy of a single TP is depicted for comparison. Relative formation energies belonging to the different numbers of TPs are related to distinct absolute energies. (□) Energy curves associated to two TPs with subtracted single TP curve contribution.

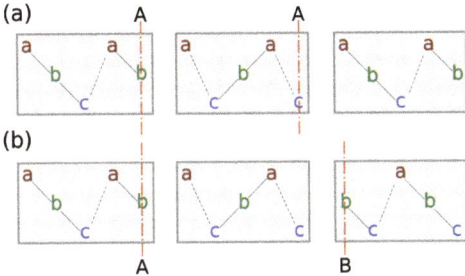

Figure 4. Twin plane orientation. There exist two mutual orientation of TPs. (a) TPs are located at the same sides of QLs (similar letters—AA resp. BB). (b) TPs occurs at the different sides of QLs (distinct letters—AB).

Considering two TPs in the multilayer structure consisting of 10 QLs, we find that the dependence related to a single TP, discussed in detail in the Appendix (Figure A1), is changed almost only for adjacent TPs, where an extra interaction energy appears (Figure 3). However, for such a small multilayer structure, it is not convenient to study TP interactions because of the strong interplay with the interfaces, which is shown in the Appendix (Figure A1a). It is reflected in the bending of calculated energy dependencies (Figure 3) caused by non-negligible energy contribution originating from the interactions between TPs and vacuum interfaces (Figure A1).

Therefore, we introduce a larger structure, consisting of 20 Bi_2Se_3 QLs, where positions of two border TPs are fixed and the third one is able to move in between them (Figure 5a). This allows us to study the behavior of a TP in a more realistic situation, where it is affected primarily by other surrounding TPs rather than a surface. Interface proximity effects are thus reduced in this situation. The 3 TP calculation again shows a clearly visible repulsion of neighboring TPs (Figure 5a), especially after the subtraction of the surface-induced contribution to single TP energy (Figure A1). It reveals the occurrence of a significant interaction energy contribution appearing for TPs distant up to the length of three QLs. This suggests that TPs in a pure sample are likely spread over the sample with mutual distances which exceed at least the width of three or four QLs. Previous experiments utilizing X-ray nanobeam microscopy (D.Kriegner) [21] prove that if more TPs are observed, they are clearly several 10 nm apart. Hence, one can compare it with the width of one QL, which is about 1 nm [19]. This finding is supported by another experiment evidencing TPs separated by several QLs [20,41]. Nonetheless, one should be aware that we are comparing ground state calculations with a molecular beam epitaxy growth, which occurs far from equilibrium conditions.

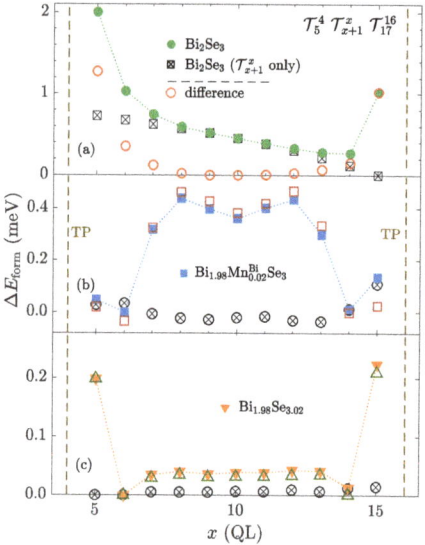

Figure 5. (●,■,▼) Relative formation energy of three TPs as a function of the position x of the middle TP. Positions of border TPs are fixed. The Bi_2Se_3 multilayer consists of 20 QLs. (**a**) pure system without any disorder. (**b**) system with homogeneous magnetic doping. (**c**) system under presence of homogeneously distributed nonmagnetic disorder. (⊗,⊠) Relative formation energy related to a single TP depicted for comparison. (○,□,△) Energy curves associated to three TPs with a subtracted single TP curve contribution. Particular relative formation energy curves are related to different absolute energies.

3.1.2. Native and Magnetic Defects

The interaction between TPs significantly changes when chemical disorder is introduced in the sample. For all studied types of doping (Mn^{Bi} shown in Figure 5b, or Se^{Bi} shown in Figure 5c), we observed a modification of the dependence of the relative formation energy on the distributions of TPs in comparison to the pure sample (Figure 5a). Due to the presence of disorder, the monotonous dependence on the distance from a certain vacuum interface disappears (compare Figures 5b,c to Figure 5a; ⊗-, ⊠-points). Moreover, the observed relative energy differences are almost one order of magnitude smaller (Figure 5b,c). It may be caused by suppressed interactions between TPs compared to the ideal case. The presence of a TP apparently does not represent such significant perturbation in disordered systems as it does in the case of pure, regular systems.

Calculations show that a system with magnetic disorder (Mn^{Bi}) prefers the gathering of TPs (Figure 5b) instead of their spreading observed in the undoped system. On the other hand, nonmagnetic disorder rather maintains a repulsion between TPs, although it is quite weak (Figure 5c) in comparison to the ideal case (Figure 5a), and it is non-negligible only for adjacent TPs. This indicates a significance of magnetism related effects, although we cannot exclude the role of different chemistry between dopant species. Therefore, we study the influence of the magnetism in more detail in the following sections.

3.2. Twin Plane Formation under Chemical Disorder

We have calculated dependencies of the relative formation energy of TPs on the concentration x of magnetic Mn^{Bi} or Fe^{Bi} and native Se^{Bi} defects to describe the influence of the defect presence on the tendency to TP formation (Figure 6).

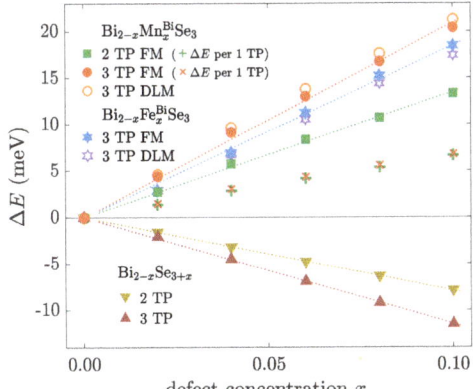

Figure 6. Relative change of the TP formation energy ΔE as a function of the concentration of defects x. Structures either with two (T_8^7, T_{13}^{12}) or three TPs (T_5^4, T_{10}^9, and T_{17}^{16}) within multilayers consisting of 20 QLs were used. Dependencies under presence of magnetic and native defects are depicted. (+) and (×) denote hypothetical relative formation energy related to a single TP. Dotted lines depict calculated linear fits.

The calculated relative formation energy, obtained for a different number of included TPs concerning also their distinct positions, almost linearly grows with the increasing concentration of magnetic defects. This proves that an increased amount of magnetic dopants x leads to the suppression of TPs in the multilayer. On the other hand, the nonmagnetic disorder (Se^{Bi}) decreases the relative formation energy of TPs in the structure. However, the appropriate dependencies exhibit linear behavior as well.

We assume that the suppression of TPs with respect to the increasing concentration of magnetic dopants corresponds to the observed tendency to gathered TPs in the case of magnetic doping (Figure 5b). We suppose that the gathering of TPs likely minimizes an induced effect on the electron structure, which arises from the interplay of TPs and disorder in connection with the magnetism. It agrees with the observation that TPs are less favorable in the magnetically doped systems (Figure 6). Analogously, the fact that the presence of nonmagnetic disorder does favor an occurrence of TPs (Figure 6) can be related to the suppressed impact of TPs on the system in that case (Figure 5c). Besides, one might note a proportionality of the relative formation energy to the number of the occuring TPs. It is confirmed by the comparison of the formation energy per single TP (Figure 6).

So far, we discussed ferromagnetically (FM) ordered magnetic dopants. Now, for a moment, we introduce a paramagnetic state represented by the disordered local moment (DLM) model, in order to decide whether the TP formation energy depends on the type of the magnetic order. In general, these two magnetic phases stand for the limiting cases of the magnetic order. One describes a perfectly ordered system, the other one an absolute disorder. One can observe that calculations exhibit only slight changes of the formation energy with a respect to the former FM order. It indicates that the formation energy likely hardly depends on the type of the magnetic order. More precisely, TPs become more favorable in the case of Fe doping. On the other hand, Mn doping illustrates an opposite behavior. We suppose that different slopes of energy dependencies induced either by Mn or by Fe dopants are likely related to different magnitudes of local exchange splitting. Calculations show that Fe atoms bear about 0.8 μ_B smaller magnitudes of magnetic moments than the Mn ones. Therefore, one might assume that the TP formation energy likely scales with the size of the change of the local exchange splitting caused by the mirrored crystal structure.

The mentioned quite large difference between the magnitudes of Fe and Mn magnetic moment stems from a distinct character of the magnetic exchange interactions (Figure 7), where they are evaluated by employing the Liechtenstein formula [30,55,56]. A comparison shows that unlike Mn related interactions, which are nearly positive except the nearest ones, the exchange interactions between Fe dopants are predominantly antiferromagnetic [57], regardless of the considered magnetic sublattices.

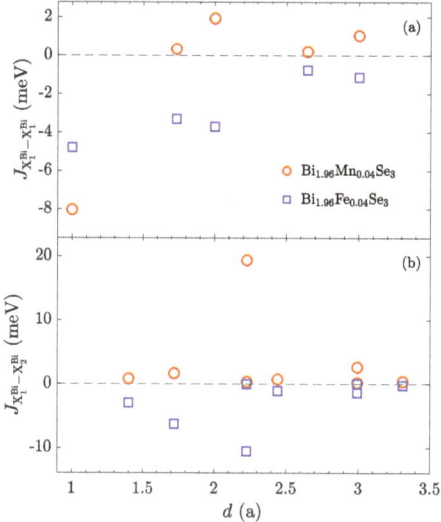

Figure 7. Exchange interactions between magnetic atoms at the substitution position within Bi_2Se_3 as a function of the distance in units of the lattice parameter a. Multilayered structures composed of 20 QLs and FM ordering are employed. Exchange interactions at central QLs are evaluated. (**a**) Exchange interaction within the same sublattice. Interactions within the atomic layer are depicted. (**b**) Exchange interaction between atoms occupying different sublattices. Interactions across the vdW gap are depicted only.

3.3. Magnetic Dopants Behavior

Next, we focused on the influence of TP on the behavior of magnetic dopants, and we calculated relative formation energies of Mn dopants as a function of the position of the dopant determined by indices of the Bi site and QL (Figure 8a). Only one Bi site in the whole structure is partly substituted by Mn. Comparing the shape of corresponding curves differing in the presence of a TP, one observes a clear variation of the relative formation energy caused by the TP. Dependencies without TPs bear a nearly symmetrical behavior, where a deviation is likely caused by an asymmetry of Bi sites concerning the QL structure. The occurrence of the TP modulates the shape of the former energy dependencies as particular sites become relatively more favored or disfavored according to their location with respect to the TP. A comparison of the formation energies (Figure 8a) with the magnitudes of induced magnetic moments on Mn dopants (Figure 8b) indicates a possible relation between the magnetism or spin splitting and the distribution of the relative formation energy of magnetic dopants. One can observe that the effectively suppressed relative formation energies, compared to the ideal structure, correspond to the weakening of induced magnetic moments, and vice versa (Figure 8b). A modulation of magnitudes of magnetic moments by stacking faults has been discussed e.g., in Pd films by means of ab-initio calculations [58]. The exceptional change of the magnitude of the magnetic moment, which is about two orders of magnitude larger than the other ones, stems from the proximity of the TP and it can be ascribed to the large charge transfer observed in the undoped structure, as described in the Appendix (Figure A2).

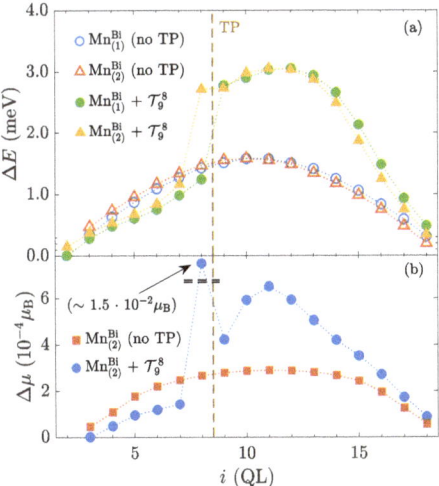

Figure 8. Magnetic doping of one layer inside the Bi_2Se_3 multilayer. (**a**) Relative formation energy of Mn^{Bi} substition defects as a function of the doped QL i. $Mn^{Bi}_{(1)}$ and $Mn^{Bi}_{(2)}$ stand for substitutions at distinct Bi sites. $Mn^{Bi}_{(1)}$ faces neighbouring QL with lower x. Dependencies with and without TP are depicted. For clarity, they are shifted to fit at the end points. (**b**) Distribution of magnitudes of Mn^{Bi} magnetic moments as a function of the position of the substitution i. Only one Bi site labeled by the index i of the appropriate QL is doped by 1% of Mn.

The described interplay of TPs and magnetic defects could explain the energetic gain observed for gathered TPs in a magnetic material (Figure 5). Close TPs lead to a smaller perturbation of the whole electronic structure. This might be deduced from the distribution of calculated magnitudes of magnetic moments in a homogeneously doped multilayer as a function of the positions of incorporated TPs (Figure 9a). We see that the closer TPs are, the smaller the overall variation of magnitudes of magnetic moments is (Figure 9c).

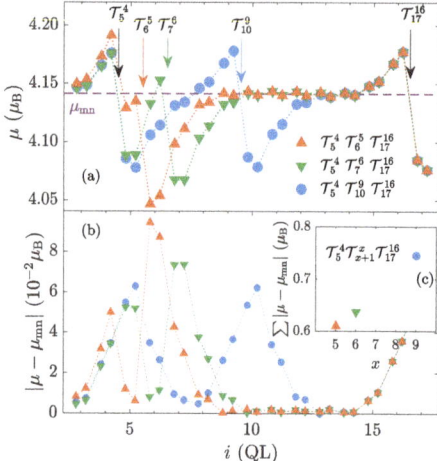

Figure 9. (a) Distributions of magnitudes of magnetic moments in homogeneously magnetically doped Bi$_{1.98}$Mn$_{0.02}$Se$_3$ for various positions of TPs. Mn^{Bi} are labeled by the index i of the appropriate QL. Arrows point to positions of introduced TPs. The outer TPs have fixed positions in all the cases (black arrows). Whereas, the inner one is being moved, the color of the arrow corresponds to the color of the function. (b) The absolute value of the change of the magnetic moment with the respect to the mean moment value μ_{mn}. (c) The sum of the magnetic moment changes from the previous subplot.

3.4. Comparison of Different TP Orientations

In the previous part, we described the simplest case consisting in the identical orientations of three TPs (AAA) (Figure 4). Now, we focus on the influence of nonidentical TPs on the preceding result. To examine it, we invert the orientation of each TP in the former three-TP structures, namely the orientations BAA, ABA, AAB are used (Figure 4), and we calculate the distribution of the formation energy.

We recall that in our approach, we are not able to compare the formation energies of the former identically oriented TPs with the case containing a single TP with the inverted orientation well. However, the mutual comparison of the new structures is feasible (Figure 10). Concerning the undoped structure, one can observe that, for the studied number of TPs, their formation energy strongly depends on the order of TP-type (Figure 10a). Considering an increasing index of QLs, it is evident that the BA order of two TPs, representing TPs at opposite QL sites (Figure 4), is more favorable than the AB one, which stands for TPs at an adjacent QL site. Namely, the BAA order, containing no AB sequence, has the lowest relative formation energy. Besides, it is clearly illustrated at "touching points", where two adjacent TPs, either A or B type, are switching (Figure 10a), and the AB order with the BA one are interchanged (Figure 11). One can assume that the energy difference originates from the mentioned asymmetrical influence of TPs on the surrounding (Figure 8), which differentiate the AB and BA order. One can notice that the type of the TP sequence can be characterized by number of the vdW gaps in between TPs with respect to the system of identical TPs. According to the Figure 11), the AB segment contains an extra vdW gap, whereas the BA segment misses one.

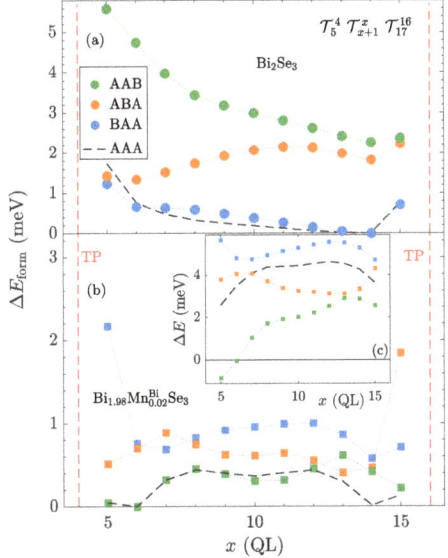

Figure 10. Dependences of the relative formation energies of three TPs on the position of the middle TP x. Different mutual orientations of the TPs are used. The orientation of TPs is labeled by letters A and B. (**a**) undoped structure. (**b**) magnetically doped structure. (**c**) change of the TP formation energy caused by presence of the magnetic defects—$Bi_{1.98}Mn_{0.02}Se_3$ with respect to the undoped case. The relative formation energy curve related to identical TPs (AAA—dashed lines) depicted in panels (**a**,**b**) serves only as a shape reference.

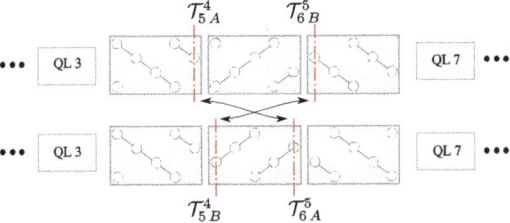

Figure 11. Sketch of the interchange of the AB TP order with the BA one in the case of gathered TPs.

Except the case of the inverted middle TP (ABA), the calculated energy curves qualitatively resemble the relative formation energy curve of identical TPs (Figure 10a). Namely, the BAA and AAA are nearly the same. The observed disfavor of the AB order likely gives rise to a slightly higher slope of the AAB energy curve in comparison to the case of identical TPs. The relative increase of the formation energy might be ascribed to elongation of the segment between A- and B-typed TPs. The shape of ABA formation energy curve can be explained in the similar way. There occurs a local maximum of the relative formation energy as a function of the position of the middle TPs. It likely originates from a complex interplay arising from the occurrence of two diametrically opposite inter-TP segments AB and BA, while their length is modified.

Considering the symmetry of the Bi_2Se_3 slab placed into the vacuum, where the BAA order is equivalent to the BBA one by a side inversion, one might assume that such immediate alternation of the TP types (ABA) is unlikely based on the calculated formation energies. Hence, it appears that the role of the TPs orientation can be regarded as marginal concerning the distribution of TPs in the undoped structure.

The relative formation energies dramatically change in the presence of the magnetic defects, similarly to the case of the identical TPs (Figure 10b). Although the energy curves are modified by the presence of distinct TPs, the local energy minima belonging to gathered TPs are kept. The presence of magnetic dopants reorders the formation energies according to the variation of the TPs orientation. It is likely related to the described interplay of TPs and the magnetic dopants (Figure 9). We show that the formation energy of TPs still grows with increased amount of the magnetic dopants, nearly irrespective of the TPs positions (compare Figure 10c and Figure 6). The existing exceptions originate from the special order of TPs, where TPs are more favorable under magnetic doping (Figure 10b). Besides, one should be aware that the calculated curves (Figure 10c) are influenced by vacuum interface-induced effects similar to the identical TPs (Figure 5). Finally, one can conclude that the TPs orientation does not cause significant qualitative changes in the TPs behavior, even in the case of magnetic disorder, as the lowest energy case almost mimics the behavior of the system with identical TPs.

3.5. Surface States

Conductive Dirac surface states are one of the most interesting properties of TIs. The appearance of TPs can strongly influence their presence, since the mirroring of the structure symmetry could represent a boundary in the structure. Hence, in this paper, we also try to simulate the influence of the presence of TP and its position on the surface states. We calculated Bloch spectral functions (BSF) in the vicinity of the Γ point, where the Dirac cone exists, on the path between high symmetrical reciprocal points M and K. In order to study the band gap and surface states, we project BSFs along the mentioned $K - \Gamma - M$ path to the energy-intensity plane in a way that the maximal intensity of the BSF over the k-path is selected for particular energy points. Then, a formation of the Dirac states is indicated by the vanishing of the energy gap and an occurrence of a strikingly high intensity at the Dirac point, where the surface states intersects (Figure 12a). The projected BSFs (PBSF) reveal that the presence of a TP in at a certain distance from the surface breaks the surface Dirac states, which exist in the unperturbed structure (Figure 12a). Our calculations showed that, for TPs which are closer than 6 QLs to the surface, a gap opens, as seen most clearly in the presence of T_4^3 (Figure 12a).

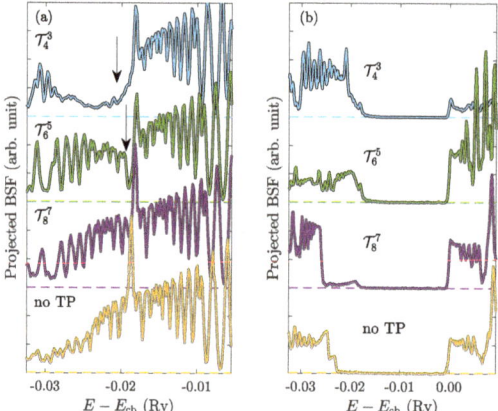

Figure 12. Projected BSFs of pure Bi_2Se_3 in the vicinity of Γ point as a function of the position of TP. Spin up and spin down channels are overlaped. Obtained surface gaps are denoted by arrows. (**a**) PBSF of the surface QL , (**b**) PBSF of the fifth QL from the surface. Energy axes are scaled to the position of conduction band edge E_{cb} at the fifth QL.

The oscillations occurring in PBSF dependencies are caused by finite energy- and k-mesh, which prevents obtaining smooth electron bands in terms of the BSF as well as a narrow k-window, which cuts energy bands. Energy scales are related to the position of the well defined conduction band edge (E_{cb}) at an inner QL (Figure 12b). As was mentioned, the observed canceling of surface states and gap opening likely arise from the proximity of two interfaces, which leads to a destructive interference [6]. Comparing results obtained for TP below the seventh QL with the unperturbed system, we found almost no difference as the Dirac cone is recovered. Similarly, one can mention a modulation of the bulk band gap width in the vicinity of a TP (Figure 12b), where a band gap width changes due to the presence of a boundary.

4. Conclusions

We have studied the behavior of TPs in the pure layered Bi_2Se_3 system, as well as in Bi_2Se_3 under the presence of magnetic and nonmagnetic disorder by first principles calculations. Our results show that interactions between TPs in the pure Bi_2Se_3 become negligible for distances above three QLs. However, for smaller distances, a significant increase of the TP formation energy was observed, in agreement with the experimentally observed spatial separation of TPs in real samples.

The distribution of TPs and their interplay significantly changes in the presence of chemical disorder. The presence of nonmagnetic disorder weakens the influence of TPs on the electron structure, and therefore the interactions between TPs are significantly smaller. However, the occurrence of magnetic defects modified the behavior of TPs significantly. Adjacent TPs become energetically more favorable, which corresponds to the dependence of the relative formation energy of TPs on the concentration of magnetic doping. It reflects a suppression of the TPs formation in magnetically doped structures, unlike nonmagnetic Se^{Bi} antisites. The gathering of TPs leads to a smaller total perturbation of the electron structure and hence might be comprehended as a tendency to TPs annihilation. A thorough analysis indicates that the observed mismatch between the magnetic doping and the presence of TPs consists in the influence of TPs on the spin splitting of magnetic atoms.

On the other hand, the variation of spin splitting, caused by TPs, influences the site preference of magnetic defects. Mn generally does not prefer to occupy sites right at the twin boundary according to our calculations. Such behavior is indicated also in experiments, since no metallic Mn-Mn bonds were observed, although they would probably arise if clustering of Mn at these boundaries was present [54].

Author Contributions: Funding acquisition, J.H.; Investigation, J.Š.; Project administration, K.C. and J.H.; Supervision, K.C.; Visualization, J.Š.; Writing—original draft, J.Š. and K.C.; Writing—review and editing, D.K. and J.H. All authors have read and agreed to the published version of the manuscript.

Funding: This research was funded the Czech Science Foundation (Grant No. 19-13659S).

Acknowledgments: Access to computing and storage facilities owned by parties and projects contributing to the National Grid Infrastructure MetaCentrum provided under the programme "Projects of Large Research, Development, and Innovations Infrastructures" (CESNET LM2015042), is greatly appreciated. This work was supported by The Ministry of Education, Youth and Sports from the Large Infrastructures for Research, Experimental Development and Innovations project "e-Infrastructure CZ – LM2018140".

Conflicts of Interest: The authors declare no conflict of interest.

Appendix A. Single TP Distribution

To check the behavior of a single TP in a pure multilayer structure, and to discuss the vacuum interface caused effect, a distribution of the single TP in a multilayer consisting of 10 QLs is studied. Thus, an asymmetric dependence of the relative TP formation energy as the function of the TP position was obtained (Figure A1a). According to the Figure 2, a TP occurs at Se sites in the vicinity of vdW gap. Generally, one is used to dividing a system to twin domains adjacent to the twin boundary. However, according to the structure composed of QLs, we hypothetically split the present system into two domains separated by a vdW gap. It seems to be more convenient to deal with entire QLs in energy comparisons. Nevertheless, one has to be aware that one of the domains contains a mirror layer. Since we used layer stacking according to Figure 2, a TP is located in the domain bearing smaller QL indices. The relative formation energy (Figure A1a) bears a monotonous dependence, which favors a width maximization of the domain, where the mirror layer belongs. It suggests a significant interplay of the TP with the surface interfaces, which depends on the orientation of the TP.

Naturally, the occurrence of a TP in the structure causes a charge transfer compared to the unperturbed system. Calculations show (Figure A2) that the electron density changes primarily in the QL possessing a mirror plane of the Bi_2Se_3 layer stacking. Especially, the charge is depleted from the vdW gap and it flows to the Bi layer adjacent to the TP located at Se sites. Besides, a TP causes charge oscillations, which spread out of the TP. Evaluating the absolute charge transfer $\sum |\Delta \rho_t|$ per a domain based on the charge transfer distribution, belonging to the system with symmetrical domain sizes (Figure A2); one finds that for the present way of stacking (Figure 2) the charge modulation is larger for the domain, consisting of QLs with smaller indices. The difference is about one order of magnitude, which implies a different impact on domains. Having also considered vacuum interfaces, which represent another structure fault, the system naturally should tend to a suppression of the energetically more demanding perturbations by their separation (Figure A1a).

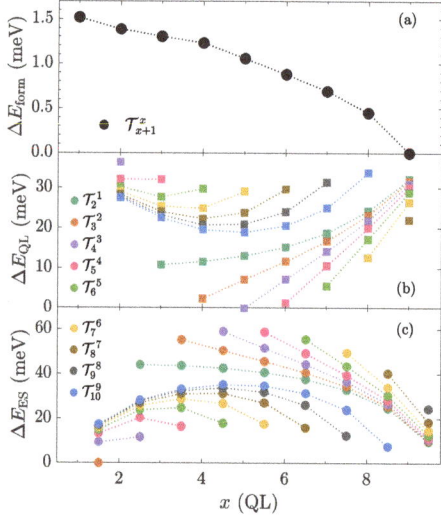

Figure A1. (a) Single TP relative formation energy ΔE_form as a function of the TP position x in the Bi_2Se_3 multilayer. (b) Relative energy contribution ΔE_QL of the particular QL x. (c) Relative energy contribution of ES as a function of their position. Bi_2Se_3 multilayer consists of 10 QLs. Concerning plots (b) and (c), dependencies belonging to several locations of TPs are depicted, where data points in the vicinity of TPs are excluded for clarity.

Let us consider QLs and ESs separately and evaluate their relative total energy contributions. One observes that a clear discrepancy occurs between the two presented domains, regardless of the TP position. Relative magnitudes of total energies ΔE_QL related to particular QLs x as a function of the position of the TP display that the domain containing the TP, namely QLs with the smaller indices, possesses higher ΔE_QL and the closer a TP is to vacuum, the higher the magnitude of ΔE_QL. For brevity, we exclude QLs in the vicinity of the TP, since their relative energy changes are of a different scale. Similarly, relative energy contributions of ESs (ΔE_ES) are also influenced by the position of the TP (Figure A1c). However, they follow an opposite evolution compared to the energy contribution of QLs, likely because of an opposite impact of the stacking fault. The charge transfer of the related QLs and ESs differs in the sign. The formation energy curve (Figure A1a) includes both contributions, ΔE_QL and ΔE_ES. However, we observe that the shape of the formation energy curve ΔE_form (Figure A1a) is mostly determined by ES's contribution ΔE_ES (Figure A1c)

It is worth studying the influence of a single TP on surrounding atomic layers in a pristine structure, as it describes the bare effect of TP. It shows that the perturbation caused by a TP is hardly local. Dependencies of the charge transfer (Figure A2) or the QL resolved total energy (Figure A1c) indicate that the charge as well as energy modulation spread over few QL. Moreover, a strong interplay with the Bi_2Se_3 boundaries is visible, as it likely yields the shape of the E_{form} dependence (Figure A1a).

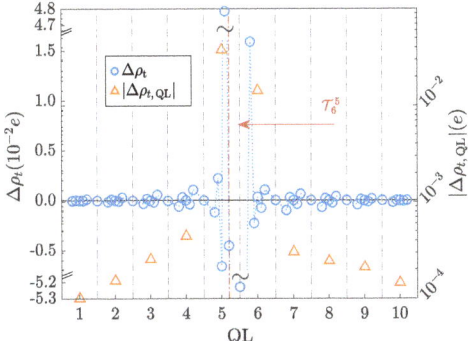

Figure A2. (left axis) Distribution of the charge transfer $\Delta\rho_t$ caused by the presence of a twin plane ($\mathcal{T}_6^5$) in the multilayer composed of 10 Bi_2Se_3 QLs. Each data point denotes a particular atomic layer or an ES representing the vdW gap. (right axis) Dependence of the variance of the charge transfer at particular QLs. Dashed lines separate QLs and denote position of the vdW gap.

References

1. Bansil, A.; Lin, H.; Das, T. Colloquium: Topological band theory. *Rev. Mod. Phys.* **2016**, *88*, 021004. [CrossRef]
2. Xia, Y.; Qian, D.; Hsieh, D.; Wray, L.; Pal, A.; Lin, H.; Bansil, A.; Grauer, D.; Hor, Y.S.; Cava, R.J.; et al. Observation of a large-gap topological-insulator class with a single Dirac cone on the surface. *Nat. Phys.* **2009**, *5*, 398–402. [CrossRef]
3. Qi, X.L.; Zhang, S.C. Topological insulators and superconductors. *Rev. Mod. Phys.* **2011**, *83*, 1057–1110. [CrossRef]
4. Hsieh, D.; Xia, Y.; Qian, D.; Wray, L.; Meier, F.; Dil, J.H.; Osterwalder, J.; Patthey, L.; Fedorov, A.V.; Lin, H.; et al. Observation of Time-Reversal-Protected Single-Dirac-Cone Topological-Insulator States in Bi_2Te_3 and Sb_2Te_3. *Phys. Rev. Lett.* **2009**, *103*, 146401. [CrossRef] [PubMed]
5. Hasan, M.Z.; Kane, C.L. Colloquium: Topological insulators. *Rev. Mod. Phys.* **2010**, *82*, 3045–3067. [CrossRef]
6. Zhang, Y.; He, K.; Chang, C.Z.; Song, C.L.; Wang, L.L.; Chen, X.; Jia, J.F.; Fang, Z.; Dai, X.; Shan, W.Y.; et al. Crossover of the three-dimensional topological insulator Bi_2Se_3 to the two-dimensional limit. *Nat. Phys.* **2010**, *6*, 584–588. [CrossRef]
7. Cayssol, J. Introduction to Dirac materials and topological insulators. *C. R. Phys.* **2013**, *14*, 760–778. [CrossRef]
8. Zhang, H.; Liu, C.X.; Qi, X.L.; Dai, X.; Fang, Z.; Zhang, S.C. Topological insulators in Bi_2Se_3, Bi_2Te_3 and Sb_2Te_3 with a single Dirac cone on the surface. *Nat. Phys.* **2009**, *5*, 438–442. [CrossRef]
9. Freitas, R.R.Q.; de Brito Mota, F.; Rivelino, R.; de Castilho, C.M.C.; Kakanakova-Georgieva, A.; Gueorguiev, G.K. Tuning band inversion symmetry of buckled III-Bi sheets by halogenation. *Nanotechnology* **2016**, *27*, 055704. [CrossRef]
10. Bernevig, B.; Hughes, T. *Topological Insulators and Topological Superconductors*; Princeton University Press: Princeton, NJ, USA, 2013.
11. Ortmann, F.; Roche, S.; Valenzuela, S. *Topological Insulators: Fundamentals and Perspectives*; Wiley-VCH: Weinheim, Germany, 2015.
12. König, M.; Wiedmann, S.; Brüne, C.; Roth, A.; Buhmann, H.; Molenkamp, L.W.; Qi, X.L.; Zhang, S.C. Quantum Spin Hall Insulator State in HgTe Quantum Wells. *Science* **2007**, *318*, 766–770. [CrossRef] [PubMed]

13. Wray, L.A.; Xu, S.Y.; Xia, Y.; Hsieh, D.; Fedorov, A.V.; Hor, Y.S.; Cava, R.J.; Bansil, A.; Lin, H.; Hasan, M.Z. A topological insulator surface under strong Coulomb, magnetic and disorder perturbations. *Nat. Phys.* **2010**, *7*, 32–37. [CrossRef]
14. Xu, S.Y.; Neupane, M.; Liu, C.; Zhang, D.; Richardella, A.; Andrew Wray, L.; Alidoust, N.; Leandersson, M.; Balasubramanian, T.; Sánchez-Barriga, J.; et al. Hedgehog spin texture and Berry's phase tuning in a magnetic topological insulator. *Nat. Phys.* **2012**, *8*, 616–622. [CrossRef]
15. Chen, Y.L.; Chu, J.H.; Analytis, J.G.; Liu, Z.K.; Igarashi, K.; Kuo, H.H.; Qi, X.L.; Mo, S.K.; Moore, R.G.; Lu, D.H.; et al. Massive Dirac Fermion on the Surface of a Magnetically Doped Topological Insulator. *Science* **2010**, *329*, 659–662. [CrossRef] [PubMed]
16. Chang, C.Z.; Zhang, J.; Liu, M.; Zhang, Z.; Feng, X.; Li, K.; Wang, L.L.; Chen, X.; Dai, X.; Fang, Z.; et al. Thin Films of Magnetically Doped Topological Insulator with Carrier-Independent Long-Range Ferromagnetic Order. *Adv. Mater.* **2013**, *25*, 1065–1070. [CrossRef] [PubMed]
17. Carva, K.; Kudrnovský, J.; Máca, F.; Drchal, V.; Turek, I.; Baláž, P.; Tkáč, V.; Holý, V.; Sechovský, V.; Honolka, J. Electronic and transport properties of the Mn-doped topological insulator Bi_2Te_3: A first-principles study. *Phys. Rev. B* **2016**, *93*, 214409. [CrossRef]
18. Máca, F.; Kudrnovský, J.; Baláž, P.; Drchal, V.; Carva, K.; Turek, I. Tetragonal CuMnAs alloy: Role of defects. *J. Magn. Magn. Mater.* **2019**, *474*, 467–471. [CrossRef]
19. Zhang, W.; Yu, R.; Zhang, H.J.; Dai, X.; Fang, Z. First-principles studies of the three-dimensional strong topological insulators Bi_2Te_3, Bi_2Se_3 and Sb_2Te_3. *New J. Phys.* **2010**, *12*, 065013. [CrossRef]
20. Lee, Y.; Punugupati, S.; Wu, F.; Jin, Z.; Narayan, J.; Schwartz, J. Evidence for topological surface states in epitaxial Bi_2Se_3 thin film grown by pulsed laser deposition through magneto-transport measurements. *Curr. Opin. Solid State Mater. Sci.* **2014**, *18*, 279–285. [CrossRef]
21. Kriegner, D.; Harcuba, P.; Veselý, J.; Lesnik, A.; Bauer, G.; Springholz, G.; Holý, V. Twin domain imaging in topological insulator Bi_2Te_3 and Bi_2Se_3 epitaxial thin films by scanning X-ray nanobeam microscopy and electron backscatter diffraction. *J. Appl. Crystallogr.* **2017**, *50*, 369–377. [CrossRef]
22. Eremeev, S.V.; Vergniory, M.G.; Menshchikova, T.V.; Shaposhnikov, A.A.; Chulkov, E.V. The effect of van der Waal's gap expansions on the surface electronic structure of layered topological insulators. *New J. Phys.* **2012**, *14*, 113030. [CrossRef]
23. Zhang, J.M.; Ming, W.; Huang, Z.; Liu, G.B.; Kou, X.; Fan, Y.; Wang, K.L.; Yao, Y. Stability, electronic, and magnetic properties of the magnetically doped topological insulators Bi_2Se_3, Bi_2Te_3, and Sb_2Te_3. *Phys. Rev. B* **2013**, *88*, 235131. [CrossRef]
24. Hor, Y.S.; Roushan, P.; Beidenkopf, H.; Seo, J.; Qu, D.; Checkelsky, J.G.; Wray, L.A.; Hsieh, D.; Xia, Y.; Xu, S.Y.; et al. Development of ferromagnetism in the doped topological insulator $Bi_{2-x}Mn_xTe_3$. *Phys. Rev. B* **2010**, *81*, 195203. [CrossRef]
25. Zhang, J.M.; Zhu, W.; Zhang, Y.; Xiao, D.; Yao, Y. Tailoring Magnetic Doping in the Topological Insulator Bi_2Se_3. *Phys. Rev. Lett.* **2012**, *109*, 266405. [CrossRef] [PubMed]
26. Ghasemi, A.; Kepaptsoglou, D.; Figueroa, A.I.; Naydenov, G.A.; Hasnip, P.J.; Probert, M.I.J.; Ramasse, Q.; van der Laan, G.; Hesjedal, T.; Lazarov, V.K. Experimental and density functional study of Mn doped Bi_2Te_3 topological insulator. *APL Mater.* **2016**, *4*, 126103. [CrossRef]
27. Figueroa, A.I.; van der Laan, G.; Collins-McIntyre, L.J.; Cibin, G.; Dent, A.J.; Hesjedal, T. Local Structure and Bonding of Transition Metal Dopants in Bi_2Se_3 Topological Insulator Thin Films. *J. Phys. Chem. C* **2015**, *119*, 17344–17351. [CrossRef]
28. Rienks, E.D.L.; Wimmer, S.; Sánchez-Barriga, J.; Caha, O.; Mandal, P.S.; Ruzicka, J.; Ney, A.; Steiner, H.; Volobuev, V.V.; Groiss, H.; et al. Large magnetic gap at the Dirac point in $Bi_2Te_3/MnBi_2Te_4$ heterostructures. *Nature* **2019**, *576*, 423–428. [CrossRef]
29. Růžička, J.; Caha, O.; Holý, V.; Steiner, H.; Volobuiev, V.; Ney, A.; Bauer, G.; Duchoň, T.; Veltruská, K.; Khalakhan, I.; et al. Structural and electronic properties of manganese-doped Bi_2Te_3 epitaxial layers. *New J. Phys.* **2015**, *17*, 013028. [CrossRef]
30. Carva, K.; Baláž, P.; Šebesta, J.; Turek, I.; Kudrnovský, J.; Máca, F.; Drchal, V.; Chico, J.; Sechovský, V.; Honolka, J. Magnetic properties of Mn-doped Bi_2Se_3 topological insulators: Ab initio calculations. *Phys. Rev. B* **2020**, *101*, 054428. [CrossRef]
31. Ptok, A.; Kapcia, K.J.; Ciechan, A. Electronic properties of Bi_2Se_3 dopped by 3d transition metal (Mn, Fe, Co, or Ni) ions. *J. Phys. Condens. Matter* **2020**. [CrossRef]

32. Wei, X.; Zhang, J.; Zhao, B.; Zhu, Y.; Yang, Z. Ferromagnetism in Fe-doped Bi$_2$Se$_3$ topological insulators with Se vacancies. *Phys. Lett. A* **2015**, *379*, 417–420. [CrossRef]
33. Hor, Y.S.; Richardella, A.; Roushan, P.; Xia, Y.; Checkelsky, J.G.; Yazdani, A.; Hasan, M.Z.; Ong, N.P.; Cava, R.J. p-type Bi$_2$Se$_3$ for topological insulator and low-temperature thermoelectric applications. *Phys. Rev. B* **2009**, *79*, 195208. [CrossRef]
34. Scanlon, D.O.; King, P.D.C.; Singh, R.P.; de la Torre, A.; Walker, S.M.; Balakrishnan, G.; Baumberger, F.; Catlow, C.R.A. Controlling Bulk Conductivity in Topological Insulators: Key Role of Anti-Site Defects. *Adv. Mater.* **2012**, *24*, 2154–2158. [CrossRef] [PubMed]
35. Wolos, A.; Drabinska, A.; Borysiuk, J.; Sobczak, K.; Kaminska, M.; Hruban, A.; Strzelecka, S.G.; Materna, A.; Piersa, M.; Romaniec, M.; et al. High-spin configuration of Mn in Bi$_2$Se$_3$ three-dimensional topological insulator. *J. Magn. Magn. Mater.* **2016**, *419*, 301–308. [CrossRef]
36. Huang, F.T.; Chu, M.W.; Kung, H.H.; Lee, W.L.; Sankar, R.; Liou, S.C.; Wu, K.K.; Kuo, Y.K.; Chou, F.C. Nonstoichiometric doping and Bi antisite defect in single crystal Bi$_2$Se$_3$. *Phys. Rev. B* **2012**, *86*, 081104. [CrossRef]
37. Miao, L.; Xu, Y.; Zhang, W.; Older, D.; Breitweiser, S.A.; Kotta, E.; He, H.; Suzuki, T.; Denlinger, J.D.; Biswas, R.R.; et al. Observation of a topological insulator Dirac cone reshaped by non-magnetic impurity resonance. *NPJ Quantum Mater.* **2018**, *3*, 29. [CrossRef]
38. Sánchez-Barriga, J.; Varykhalov, A.; Springholz, G.; Steiner, H.; Kirchschlager, R.; Bauer, G.; Caha, O.; Schierle, E.; Weschke, E.; Ünal, A.A.; et al. Nonmagnetic band gap at the Dirac point of the magnetic topological insulator (Bi$_{(1-x)}$Mn$_x$)$_2$Se$_3$. *Nat. Commun.* **2016**, *7*, 10559. [CrossRef]
39. Medlin, D.L.; Yang, N.Y.C. Interfacial Step Structure at a (0001) Basal Twin in Bi$_2$Te$_3$. *J. Electron. Mater.* **2012**, *41*, 1456–1464. [CrossRef]
40. Medlin, D.L.; Ramasse, Q.M.; Spataru, C.D.; Yang, N.Y.C. Structure of the (0001) basal twin boundary in Bi$_2$Te$_3$. *J. Appl. Phys.* **2010**, *108*, 043517. [CrossRef]
41. Tarakina, N.V.; Schreyeck, S.; Luysberg, M.; Grauer, S.; Schumacher, C.; Karczewski, G.; Brunner, K.; Gould, C.; Buhmann, H.; Dunin-Borkowski, R.E.; et al. Suppressing Twin Formation in Bi$_2$Se$_3$ Thin Films. *Adv. Mater. Interfaces* **2014**, *1*, 1400134. [CrossRef]
42. Levy, I.; Garcia, T.A.; Shafique, S.; Tamargo, M.C. Reduced twinning and surface roughness of Bi$_2$Se$_3$ and Bi$_2$Te$_3$ layers grown by molecular beam epitaxy on sapphire substrates. *J. Vac. Sci. Technol. B* **2018**, *36*, 02D107. [CrossRef]
43. Aramberri, H.; Cerdá, J.I.; Muñoz, M.C. Tunable Dirac Electron and Hole Self-Doping of Topological Insulators Induced by Stacking Defects. *Nano Lett.* **2015**, *15*, 3840–3844. [CrossRef] [PubMed]
44. Skriver, H.L. *The LMTO Method: Muffin-Tin Orbitals and Electronic Structure*; Springer: Berlin, Germnay, 2012.
45. Turek, I.; Drchal, V.; Kudrnovsky, J.; Sob, M.; Weinberger, P. *Electronic Structure of Disordered Alloys, Surfaces and Interfaces*; Kluwer: Boston, MA, USA, 1997.
46. Vosko, S.H.; Wilk, L.; Nusair, M. Accurate spin-dependent electron liquid correlation energies for local spin density calculations: A critical analysis. *Can. J. Phys.* **1980**, *58*, 1200–1211. [CrossRef]
47. Freitas, R.R.Q.; de Brito Mota, F.; Rivelino, R.; de Castilho, C.M.C.; Kakanakova-Georgieva, A.; Gueorguiev, G.K. Spin-orbit-induced gap modification in buckled honeycomb XBi and XBi$_3$ (X = B, Al, Ga, and In) sheets. *J. Phys. Condens. Matter* **2015**, *27*, 485306. [CrossRef] [PubMed]
48. Korzhavyi, P.A.; Ruban, A.V.; Abrikosov, I.A.; Skriver, H.L. Madelung energy for random metallic alloys in the coherent potential approximation. *Phys. Rev. B* **1995**, *51*, 5773–5780. [CrossRef]
49. Velický, B.; Kirkpatrick, S.; Ehrenreich, H. Single-Site Approximations in the Electronic Theory of Simple Binary Alloys. *Phys. Rev.* **1968**, *175*, 747–766. [CrossRef]
50. Kudrnovský, J.; Drchal, V.; Blaas, C.; Weinberger, P.; Turek, I.; Bruno, P. Ab initio theory of perpendicular magnetotransport in metallic multilayers. *Phys. Rev. B* **2000**, *62*, 15084–15095. [CrossRef]
51. Turek, I.; Kudrnovský, J.; Šob, M.; Drchal, V.; Weinberger, P. Ferromagnetism of Imperfect Ultrathin Ru and Rh Films on a Ag(001) Substrate. *Phys. Rev. Lett.* **1995**, *74*, 2551–2554. [CrossRef]
52. Kudrnovský, J.; Drchal, V.; Turek, I.; Dederichs, P.; Weinberger, P.; Bruno, P. Ab initio theory of perpendicular transport in layered magnetic systems. *J. Magnetism Magn. Mater.* **2002**, *240*, 177–179. [CrossRef]
53. Spedding, F.H.; Daane, A.H.; Herrmann, K.W. The crystal structures and lattice parameters of high-purity scandium, yttrium and the rare earth metals. *Acta Crystallogr.* **1956**, *9*, 559–563. [CrossRef]

54. Vališka, M.; Warmuth, J.; Michiardi, M.; Vondráček, M.; Ngankeu, A.S.; Holý, V.; Sechovský, V.; Springholz, G.; Bianchi, M.; Wiebe, J.; et al. Topological insulator homojunctions including magnetic layers: The example of n-p type (n-QLs Bi_2Se_3/Mn-Bi_2Se_3) heterostructures. *Appl. Phys. Lett.* **2016**, *108*, 262402. [CrossRef]
55. Liechtenstein, A.; Katsnelson, M.; Antropov, V.; Gubanov, V. Local spin density functional approach to the theory of exchange interactions in ferromagnetic metals and alloys. *J. Magn. Magn. Mater.* **1987**, *67*, 65–74. [CrossRef]
56. Turek, I.; Kudrnovský, J.; Drchal, V.; Bruno, P. Exchange interactions, spin waves, and transition temperatures in itinerant magnets. *Philos. Mag.* **2006**, *86*, 1713–1752. [CrossRef]
57. Polyakov, A.; Meyerheim, H.L.; Crozier, E.D.; Gordon, R.A.; Mohseni, K.; Roy, S.; Ernst, A.; Vergniory, M.G.; Zubizarreta, X.; Otrokov, M.M.; et al. Surface alloying and iron selenide formation in Fe/Bi_2Se_3(0001) observed by x-ray absorption fine structure experiments. *Phys. Rev. B* **2015**, *92*, 045423. [CrossRef]
58. Alexandre, S.S.; Anglada, E.; Soler, J.M.; Yndurain, F. Magnetism of two-dimensional defects in Pd: Stacking faults, twin boundaries, and surfaces. *Phys. Rev. B* **2006**, *74*, 054405. [CrossRef]

Publisher's Note: MDPI stays neutral with regard to jurisdictional claims in published maps and institutional affiliations.

© 2020 by the authors. Licensee MDPI, Basel, Switzerland. This article is an open access article distributed under the terms and conditions of the Creative Commons Attribution (CC BY) license (http://creativecommons.org/licenses/by/4.0/).

Article

Theoretical Investigation of Azobenzene-Based Photochromic Dyes for Dye-Sensitized Solar Cells

Md Al Mamunur Rashid [1], Dini Hayati [1], Kyungwon Kwak [2,*] and Jongin Hong [1,*]

1. Department of Chemistry, Chung-Ang University, Seoul 06974, Korea; ndcmamun@korea.ac.kr (M.A.M.R.); dinihayati300194@gmail.com (D.H.)
2. Center for Molecular Spectroscopy and Dynamics, Institute for Basic Science (IBS) & Department of Chemistry, Korea University, Seoul 02841, Korea
* Correspondence: kkwak@korea.ac.kr (K.K.); hongj@cau.ac.kr (J.H.); Tel.: +82-2-820-5869 (J.H)

Received: 26 February 2020; Accepted: 16 April 2020; Published: 9 May 2020

Abstract: Two donor-π-spacer-acceptor (D-π-A) organic dyes were designed as photochromic dyes with the same π-spacer and acceptor but different donors, based on their electron-donating strength. Various structural, electronic, and optical properties, chemical reactivity parameters, and certain crucial factors that affect short-circuit current density (J_{sc}) and open circuit voltage (V_{oc}) were investigated computationally using density functional theory and time-dependent density functional theory. The *trans-cis* isomerization of these azobenzene-based dyes and its effect on their properties was studied in detail. Furthermore, the dye-(TiO$_2$)$_9$ anatase nanoparticle system was simulated to understand the electronic structure of the interface. Based on the results, we justified how the *trans-cis* isomerization and different donor groups influence the physical properties as well as the photovoltaic performance of the resultant dye-sensitized solar cells (DSSCs). These theoretical calculations can be used for the rapid screening of promising dyes and their optimization for photochromic DSSCs.

Keywords: dye-sensitized solar cells; azobenzene; density functional theory

1. Introduction

To meet the ever-increasing global energy demands, the utilization of solar energy—a clean, renewable, and naturally abundant energy resource—has attracted considerable attention in recent decades. Accordingly, photovoltaic devices (or solar cells) have been extensively developed to meet this energy demand. Dye-sensitized solar cells (DSSCs) have been widely investigated as a promising candidate for low-cost photovoltaic cells in the past two decades because of their distinctive features, including shape flexibility, transparency, better performance under prolonged low-light conditions, thermal dual stress, different solar incident angles, easy material synthesis, low weight, and cost-effectiveness. Moreover, new functional materials have been designed to increase the solar-to-electrical energy conversion efficiency of DSSCs [1,2]. In the public sector, DSSCs are used in flat and curved building skins for building-integrated photovoltaics because of their transparency and aesthetic value. Although numerous studies have been conducted based on device physics, material innovation, and commercialization to achieve high performance and long-term fidelity of DSSCs [3], they are still deficient in various aspects.

The photosensitizer is the core of a DSSC that absorbs solar radiation over a broad spectral range. Moreover, it contains functional groups, which aid in adsorption on the TiO$_2$ surface and injection of electrons into the conduction band (CB) of TiO$_2$ after solar light excitation. Organic dyes are attracting increased attention not only as alternative photosensitizers, but also as promising photofunctional materials for optical devices and photovoltaic cells because of their low cost, environment friendliness, and high molecular extinction coefficients [4]. Metal-free organic dyes, which commonly feature a push-pull architecture like dipolar donor–π-bridge–acceptor (D–π–A) frameworks, are being studied for use in DSSCs more than Ru-based dyes. This is because metal-free organic dyes have attractive attributes, such as efficient intramolecular charge transfer (ICT), a wider variety of structural designs, easy fabrication, raw material abundance, various synthetic protocols, good flexibility for molecular tailoring, tunable spectral properties, high efficiency, cost-effectiveness, and applicability as organic optoelectronic materials [5,6]; consequently, their commercial application is promising. Because of these features, recent research has focused on designing new metal-free organic dyes to further improve the performance of DSSCs. In the D–π–A structure, the donor unit plays an important role in not only tuning and modifying the absorption spectra but also controlling the molecular energy levels and intramolecular charge separation. Thus, several studies have been conducted to investigate the effect of changing the donor units on the absorption characteristics of the dyes and DSSC performance [7,8]. Although triphenylamine, dialkylamine, and diphenylamine moieties are commonly used as electron donors [4,9], only a few studies have systematically investigated the molecular origin of the DSSC performance modulated by these donor groups.

Azobenzene dyes are organic compounds that contain the photoreactive -N=N- group, which undergoes reversible *trans-cis-trans* isomerization when irradiated by sunlight. Therefore, these compounds are used in photoresponsive material systems as phototriggers [10]. Azobenzene photochemistry has also been observed in numerous constricted and/or interfacial environments, such as molecular or liquid crystals for molecular level photoswitching, or embedded within cyclodextrins, polymers, and metal-organic frameworks [11,12]. Recently, D–π–A-type azobenzene derivatives have generated considerable interest because of the presence of both, electron-donating and electron-accepting groups, on the π-conjugated system of the azo chromophore. Several studies have been conducted on conjugated π-spacers, such as acetylene, vinyl, and phenyl [8,13]. However, in metal-free organic dyes, the effect of using azobenzene as a π-spacer in the D–π–A structure has not been widely studied; examples of the effective inclusion of azobenzene dyes into DSSCs are rare, and the correlation between the molecular arrangement of these dyes and DSSC properties has not been studied extensively [9].

Quantum chemical methods have been employed in recent decades as a sustainable approach for elucidating the relationship between molecular geometries and dye characteristics, thus offering a reliable theoretical platform for the rapid screening of efficient dyes prior to expensive and time-consuming syntheses. Density functional theory (DFT) and time-dependent density functional theory (TDDFT) have been extensively used to investigate the electronic and optical properties of virtual photosensitizers in the ground and excited states for the development of DSSCs [14,15]. Therefore, the theoretical predictions based on DFT calculations are promising, as they correlate well with the experimental data on DSSCs [16]. Numerous research groups have successfully calculated the photoelectric properties of organic dyes using quantum chemical methods. Donor modifications can improve the light-harvesting efficiency (LHE) and electron injection ability, which contribute to the solar cell efficiency [8]. The use of a bulky donor moiety leads to a high open circuit voltage, longer electron lifetime, and slower back-transfer of electrons, resulting in higher photovoltaic performance [17]. The role of donor moieties in the photoinjection mechanism has also been investigated for a series of D–π–A-structured dyes adsorbed on a $(TiO_2)_{15}$ anatase cluster in the DFT framework using various functionals [18]. Novir et al. investigated the properties of numerous azobenzene-based dyes with different electron-donating groups and reported that the donor groups did not have any significant effect on their optical properties, such as LHE and exciton binding energy [19].

In this study, two photochromic azobenzene-based dyes were selected as sensitizers to investigate the various properties of DSSCs to determine the relationship between the molecular structure and photoelectric properties using reliable quantum chemical calculation methods. The objective of this study was to understand the effect of different donor groups (dimethylamine and diphenylamine) on the photophysical properties of the two azo dyes and the photovoltaic performances of the resultant DSSCs. For in-depth analysis via DFT and TDDFT, the structural, electronic, optical properties, including chemical reactivity parameters and some crucial factor relating to short circuit current density (J_{SC}) and open circuit voltage (V_{OC}) of the two dyes were determined after their adsorption on a TiO$_2$ surface. The elaborate DFT analyses presented herein can provide a better understanding of the photoelectrical properties of the two azo dyes for photochromic DSSCs.

2. Methods

The ground-state geometries of all the dyes before and after binding onto the TiO$_2$ surface were fully optimized using N,N-dimethylformamide (DMF) solvent ($\varepsilon = 37.5$) without symmetry constriction. Frequency calculations were performed to confirm that all the optimized geometries were stationary minima points. The calculations were carried out using DFT at the B3LYP level with the 6-311G(d,p) basis set for C, H, O, and N atoms and the LANL2DZ basis set for the Ti atom [20], considering the relativistic effect of heavy atoms. The excitation energies, oscillator strengths, and UV-Visible absorption spectra of all the dyes before and after binding to TiO$_2$ in the DMF solvent were simulated using TDDFT with CAM-B3LYP [21] functionals and the 6-311++G(d,p) basis set for non-metal atoms, and the LANL2DZ basis set for the Ti atom on the basis of the optimized ground-state geometries. The effective core potential (ECP) for sixty valence electrons of the dyes adsorbed on the TiO$_2$ surface was applied for the DFT and TDDFT calculations. The conductor-like polarized continuum model (C-PCM) method [22] was applied within the self-consistent reaction field theory to simulate the solvent effects throughout the study. Natural bond orbital (NBO) analysis was performed by calculating the orbital populations for the ground state and excited state using the NBO 5.0 program [23]. All calculations were performed using the Gaussian 16 package [24].

3. Results and Discussion

3.1. Isolated Dyes and Dye/TiO$_2$ Complexes

In this study, two D–π–A organic dyes were designed containing two electron-donating moieties, namely, dimethylamine and diphenylamine, an azobenzene-benzene moiety as the π-spacer, and cyanoacrylic acid as the anchoring group, as shown in Figure 1. The azo group, which showed reversible *cis-trans* photoisomerization and allowed geometrical change of the π-conjugation backbone under light and heat, led to the *trans* and *cis* structures of the two studied dyes (Figure 1b). In this study, the *trans* structures are named E-DMAC and E-DPAC and the *cis* structures are named Z-DMAC and Z-DPAC. Here, DMAC and DPAC contained methyl and phenyl moieties in their donor moieties, respectively. To provide more realistic information about the dye adsorption on the semiconductor surface in terms of electronic structure and optical properties, the dyes adsorbed on the TiO$_2$ surface were also studied and are referred to as dye/TiO$_2$ in this study. Figure 1c shows the optimized structures of the dye/TiO$_2$ complexes for both dyes. In the dye/TiO$_2$ complexes, the adsorption of dyes through carboxylic acid can occur via either physisorption or chemisorption. The carboxylic acid can bind to the TiO$_2$ surface by several anchoring modes, such as monodentate bridging, bidentate bridging, and bidentate chelating [25,26]. Because of the controversies surrounding the exact anchoring modes for the binding of dyes on TiO$_2$ nanoparticles, the studied dyes were optimized considering all the three anchoring modes, and the findings revealed that the bidentate chelating anchoring mode was the most stable form for these dyes for both the *cis* and *trans* isomers. To simulate the dye/TiO$_2$ complexes, the initial geometry of the (TiO$_2$)$_9$ anatase cluster was obtained from the previous study [26], which was large enough to reproduce the electronic and optical properties of the nanocomposites [27].

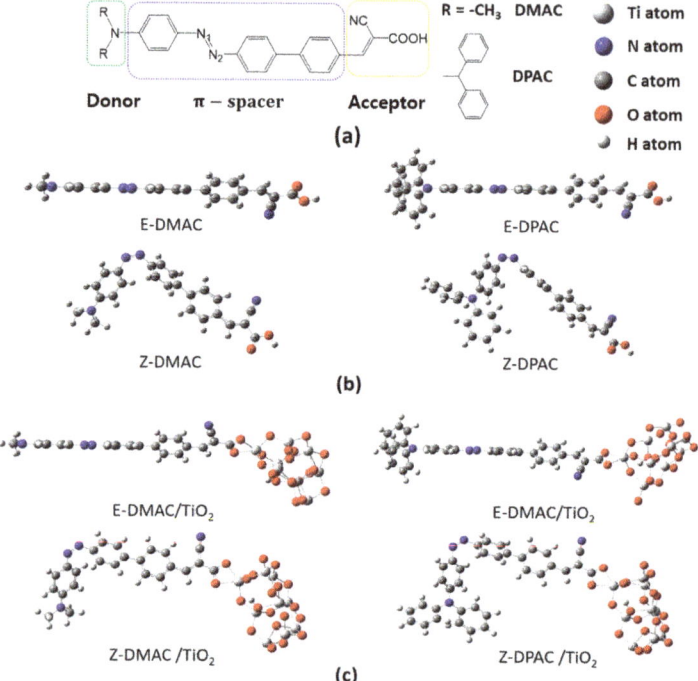

Figure 1. (a) Molecular structure of the dyes, and optimized geometries for *trans* (E) and *cis* (Z) structures of 2-cyano-3-(4'-(4-(dimethylamino)phenyl)diazenyl)-[1,1'-biphenyl]-4-yl)acrylic acid (DMAC) and 2-cyano-3-(4'-(4-(diphenylamino)phenyl)diazenyl)-[1,1'-biphenyl]-4-yl)acrylic acid (DPAC) as (b) isolated dyes and (c) dye/TiO$_2$ complexes. The titanium, nitrogen, carbon, oxygen, and hydrogen atoms are shown in the legend.

3.2. FT-IR Spectroscopic Analysis

The simulated FT-IR spectra of the two isolated dyes and dye/TiO$_2$ complexes in the range of 300–4000 cm^{-1} are shown in Figure S1. IR peaks with high intensity were observed mainly in the regions 1100–1900 cm^{-1} and 3000–3800 cm^{-1} for the *cis* and *trans* isomers of the isolated DMAC dyes. The characteristic peak at 3750 cm^{-1} arose from the stretching vibration of O-H in the carboxyl unit. Compared with the FT-IR spectrum of the dye/TiO$_2$ complexes, the O-H stretching vibration was weaker, which indicated that the O-H bond in the carboxyl unit of the DMAC dyes had ruptured. Consequently, the characteristic peak corresponding to the stretching vibration of the Ti-O bond appeared at ~470–490 cm^{-1} (Figure S1a), which indicated the formation of a Ti-O bond and the adsorption of the dye on the TiO$_2$ surface. Similarly, in the FT-IR spectra of the isolated DPAC dyes, intense IR peaks were observed in the range of 1000–1900 cm^{-1} and 3000–3800 cm^{-1} for both, the *cis* and *trans* isomers (Figure S1b). The peak at 3752 cm^{-1}, originating from the stretching vibration of the O-H bond in the carboxyl unit, disappeared in the FT-IR spectra of the dye/TiO$_2$ complexes. A peak appeared at ~487 cm^{-1} in the FT-IR spectra of the dye/TiO$_2$ complexes, which was attributed to the stretching vibration of the Ti-O bond. The results indicated that both, the DMAC and DPAC dyes, were adsorbed on the TiO$_2$ film in their *cis* and *trans* forms.

3.3. Adsorption Energy

The strength of the interaction energy between the dye and the TiO$_2$ surface was considered as the adsorption energy, which affected the rate of electron injection. In DSSCs, a high adsorption energy indicates a higher electronic coupling strength between the anchoring group and TiO$_2$ surface, which results in higher J_{SC} as well as electron transfer rate. The optimized structures of the DMAC/(TiO$_2$)$_9$ and DPAC/(TiO$_2$)$_9$ complexes are shown in Figure 2. It was evident that the photosensitizers were adsorbed almost perpendicularly onto the TiO$_2$ surface with the formation of two Ti-O bonds in the bidentate chelating anchoring mode. The calculated bond distances between the Ti and O atoms of the carboxylic acid of the dyes were in the range of 2.07–2.09 Å, which resulted in a strong interaction between the dyes and the TiO$_2$ surface. The adsorption energies of the dyes decreased in the order of E-DMAC > Z-DMAC > E-DPAC > Z-DPAC, which implied that the investigated dyes were strongly adsorbed on the TiO$_2$ surface. The DMAC dye/TiO$_2$ complexes showed a higher adsorption energy than the DPAC dye/TiO$_2$ complexes, which increased the electron transfer rate and improved the J_{SC} and photovoltaic performance of the DMAC dyes.

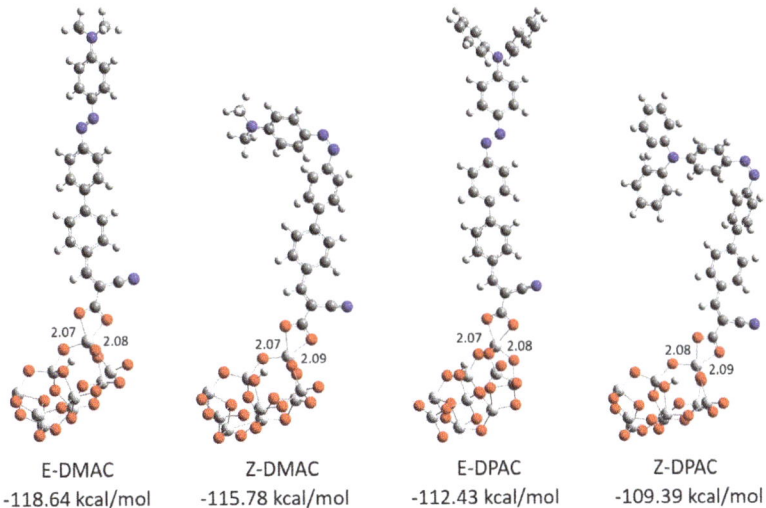

Figure 2. Optimized bidentate chelating mode and adsorption energies of DMAC and DPAC dyes on a (TiO$_2$)$_9$ anatase cluster calculated at the B3LYP level using the 6-31G(d,p) basis sets for non-metals and LANL2DZ basis sets with ECP for the Ti atom.

3.4. Structural Analysis

The degree of conjugation of the dyes affects their absorption spectra. Figure 1 shows that the *trans* dyes were fully conjugated as well as extremely coplanar compared to the twisted *cis* structures throughout the donor, π-bridge, and acceptor groups. Because of the strong π-conjugation, the planar *trans* dyes suppressed the rotational disorder and transferred more charge from the donor to the acceptor compared to the distorted *cis* dyes. The angle between the two arene rings of the azo group changed dramatically from 0° to ~78° upon *trans*-to-*cis* photoisomerization of the isolated dyes and dye/TiO$_2$ complexes. The dihedral angles between the benzene of the azo moiety and the right part benzene of the π-spacer moiety were ~32.5° owing to the steric hindrance between the hydrogens of the adjacent benzene moieties. The DPAC dyes had a distorted three-dimensional structure with a dihedral angle of ~50° between the phenyl rings owing to the internal steric hindrance among the phenyl rings. The distorted structure was beneficial for inhibiting dye aggregation on the semiconductor. To understand the relationship between the geometric properties and electron-donating strength of the dyes, the selected four bond lengths and the dihedral angle of the azobenzene moiety are summarized in Table 1. The calculated bond lengths were between the bond lengths of single and double bonds (N-C: 1.471 Å, N=C: 1.273 Å, and N=N: 1.247 Å) [28–30], which indicated that the charge was delocalized over the entire molecule. Interestingly, the bond length of the azo group (-N=N-), which is an important indicator of ICT in azo dyes, was longer in the *trans* dyes than in the corresponding *cis* dyes for both DMAC and DPAC moieties, while all the C-N bonds of the *trans* dyes were shorter than those of the *cis* dyes. As the electron-donating strength of the donor group increased from DMAC to DPAC dyes, the C-N distances increased; however, the N=N distances decreased in the respective *trans* and *cis* isomers. After binding to the TiO$_2$ surface (dye/TiO$_2$ complexes), similar trends were observed for both the dyes. The N=N bonds of the *trans* dye/TiO$_2$ complexes were longer, while the C-N bonds were shorter than those of the *cis* dye/TiO$_2$ complexes. Thus, even with a large displacement from the *trans* to *cis* form, the alternation of bond lengths was observed to be a function of the electron-donating strength. This result suggested that the electron-donating strength affected the geometric properties, which were related to the electronic structures, charge transfer, and optical properties. However, minimal changes were observed in the dihedral angles of the *cis* and *trans* forms of the DMAC and DPAC dyes before and after binding to TiO$_2$, indicating that the adsorption on TiO$_2$ did not affect the dihedral angles of the azo moiety. It is assumed that the degree of π-conjugation in the azo group could be maintained during the *trans-cis* photoisomerization even though the *cis* isomers had a distorted non-planar structure around the azo group.

Table 1. Structural parameters of DMAC and DPAC as isolated dyes and dye/TiO$_2$ complexes. A schematic representation of the dye is shown below.

Dye	Bond	Angle (Isolated Dye)	Angle (Dye/TiO$_2$)
E-DMAC	$N_1=N_2$	1.268	1.272
	$C_1-N_1=N_2-C_2$	179.86	178.68
	$N_3=C_6$	1.375	1.368
	$C_1=N_1$	1.397	1.394
	$C_2=N_2$	1.412	1.412
Z-DMAC	$N_1=N_2$	1.255	1.261
	$C_1-N_1=N_2-C_2$	−11.49	−11.85
	$N_3=C_6$	1.379	1.372
	$C_1=N_1$	1.42	1.412
	$C_2=N_2$	1.425	1.426
E-DPAC	$N_1=N_2$	1.267	1.269
	$C_1-N_1=N_2-C_2$	179.90	−179.99
	$N_3=C_6$	1.404	1.399
	$C_1=N_1$	1.402	1.400
	$C_2=N_2$	1.414	1.415
Z-DPAC	$N_1=N_2$	1.254	1.257
	$C_1-N_1=N_2-C_2$	−12.167	−10.94
	$N_3=C_6$	1.407	1.408
	$C_1=N_1$	1.424	1.422
	$C_2=N_2$	1.429	1.431

3.5. Cation-to-TiO$_2$ Surface Distance

In DSSCs, the undesirable recombination processes are closely related to the contact distance between the cation and semiconductor surface. If the contact distance is small, there is a possibility of electron back-transfer to either the cation or electrolyte during binding to TiO$_2$. Because of a smaller cation-to-TiO$_2$ distance, the *cis* dyes were expected to exhibit greater recombination while being adsorbed on the TiO$_2$ surface, which would lead to lower J_{SC} and V_{OC} as compared with those of the *trans* dyes. The contact distance between the cation and TiO$_2$ surface is shown in Figure S2. In the case of DMAC dyes (Figure S2a), the cation-to-TiO$_2$ distance for the *cis* dye (15.65 Å) was two-thirds of that of the *trans* dye (21.75 Å). A similar trend was observed in the case of DPAC dyes (Figure S2b), where the cation-to-TiO$_2$ contact distance for the *cis* dye (13.78 Å) was two-thirds of that of the *trans* dye (22.72 Å). This indicated that the J_{SC} and V_{OC} of the *trans* isomers were higher than those of the *cis* isomers for both, DMAC and DPAC dyes.

3.6. Molecular Orbitals

The frontier molecular orbitals (FMOs) of a molecule can be used to predict its optical and electronic properties. For a better understanding of the electron distribution and the relationship between the electronic structure and electron transition characteristics, the qualitative representation of ICT, i.e., the electron density distributions of the selected FMOs of the two dyes for the *trans* and *cis* isomers are shown in Figure 3.

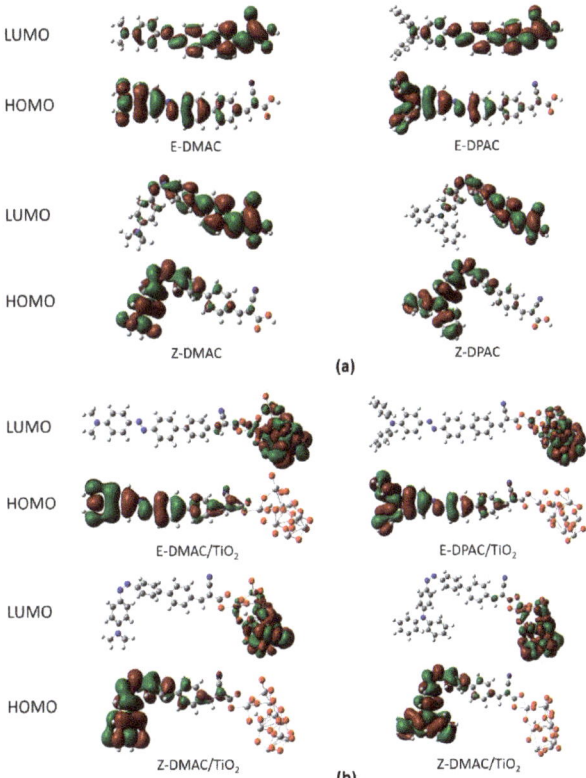

Figure 3. Frontier molecular orbitals of *trans* and *cis* isomers of DMAC and DPAC as (a) isolated dyes and (b) dye/TiO$_2$ complexes.

For both, the DMAC and DPAC dyes (Figure 3a), the electron densities of the HOMOs were extended to the donor up to the azobenzene moiety of the π-spacer, whereas the electron densities of the LUMOs were mainly delocalized along the right part of the π-spacer to the cyanoacrylic acid moiety. The electron distribution of the molecular orbitals confirmed that electron injection occurred from the diarylamine unit (D) to the cyanoacrylic acid unit (A). This was beneficial for the photon-driven ICT process and led to a charge transfer from the donor to the acceptor. ICT is facilitated if the electron density distribution of the HOMO is located near the electron donor, while that of the LUMO is delocalized around an anchoring group, ready for electron injection into the CB of the TiO$_2$ semiconductor. Interestingly, the *trans-cis* conformation did not affect the HOMO-LUMO electron distribution significantly, which suggested that azobenzene was a good π-spacer for ICT under illumination. Additionally, ICT was maintained even with an evident structural change. Therefore, it was evident that both the *trans* and *cis* forms would serve as a photosensitizer in DSSCs. The electron densities of the FMOs of the dye/TiO$_2$ complexes are shown in Figure 3b. The electron densities of the HOMOs for the trans and cis dye/TiO$_2$ complexes were distributed from the donor to the π-spacer, similar to the isolated dyes, whereas the electron densities of the LUMOs of the dye/TiO$_2$ complexes were almost entirely concentrated on TiO$_2$, which indicated that the LUMO located close to the cyanoacrylic acid anchoring group enhanced the orbital overlap with the 3d orbitals of Ti. As a result, the excited electrons were easily injected into TiO$_2$ via the anchoring unit, leading to an increase in J_{SC}. In summary, the study of FMOs suggested that both the dyes showed large ICTs, and consequently, a strong electronic coupling with the TiO$_2$ surface.

3.7. UV-Visible Spectroscopic Analysis

The maximum absorption wavelengths (λ_{max}), oscillator strength (*f*), excited state transition characteristics, nature of the most relevant transitions of the electronic absorption bands, and LHE are summarized in Table 2. The simulated UV-Vis absorption spectra of the DMAC and DPAC dyes in DMF solvent obtained from the TDDFT calculations for the isolated dyes and dye/TiO$_2$ complexes are shown in Figure 4. The red and black colors represent the DMAC and DPAC dyes, respectively. The solid and dotted lines represent the *trans* and *cis* dyes, respectively. Both DMAC and DPAC dyes exhibited a broad absorption band and a high molar extinction coefficient, which resulted in the highest sunlight absorption ability. For the isolated dyes (Figure 4a), the *trans* isomers showed a relatively strong absorption at 400–525 nm, with the maximum absorption peaks of the DMAC and DPAC dyes appearing at 430 nm and 440 nm, respectively (Table 2). These strong absorption bands corresponded to the π–π* transition of the FMOs. The absorption ranges of the two dyes were mainly spread over the visible region, thus ensuring effective solar energy usage. Interestingly, two absorption bands were observed for the *cis* dyes. The strong absorption band at ~341–347 nm was possibly due to the π–π* transition, while the weak band at ~457–471 nm could be attributed to the n–π* transition for both *cis* dyes. The spectral difference between the *trans* and *cis* isomers would impart different colors in the DSSC. As the photoirradiation proceeded, the intensity of the *trans* dyes in the 400–500 nm region decreased and that of the *cis* dyes in the 300–400 nm region increased. For the isolated dyes, the major electron transition involved the HOMO, HOMO−1, LUMO, and LUMO+1 orbitals. The change in the electron density between the molecular orbitals (Figure 2) showed that the electron moved from the donor to the acceptor unit, which is an ICT and conducive to a high J_{SC}. The transition from HOMO/LUMO corresponding to the π–π* transition was the main contributor to the lowest electronic excitation in the *trans* dyes, although transitions from the HOMO/LUMO+1 orbital also contributed to this excitation. In the case of *cis* dyes, the transition from HOMO/LUMO, representing the π–π* transition, contributed to the strong absorption for both, the DMAC and DPAC dyes. The weak absorption by the *cis* dyes was primarily related to HOMO/LUMO+1 of the occupied orbitals corresponding to the n–π* transition, which was due to the presence of unshared electron pairs of the nitrogen atoms. The coplanar structure of the azobenzene unit in the *trans* dyes prevented the n–π* transition, while the n–π* transition in the *cis* dyes resulted from the interaction between the azo bond (N=N) and the π-conjugated system. The transition properties of the dyes adsorbed on the (TiO$_2$)$_9$ cluster based on the optimized ground-state structures were investigated using the CAM-B3LYP/6-311++G(d,p) method. The isolated dyes and the dye/TiO$_2$ complexes exhibited almost similar UV-Vis absorption spectra (Figure 4b). After binding to TiO$_2$, the dyes showed a red shift in the maximum absorption wavelengths as compared with those of the isolated dyes. The absorption peaks of the *trans* dye/TiO$_2$ complexes showed a red shift of 10–12 nm compared with that of the isolated *trans* dyes, which corresponded mainly to the HOMO/LUMO transition (Table 2). The strong absorption band of the *cis* dye/TiO$_2$ complexes, which also corresponded to the HOMO/LUMO transition, showed red shifts of 17 nm (for DMAC dye) and 9 nm (for DPAC dye) compared to those of the isolated dyes, respectively. The red shift of the maximum absorption wavelength of the dye after binding to TiO$_2$ could be explained on the basis of the interactions between the electron acceptor group of the dye (–COOH) and the 3d orbitals of the Ti atom, which resulted in a decrease in the LUMO energies as compared to the isolated dyes. The UV-Vis absorption spectra also revealed the mechanism of photoinjection from the dye to the semiconductor. Compared to the UV-Vis spectrum of the isolated dye, the appearance of a new band in the spectrum of the dye/TiO$_2$ complex indicates that it shows a Type II (direct) mechanism [31], whereas the absence of a new band suggests that it exhibits a Type I (indirect) mechanism [32]. As can be seen in Figure 4, both the DMAC and DPAC dyes exhibited a Type I (indirect) injection route during binding to the TiO$_2$ surface.

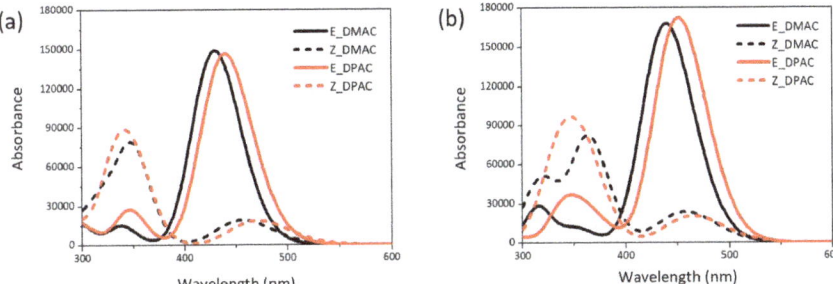

Figure 4. UV-Vis absorption spectra of *cis* and *trans* isomers of DMAC and DPAC as (**a**) isolated dyes and (**b**) dye/TiO$_2$ complexes.

Table 2. Maximum absorption wavelengths (λ_{max}), oscillator strengths (f), excited state transition characteristics, nature of the transitions for the most relevant transitions of the electronic absorption bands, and light-harvesting efficiencies (LHEs) of the dyes.

Dye	Excited State Character	Transition Assignment (%)	Oscillator Strength, f	λ_{max}	LHE
E-DMAC	$\pi \to \pi^*$	H-L (66.6%) H-L+1 (32.3%)	2.0486	430	0.9911
Z-DMAC	$n \to \pi^*$	H-L+1 (52.9%) H−1-L (21.2%)	0.2647	457	0.4564
	$\pi \to \pi^*$	H-L (60.2%) H−1-L (18.3%)	0.9926	347	0.8983
E-DPAC	$\pi \to \pi^*$	H-L (65.9%) H-L+1 (26.1%)	1.8015	440	0.9475
Z-DPAC	$n \to \pi^*$	H-L+1 (51.4) H−1-L+1 (20.3%)	0.2583	471	0.4483
	$\pi \to \pi^*$	H-L (64.4%) H−1-L+1 (15.6%)	0.7985	341	0.8411
E-DMAC/TiO$_2$		H-L (82.6%) H-L+1 (30.1%)	2.3227	440	0.9953
Z-DMAC/TiO$_2$		H-L (53.8%)	0.3276	458	0.5297
E-DPAC/TiO$_2$		H-L (85.7%) H-L+1 (28.8%)	1.1195	364	0.9241
			2.3189	452	0.9951
Z-DPAC/TiO$_2$		H-L (64.3%)	0.2844	468	0.4805
			0.8237	350	0.8499

3.8. Energy Diagram

To investigate the electronic and transition properties of the dyes, the FMO energy levels from HOMO−2 to LUMO+2 of the isolated dyes and dye/TiO$_2$ complexes for both DMAC and DPAC were calculated using the B3LYP/6-311G(d,p) level, and the results are shown in Figure 5.

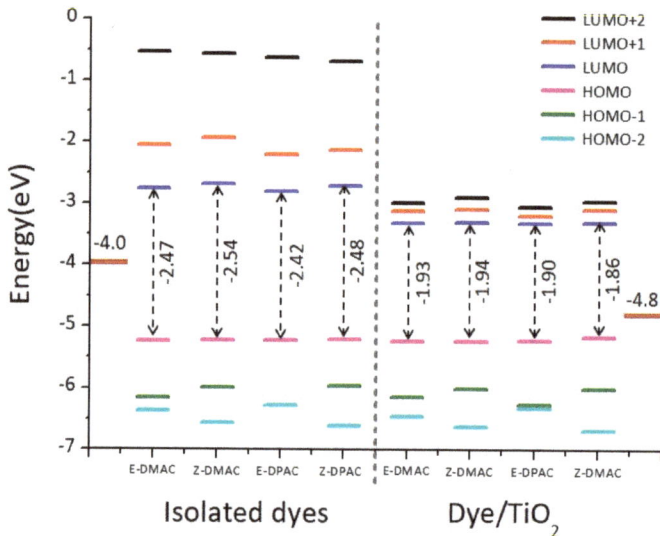

Figure 5. Molecular orbital energy diagrams of *trans* and *cis* isomers of DMAC and DPAC as isolated dyes and dye/TiO$_2$ complexes.

To design an effective dye, the HOMO and LUMO energy levels of the dyes must be below the redox potential of the I$^-$/I^{3-} electrolyte and above the CB of TiO$_2$, respectively. The measured HOMO energy levels of the isolated dyes were lower than the redox potential of I$^-$/I^{3-} (−4.80 eV) [4], which implied that the oxidized dyes could restore the electrons from the electrolyte. Similarly, the LUMO energy levels of the *trans* and *cis* dyes were above the CB of TiO$_2$ (−4.00 eV) [33], which indicated that the designed excited state dyes could quickly and efficiently inject electrons into the TiO$_2$ CB. The HOMO-LUMO energy values and their energy gaps are summarized in Table 3. The HOMO energy values of both, the *trans* and *cis* isomers of the DMAC and DPAC dyes, were similar. For LUMO, the DPAC dyes showed a higher energy than the DMAC dyes when comparing their respective isomers. The HOMO-LUMO energy gaps of the *cis* isomers were higher than those of the *trans* isomers owing to the higher LUMO level. The Z-DMAC dye exhibited the highest energy difference (2.54 eV), whereas the E-DPAC dye exhibited the lowest energy gap (2.42 eV). As the HOMO-LUMO energy gaps of the *trans* dyes were lower than those of the *cis* dyes, they absorbed more light from the visible range and showed a bathochromic shift (Table 2). A higher LUMO level increases the V_{OC}, thus enhancing the efficiency of the DSSC. Therefore, it is necessary to monitor the enhanced performance of the dye with a higher LUMO energy level. Because of a higher LUMO energy level, the *cis* dyes seemingly had a higher driving force for electron injection compared to the *trans* dyes. However, the *cis* dyes had a larger band gap, which was unfavorable for optical absorption [34]. After binding to the TiO$_2$ surface, the FMO energy levels (HOMO−2 to LUMO+2) of the dye/TiO$_2$ complexes were calculated to further investigate the electronic coupling between the FMOs and CB of TiO$_2$, which are shown in Figure 5. For the dye/TiO$_2$ complexes, all the HOMO energy levels were lower than the redox potential of the I$^-$/I^{3-} electrolyte and the LUMO energy levels were higher than the CB of TiO$_2$, which indicated a strong driving force for electron injection from the dye to the semiconductor as well as a suitable regeneration of the neutral dye. There was almost no change in the energies of the HOMO levels of the dye/TiO$_2$ complexes as compared to the isolated dyes. However, the LUMO energy levels remarkably decreased after the dyes adsorbed onto the TiO$_2$ surface because of bonding between the semiconductor CB and dye. This implied that the LUMO energy levels of these dyes were strongly coupled with TiO$_2$, which is favorable for increasing electron injection into TiO$_2$. The HOMO-LUMO energy gap decreased after the dyes adsorbed onto the TiO$_2$ surface owing to the relatively low LUMO energy level, which suggested that the adsorption of the dye on the semiconductor surface facilitated the HOMO-LUMO energy level properties crucial for favorable light absorption.

Table 3. HOMO and LUMO energy values and energy gaps, excited state lifetimes, dipole moments, exciton binding energies, and coupling constants of the isolated dyes and dye/TiO$_2$ complexes.

Dye	HOMO	LUMO	HOMO-LUMO Gap	Ex-State Lifetime, τ	Dipole Moment, D	Exciton Binding Energy, EBE	Coupling Constant, |VRP|
E-DMAC	−5.2341	−2.7576	2.477	1.43	11.61	0.41	0.6171
Z-DMAC	−5.2276	−2.6858	2.541	1.99	12.20	0.89	0.6138
E-DPAC	−5.2270	−2.8066	2.420	1.52	8.67	0.47	0.6135
Z-DPAC	−5.2034	−2.7217	2.482	2.57	9.53	0.93	0.6017
E-DMAC/TiO$_2$	−5.2398	−3.3065	1.936	1.39	22.5	0.88	0.6199
Z-DMAC/TiO$_2$	−5.2352	−3.2997	1.936	1.95	28.7	1.33	0.6176
E-DPAC/TiO$_2$	−5.2200	−3.3206	1.900	1.10	21.4	0.98	0.6101
Z-DPAC/TiO$_2$	−5.1734	−3.3051	1.868	1.74	26.2	1.42	0.5867

3.9. Electrostatic Potential

To understand the chemical reactions (such as H bonding interactions), the molecular electrostatic potential (MEP), which is closely related to the electron cloud, of the isolated dyes and dye/TiO$_2$ complexes, were calculated at the B3LYP/6-311G(d,p) level, and the results are shown in Figure 6. Generally, the MEP is used to describe the nucleophilic and electrophilic reaction sites. The different colors at the surface represent different electrostatic potential values. The red and blue areas of the MEP depict the electrophilic activity corresponding to the electron-rich areas and nucleophilic activity corresponding to the electron-deficient areas, respectively. The electrostatic potential increased in the order: red < orange < yellow < green < blue. The color code of the MEP maps ranged from −0.06 a.u. (deepest red) to 0.06 a.u. (deepest blue). The MEPs of the two isolated dyes (Figure 6a) indicated that the carboxyl H atom in all the dyes had the highest nucleophilic potential. For both dyes, the highest electrophilic potential was exhibited by the N atom of the –CN group in the *trans* structures and the –CN and azo (N=N) groups in the *cis* structures. The H and N atoms represent the strongest attraction and repulsion, respectively. For the dye/TiO$_2$ complexes (in Figure 6b), the change was less distinct when the dyes were anchored on the TiO$_2$ surface owing to the interactions between the dyes and TiO$_2$, which made the dye molecules more neutral in all the regions. However, the highest nucleophilic potential was exhibited mainly by the terminal H of the TiO$_2$ cluster, while the highest electrophilic potential was exhibited by the O atoms on the TiO$_2$ cluster for both the *cis* and *trans* isomers of the DMAC and DPAC dyes.

3.10. Charge Density Difference

To investigate the charge transfer properties of the excited state complexes, the charge difference density (CDD) between the excited and ground states of the DMAC and DPAC isolated dyes and dye/TiO$_2$ complexes were determined and are shown in Figure 7. The blue and green regions represent the depletion and accumulation of electron density upon excitation, respectively. For the isolated dyes (Figure 7a), the density depletion zones (blue) were mostly located on the donor and π-spacer regions, while the density enhancement segments (green) were mainly delocalized on the acceptor moiety, which was indicative of an ICT transfer during electron transition. The CDD plots of the dye/TiO$_2$ complexes (Figure 7b) showed that the density increment region was mostly located on the acceptor moiety, while the density depletion zone was spread over the donor moiety as well as in TiO$_2$; this implied that some of the hole and electron densities were delocalized on the dye molecule, while the rest of the electron density was localized on TiO$_2$.

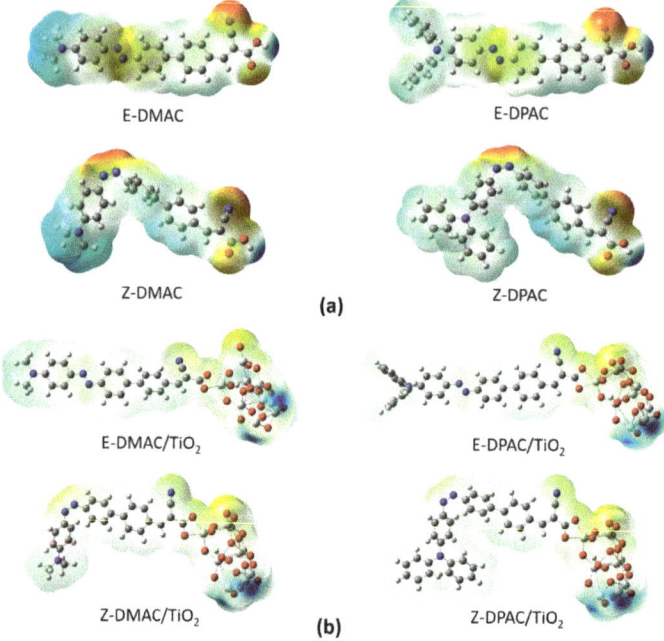

Figure 6. Molecular electrostatic potentials of *cis* and *trans* isomers of DMAC and DPAC as (**a**) isolated dyes and (**b**) dye/TiO$_2$ complexes.

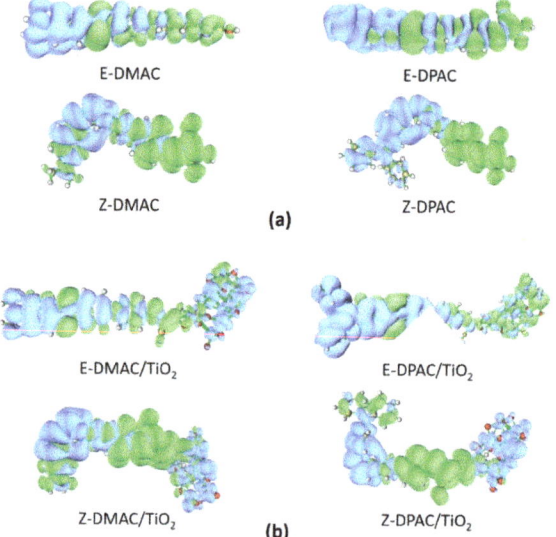

Figure 7. Electron density difference maps for *cis* and *trans* isomers of DMAC and DPAC as (**a**) isolated dyes and (**b**) dye/TiO$_2$ complexes.

3.11. NBO Analysis

Based on the optimized structure of the ground state, NBO analysis was performed to further understand the distribution of charge on the overall dye molecules and the electron transfer from the donor to the acceptor through the π-spacer to estimate the extent of ICT. The NBO population charges for the electron donor, π-spacer, and electron acceptor, which are denoted as q^{Donor}, $q^{\pi\text{-spacer}}$, and $q^{Acceptor}$, respectively, are summarized in Table 4. The most significant charge variance between the natural charges on the donor and acceptor groups is represented as Δq^{D-A}. The positive NBO values of the donor moiety indicated that they were effective electron-donating units. In contrast, the negative NBO value of the π-spacer suggested that the dye may trap electrons in the π-spacer. The negative charge of the electron acceptor could be a factor leading to electron injection from the excited dye to the TiO_2 CB. Between the DMAC and DPAC dyes, the former exhibited higher q^{donor} and Δq^{D-A} values compared to the latter. This indicated that the DMAC dyes could donate more electrons to the anchoring group compared to the DPAC dyes, thus accelerating the ICT. Moreover, the Δq^{D-A} values of the *trans* dyes were higher than those of the *cis* dyes, suggesting that the ICT ability was sensitive to the conformational changes in the π-spacer. Second order perturbation theory (SOPT) analysis of the Fock matrix on the NBO basis could determine the amount of charge transfer between the different parts of the molecule. Table S1 summarizes the NBO parameters, conjugative interaction energies (ΔE^2) between the π and π* orbitals, energy difference between the interacting NBO and matrix element (E_j-E_i), and the off-diagonal element associated with the NBO Fock matrix ($F_{(i,j)}$). Carbon atoms (C_1-C_6) and nitrogen atoms ($N_1=N_2$) were selected to investigate the electronic delocalization process. A high ΔE^2 implied more charge transfer from the donor (π) to the acceptor (π*) parts. With increasing donor size, ΔE^2 increased from the DMAC to DPAC dyes. Furthermore, the ΔE^2 of the *trans* dyes was noticeably higher than that of the *cis* dyes in the case of $\pi(C_1=C_2)$ to $\pi^*(N_1=N_2)$, which indicated that the conformational changes of the dyes also affected the ΔE^2.

Table 4. NBO analysis results for metal-free organic dyes in the ground state. Here, q^{Donor}, $q^{\pi\text{-spacer}}$, and $q^{Acceptor}$ denote the total amount of natural charges on the donor group, π-spacer, and acceptor group, respectively.

Dyes	q^{Donor}	$q^{\pi\text{-spacer}}$	$q^{Acceptor}$	Δq^{D-A}
E-DMAC	0.3078	−0.1959	−0.1119	0.4197
Z-DMAC	0.2523	−0.1395	−0.1127	0.3650
E-DPAC	0.2784	−0.1711	−0.1073	0.3858
Z-DPAC	0.2090	−0.1010	−0.1079	0.3169

3.12. Natural Transition Orbitals and Density of States

The electronic density distributions of the dyes are illustrated in Figure S3 for both the isolated dyes and dye/TiO_2 complexes determined by natural transition orbital (NTO) analysis. As ICT occurred under light illumination, it was reasonable to analyze the electronic distribution during electronic transition. NTOs can provide detailed information about the excited state transitions apart from the mixed electronic configurations because of multiple excitations among the molecular orbitals. Hole and particle transition orbitals represent the unoccupied and occupied NTOs, respectively. An eigenvalue λ denotes the fraction of the hole-particle pair contribution to the electronic transition. Importantly, the HOMO → LUMO excitation contributed mostly to the $S_0 \rightarrow S_1$ transition. As shown in Figure S3, the electron density of the hole NTOs was localized on the donor moiety and extended along the π-spacer for the E-DMAC and E-DPAC dyes, whereas the density was delocalized from the donor to the acceptor moiety for the *cis* dyes. Additionally, the electron density of the particle NTOs was delocalized mainly on the π-spacer to the acceptor moiety for all the dyes. A similar scenario was observed in the case of dye/TiO_2 complexes for both the DMAC and DPAC dyes. This indicated that photoinduced charge transfer occurs mostly in the *trans* dyes rather than in the *cis* dyes. In addition, the NTO eigenvalues (λ) of the *trans* dyes were higher than those of the *cis* dyes. During visible-light absorption, the electronic transition allowed a net electron transfer from the donor to the acceptor, and subsequently to the TiO_2 surface. In this regard, the donating capability of the donor was important for charge transfer, as additional noticeable electronic density separation required a stronger donor (Figure S3). The total density of states (TDOS) and partial density of states (PDOS) are represented in Figure S4 for the isolated dyes and dye/TiO_2 complexes. The vertical dotted line represents the HOMO energy level. For the isolated dyes (Figure S4a), the PDOS of the *p*-orbitals dominated the TDOS of the occupied orbitals, whereas the PDOS of *s*- and *p*-orbitals dominated the TDOS of the unoccupied orbitals for the DMAC and DPAC dyes. In the dye/TiO_2 complexes (Figure S4b), the PDOS of the *p*-orbitals was the main contributor to the TDOS of the unoccupied orbitals, similar to the isolated dyes. However, for the occupied orbitals, the PDOS of the *p*- and *d*-orbitals dominated the TDOS of the unoccupied orbitals in the dye/TiO_2 complexes of the two dyes.

3.13. Polarizability and Hyperpolarizability

Polarizability and hyperpolarizability characterize the response of a system in an applied electric field. They determine the strength of molecular interactions, such as long-range intermolecular induction and dispersion forces, as well as the cross sections of different scattering and collision processes of the system. Generally, a dye with a higher polarizability strongly interacts with the surrounding species and increases the local concentration of the acceptor species at the TiO_2 surface, which increases the possibility of the acceptor species penetrating the dye adsorption layer. The total static first hyperpolarizability is expressed as follows [35]:

$$\beta_{tot} = \sqrt{\beta_x^2 + \beta_y^2 + \beta_z^2} \tag{1}$$

The individual static component in the above equation is calculated from:

$$\beta_i = \beta_{iii} + \frac{1}{3}\sum_{i \neq j}(\beta_{ijj} + \beta_{jij} + \beta_{jji}) \tag{2}$$

where β_{ijk} (i, j, k = x, y, z) are the tensor components of the total static first hyperpolarizability. Owing to Kleinman symmetry, the following equation is finally obtained:

$$\beta_{tot} = \left[(\beta_{xxx} + \beta_{xyy} + \beta_{xzz})^2 + (\beta_{yyy} + \beta_{yzz} + \beta_{yxx})^2 + (\beta_{zzz} + \beta_{zxx} + \beta_{zyy})^2\right]^{1/2} \tag{3}$$

The polarizability and hyperpolarizability of the dyes are shown in Figure S5, and the values are listed in Table S2. The polarizability values of the dyes increased in the order: E-DMAC > E-DPAC > Z-DMAC > Z-DPAC. E-DMAC exhibited the highest polarizability, which implied that *trans* DMAC was a better dye. Owing to the important application of hyperpolarizability as well as its close relationship with ICT, the first hyperpolarizabilities of the two dyes were also investigated (Figure S5), the results of which are listed in Table S2. The first hyperpolarizabilities of the two dyes were in the order of Z-DMAC < Z-DPAC < E-DMAC < E-DPAC. It is noteworthy that all the components of the first hyperpolarizabilities of the two dyes were mainly along β_{xxx}, which indicated a unidirectional charge transfer from the donor to the acceptor. The β_{total} values of the *trans* dyes were considerably higher than those of the *cis* dyes, suggesting that the *trans* dyes led to more photoinduced electron transfer in the excited state. Although the first hyperpolarizability of DPAC was higher than that of DMAC, the former prevented electron transfer from the donor to the acceptor because of the non-planar structure of the donor, thereby affecting the effective electron injection from the dye molecule to the CB of the semiconductor.

3.14. Other Molecular Properties

Dyes with different dipole moments (Ds) can modify the CB of wide-bandgap semiconductors (e.g., TiO$_2$ and ZnO) and affect the nature of the interaction between the dye and the acceptor species. A strong electron-donating ability results in a higher D of the dyes, which can increase the distance between the charge centers, leading to enhanced electron delocalization. The Ds of the isolated dyes and dye/TiO$_2$ complexes are listed in Table 3. In the case of isolated dyes, the Ds of the DMAC dyes were higher than those of the DPAC dyes. Moreover, the Ds of the *cis* dyes were higher than those of the *trans* dyes, which increased the bond polarity; thus, the D vectors of the bonds cancelled each other. In the case of dye/TiO$_2$ complexes, the Ds of the DMAC dye/TiO$_2$ complexes were higher than those of the DPAC dye/TiO$_2$ complexes, with the *cis* dye /TiO$_2$ complexes showing higher Ds than the *trans* dye/TiO$_2$ complexes, similar to the isolated dyes. However, the Ds of the dye/TiO$_2$ complexes were significantly higher than those of the isolated dyes, which indicated that after their adsorption on the TiO$_2$ surface, the dyes showed greater electron delocalization (Figure 3b). Exciton binding energy (EBE) is another key factor affecting the efficiency of excitonic solar cells and is associated with charge separation in the solar cells. Dyes with high EBEs exhibited the lowest charge separation efficiency. The calculated EBEs of the two dyes are listed in Table 3. In the isolated dyes, the EBE of DMAC was lower than that of DPAC, with the *trans* dyes showing lower EBEs than the *cis* dyes in both the cases, which was a desirable outcome for photo-to-current energy conversion. The dyes with lower EBEs (*trans* dyes) generated current more efficiently from the absorbed light. In the case of the dye/TiO$_2$ complexes, the EBEs of the DPAC dyes were higher than those of the DMAC dyes, with the *cis* dye/TiO$_2$ complexes showing higher EBEs than the *trans* dye/TiO$_2$ complexes. This indicated that the *trans* dye/TiO$_2$ complexes had a higher charge separation efficiency than the *cis* dye/TiO$_2$ complexes, which was favorable for a better power conversion efficiency (PCE) of DSSCs. The coupling constant ($|V_{RP}|$), a factor that affects the rate of electron injection between the organic dyes and the semiconductor surface, could be derived from the following equation [36]:

$$|V_{RP}| = \Delta E_{RP}/2 \tag{4}$$

Equation (4) indicates that a high ΔE_{RP} will result in a high $|V_{RP}|$, which will enhance the electron injection in DSSCs. The ΔE_{RP} can be estimated as follows [37]:

$$E_{RP} = \left[E_{LUMO}^{dye} + 2E_{HOMO}^{dye}\right] - \left[E_{LUMO}^{dye} + E_{HOMO}^{dye} + E_{CB}^{TiO_2}\right] \tag{5}$$

The experimental value of $E_{CB}^{TiO_2}$ was −4.0 eV [33]. The calculated $|V_{RP}|$ values of the DMAC and DPAC dyes (listed in Table 3) decreased in the order of E-DMAC > Z-DMAC > E-DPAC > Z-DPAC. This trend implied that compared to the DPAC dyes, the DMAC dyes had a higher electron injection rate and the largest number of electrons in the CB, which led to a higher V_{OC}. A similar phenomenon was observed in the case of the dye/TiO$_2$ complexes. E-DMAC/TiO$_2$ showed the highest $|V_{RP}|$, whereas Z-DPAC/TiO$_2$ showed the lowest $|V_{RP}|$.

3.15. Excited State Lifetime

The efficiency of electron injection to TiO$_2$ can be determined by the excited state lifetime. Electron injection from the excited dye to the semiconductor was very fast, which suggested that increasing the concentration of the acceptor on the TiO$_2$ surface would increase the possibility of the acceptor species penetrating the adsorbed dye layer, thus leading to electron recombination following a short electron lifetime. This process would minimize the photovoltage and lower the charge collection efficiency, thereby reducing the J_{SC} and PCE. After electron injection, the dye was in a cationic state until regeneration occurred. It has been reported that the considerable reduction in the electron lifetime in porphyrin-based DSSCs is the main reason for their lower V_{OC} compared to that of the Ru sensitizer

N719 [38]. The longer the excited state lifetime, the longer the dyes remained in the cationic form, which favored charge transfer. The excited state lifetime of the dye was estimated as follows [39]:

$$\tau = 1.499/fE^2 \qquad (6)$$

where E is the excitation energy (cm^{-1}) of the different electronic states and f is the oscillator strength corresponding to the electronic state. To calculate the excited state lifetimes, the ground-state geometries of the DMAC and DPAC dyes were optimized in their first excited singlet electronic state with the CAM-B3LYP/6-311+G(d,p) level of theory for the isolated dyes and dye/TiO_2 complexes, considering the lowest excitation energy and the corresponding oscillator strength. The calculated excited state lifetimes of the two dyes are listed in Table 3. In the case of isolated dyes, the excited state lifetimes of the *trans* DPAC dyes were higher than those of their corresponding DMAC dyes and vice versa, respectively, which implied that the DPAC dyes remained stable in the cationic state for a longer time. In the case of the dye/TiO_2 complexes, interestingly, the opposite scenario was observed. After binding onto the TiO_2 surface, *trans* DPAC/TiO_2 exhibited a lower excited state lifetime compared to DMAC/TiO_2. A similar observation was made in the case of the *cis* dye/TiO_2 complexes. This indicated that after adsorbing onto the TiO_2 surface, the DMAC dyes remained in their cationic form for a longer time and allowed a greater charge transfer. This retarded the charge recombination process, which was favorable for a high V_{OC} and better PCE of DSSCs.

3.16. Chemical Reactivity Parameters

Based on the optimized neutral and ionic structures, the chemical reactivity parameters, namely, chemical hardness (η), electron affinity (EA), ionization potential (IP), electrophilicity power (ω), and electron-accepting power (ω^+), were investigated to further explain the molecular properties of the dyes; these parameters are listed in Table S3. The ω value represents the stabilization energy of the dyes. These ω values of the DMAC dyes were higher than those of the DPAC dyes and increased in the order of Z-DPAC < Z-DMAC < E-DPAC < E-DMAC. Thus, the ω values of the *trans* dyes were higher than those of the *cis* dyes, which implied that the former showed a higher energetic stability by attracting the electrons from the environment. The capability to accept an electron from a donor is measured by EA, which can be represented as ω^+. A higher value of ω^+ is desirable to achieve a high J_{SC}. The ω^+ values of the dyes decreased in the order of Z-DPAC < Z-DMAC < E-DPAC < E-DMAC, which indicated that the *trans* DMAC dye had the highest electron-withdrawing ability, and therefore, a higher ability to attract electrons from the acceptor moiety of the dye. Charge injection and balance affect the performance of the DSSC devices. IP and EA represent the energy barriers of both holes and electrons. The IP and EA of the two molecules were calculated by DFT, and these results are listed in Table S3. The IP and EA of the *trans* dyes were respectively lower and higher than those of the *cis* dyes, which promoted the hole-creating and electron-accepting abilities, respectively. Besides, the IP and EA of the DMAC dyes were respectively lower and higher than those of the DPAC dyes. Hence, E-DMAC had better hole-creating and electron-accepting abilities. The η value represents the resistance of the dyes to ICT in solar cells. A lower η and higher ω lead to a lower resistance to ICT and a better J_{SC}, resulting in a higher PCE. Therefore, to increase charge transfer and separation, dyes should have a lower η. The η values of the *trans* dyes were lower than those of the *cis* dyes (Table S3), which suggested that the *trans* dyes would show better efficiency for DSSCs. In addition, the η value of DMAC was lower than that of DPAC; thus, E-DMAC exhibited a lower resistance to ICT, leading to a higher J_{SC}. The chemical reactivity parameters were also measured for the dye/TiO_2 complexes (Table S3). It was observed that the ω and ω^+ of the DMAC dye/TiO_2 complexes were higher than those of the DPAC dye/TiO_2 complexes for both *trans* and *cis* isomers. Compared to the isolated dyes, the IP and EA of the DMAC dye/TiO_2 complexes were respectively lower and higher than those of the DPAC dye/TiO_2 complexes for both *trans* and *cis* dyes. Moreover, the η values of the DMAC dye/TiO_2 complexes were lower than those of the DPAC dye/TiO_2 complexes. It was observed that the dye/TiO_2

complexes showed a similar behavior as that of the isolated dyes. However, the chemical reactivity parameters shown in Table S3 indicate a better performance of the dye/TiO$_2$ complexes, in which the dyes are bound to the TiO$_2$ surface, compared to the isolated dyes. Based on these chemical reactivity parameters, the DMAC dyes are expected to show better ICT, higher J_{SC}, and higher PCE for DSSCs.

3.17. Factors Affecting Short-Circuit Current Density

In DSSCs, the sunlight-to-electricity conversion efficiency (n) of solar cell devices is determined by the V_{OC}, J_{SC}, and fill factor (FF), divided by the incident solar power (P_{inc}) [40]:

$$n = \frac{(V_{OC})(J_{SC})(FF)}{P_{inc}} \quad (7)$$

According to Equation (7), the product of V_{OC} and J_{SC} should be optimized to improve the efficiency (n). In DSSCs, J_{SC} can be expressed as [40]:

$$J_{SC} = e \int LHE(\lambda) \phi_{inject} \eta_{collect} d\lambda, \quad (8)$$

where LHE(λ) is the light-harvesting efficiency at a given wavelength, ϕ_{inject} is the electron injection efficiency, and $\eta_{collect}$ is the charge collection efficiency. All the components of DSSCs are only different for the dyes; hence, $\eta_{collect}$ can be assumed a constant. LHE(λ) can be expressed as [41]:

$$LHE = 1 - 10^{-f}, \quad (9)$$

where f represents the oscillator strength of the dyes corresponding to λ_{max}. Generally, a higher LHE, caused by the higher f, increases the light capturing ability and improves the efficiency of the DSSC. Dyes with a small energy gap are beneficial for achieving a red shift in the maximum absorption peak and a relatively high LHE. The LHEs of the isolated dyes and dye/TiO$_2$ complexes were calculated and are given in Table 2. The f values of the *trans* dyes (Table 2) were higher than that of the *cis* dyes for both, isolated dyes and dye/TiO$_2$ complexes, which suggested that the LHE of the *trans* dyes were greater than those of the *cis* dyes. The LHE should be as high as possible to maximize the J_{SC}. In the case of isolated dyes, the LHE values for the π–π^* transition were higher than those for the n–π^* transition, which indicates that the former transition was favorable for LHE for both *trans* and *cis* dyes. Moreover, changing the donor moiety in both *trans* and *cis* dyes affected the f and LHE, which implied that the LHE was affected by both, the conformational change of the azobenzene bridge structures and the electron-donating strength of the donor group. The LHE of E-DMAC was the highest among all the dyes for the isolated dyes and dye/TiO$_2$ complexes, which indicated that DMAC could absorb more photons, leading to a higher J_{SC}. ϕ_{inject} was related to the injection driving force (ΔG_{inject}) of the electrons injected from the excited dyes to the semiconductor substrate. According to Preat's method [42], ΔG_{inject} can be estimated as follows:

$$\Delta G_{inject} = E^{dye^*} - E_{CB}, \quad (10)$$

where E^{dye^*} is the oxidation potential of the dye in the excited state, and E_{CB} is the CB edge of the semiconductor (−4.00 eV) [33]. E^{dye^*} can be estimated as follows [43]:

$$E^{dye^*} = E^{dye} - E_{0-0}, \quad (11)$$

where E^{dye} is the redox potential of the ground state of the dye and E_{0-0} is the vertical transition energy associated with λ_{max}. Note that this relation is only valid if the entropy change during the light absorption process can be neglected. Hence, higher LHE and ΔG_{inject} are beneficial for increasing the J_{SC}. The ΔG_{inject}, E^{dye}, E^{dye^*}, and E_{0-0} for the two dyes were computed and are listed in Table 5.

The ΔG_{inject} values of all the dyes were more negative than that of the TiO$_2$ CB edge, which indicates that the excited state dyes lie above the TiO$_2$ CB, thus promoting electron injection from the excited sensitizer to the TiO$_2$ CB. The absolute values of ΔG_{inject} for both the dyes were considerably higher than 0.2 eV; thus, all the dyes showed a sufficient driving force to inject electrons into TiO$_2$ [44]. The ΔG_{inject} values for the *trans* dyes were more negative than those of the *cis* dyes, which suggested that the *trans* dyes would exhibit faster electron injection and a higher J_{SC} compared to the *cis* dyes. However, an excessively high value of ΔG_{inject} can cause energy redundancy, thus leading to a smaller V_{OC}. The DPAC dyes therefore had a lower V_{OC} than the DMAC dyes despite having a higher ΔG_{inject}. Similar to the isolated dyes, the ΔG_{inject} values of the DMAC dye/TiO$_2$ complexes were lower than those of the DPAC dye/TiO$_2$ complexes, with the *cis* dye/TiO$_2$ complexes showing lower negative ΔG_{inject} values than the *trans* dye/TiO$_2$ complexes. This implied that the *trans* dyes would exhibit a faster electron injection. The regeneration efficiency (η_{reg}), another important factor that affects the J_{SC}, is determined by the driving force of dye regeneration (ΔG_{reg}). ΔG_{reg} can be expressed as follows [45]:

$$\Delta G_{reg} = E_{redox} - E_{dye} \quad (12)$$

Table 5. Electron injection free energy (ΔG_{inject}), ground (E^{dye}) and excited (E^{dye*}) state oxidation potentials, vertical transition energy (E_{0-0}), total regeneration energy (ΔG_{reg}), and dipole moment perpendicular to the surface of TiO$_2$ (μ_{normal}) of DMAC and DPAC as isolated dyes and dye/TiO$_2$ complexes.

Dye	$-\Delta G_{inject}$	E_{dye}	E_{dye*}	E_{0-0}	ΔG_{reg}	μ_{normal}	eV_{OC}
E-DMAC	−1.649	5.234	2.351	2.883	0.434	12.2	1.243
Z-DMAC	−1.399	5.228	2.602	2.626	0.428	11.6	1.314
E-DPAC	−1.661	5.227	2.339	2.888	0.427	9.5	1.193
Z-DPAC	−1.351	5.203	2.649	2.555	0.403	8.7	1.278
E-DMAC/TiO$_2$	−1.578	5.239	2.422	2.817	0.439	24.1	0.694
Z-DMAC/TiO$_2$	−1.365	5.235	2.636	2.601	0.435	19.9	0.700
E-DPAC/TiO$_2$	−1.663	5.220	2.337	2.883	0.420	21.1	0.680
Z-DPAC/TiO$_2$	−1.389	5.173	2.611	2.562	0.373	17.4	0.691

The ΔG_{reg} of the isolated dyes and dye/TiO$_2$ complexes are listed in Table 5. The ΔG_{reg} values of the DMAC dyes were higher than those of the DPAC dyes, which would result in a higher V_{OC} of the former. Additionally, the ΔG_{reg} values of the *trans* dyes were higher than those of the *cis* dyes. The dye/TiO$_2$ complexes showed a similar trend for ΔG_{reg} values as that of the isolated dyes. The ΔG_{reg} values of the DMAC dye/TiO$_2$ complexes were higher than those of the DPAC dye/TiO$_2$ complexes, whereas the ΔG_{reg} values of the *trans* dyes were higher than those of the *cis* dyes after adsorption onto the TiO$_2$ surface.

3.18. Factors Affecting Open Circuit Voltage

In DSSCs, the V_{OC} can be expressed by the following equation [46]:

$$V_{OC} = \frac{E_{CB} + E_{CB}}{q} + \frac{k_b T}{q} \ln\left(\frac{n_c}{N_{CB}}\right) - \frac{E_{redox}}{q}, \quad (13)$$

where q is the unit charge, $k_b T$ is the thermal energy, n_c is the number of electrons in the CB, N_{CB} is the density of accessible states in the CB, and E_{redox} is the electrolyte Fermi level. ΔE_{CB} denotes the shift in E_{CB} when the dyes are adsorbed on the substrate and is defined as follows [47]:

$$E_{CB} = -\frac{\mu_{normal}}{\varepsilon_0 \varepsilon}, \quad (14)$$

where μ_{normal} is the dipole moment of an individual dye perpendicular to the surface of the semiconductor substrate; γ is the surface concentration of dyes; and ε_0 and ε represent the vacuum permittivity and dielectric permittivity, respectively. Thus, μ_{normal} is a key factor in determining V_{OC}. To analyze the relationship with the LUMO, V_{OC} can be expressed by the following formula [48]:

$$eV_{OC} = E_{LUMO} - E_{CB} \tag{15}$$

To obtain a higher eV_{OC}, the E_{LUMO} should be as high as possible. The μ_{normal} and eV_{OC} values were calculated and are given in Table 5 for the isolated dyes and dye/TiO$_2$ complexes. The μ_{normal} values of the DMAC dyes were higher than those of the DPAC dyes, while the μ_{normal} values of the *trans* dyes were higher than those of the *cis* dyes for both DMAC and DPAC dyes. The eV_{OC} values of the two dyes decreased in the order of Z-DMAC > Z-DPAC > E-DMAC > E-DPAC, which indicated that the DMAC dyes had higher eV_{OC} compared to the corresponding isomers of the DPAC dyes. Interestingly, the eV_{OC} values of the *trans* dyes were lower than those of the *cis* dyes owing to the lower energy level of the LUMO. Although the *cis* dyes showed a higher eV_{OC}, there was a possibility of electron back-transfer because of the short distance between the cation and TiO$_2$ surface in the *cis* structure (Figure S2), which lowered the actual V_{OC}. After binding onto the TiO$_2$ surface, the μ_{normal} values of the dye/TiO$_2$ complexes increased to approximately twice of those of the isolated dyes. Moreover, the μ_{normal} values of the *trans* dye/TiO$_2$ complexes were higher than those of the *cis* dye/TiO$_2$ complexes. Because of the increase in the μ_{normal}, values, the eV_{OC} values decreased. It was found that the eV_{OC} values of the dye/TiO$_2$ complexes were approximately half those of the isolated dyes. This suggests that after adsorbing onto the TiO$_2$ surface, the dyes showed better μ_{normal} and eV_{OC} values compared with the isolated dyes, which improved both V_{OC} and n. However, the dye/TiO$_2$ complexes exhibited no distinct change in eV_{OC} because the LUMO energy levels were very similar for all the dye/TiO$_2$ complexes (Table 3). Equation (13) provides only an ideal value for V_{OC}. However, the real V_{OC} of a DSSC is generally lower than the theoretical limit because of a backward reaction between the electrons and the redox electrolyte [49]. If the photogenerated electrons are not rapidly transferred to the conducting substrate, the facile recombination of the electrons and oxidized ionic species of the electrolyte will result in a downward photovoltage. Another factor that influences the efficiency of DSSCs is the reorganization energy (λ), which can represent the charge transfer characteristics based on the Marcus electron transfer theory [50]. To enhance the J_{SC}, the LHE and ϕ_{inject} need to be increased, while λ needs to be decreased. For fast electron transfer, the λ of the sensitizers must be low. The λ can also affect the kinetics of electron injection (K_{inject}), which can be described as follows [51]:

$$k_{inject} = Ae^{\left[\frac{-\lambda}{4k_BT}\right]}, \tag{16}$$

where A is a pre-exponential factor that depends on the strength of the electronic coupling between the dye and the surface, k_B is the Boltzmann constant, and T is the temperature. The λ can be divided into intermolecular and intramolecular recombination energies. The intermolecular recombination energy has no distinct effect on ICT. The energy of the neutral, cationic, and anionic molecules can be used to calculate the reorganization energy. Hence, the intramolecular recombination energy for hole/electron (λ_h/λ_e) transfer can be estimated as follows [52]:

$$\lambda_h = \left(E_0^+ - E_+\right) + \left(E_+^0 - E_0\right) = IP - HEP \tag{17}$$

and:

$$\lambda_e = \left(E_0^- - E_-\right) + \left(E_-^0 - E_0\right) = EA - EEP, \tag{18}$$

where E_0 represents the energy of the neutral molecule in the ground state, E_0^+/E_0^- represents the energy of the cation/anion with the geometry of the neutral molecule, and E_+^0/E_-^0 represents the energy of the neutral molecule with the geometry of the cationic/anionic state. HEP and EEP are the hole and electron extraction potentials, respectively. The λ values of all the dyes were calculated and the results are presented in Table S4. The total reorganization energies, λ_i (summation of λ_h^+ and λ_e^-), of the DMAC dyes were lower than those of the DPAC dyes, which implies that the DMAC dyes would exhibit faster electron transfer, higher J_{SC}, and consequently, better PCE. Furthermore, the λ_i values of the *trans* dyes were lower than those of the *cis* dyes. Thus, *trans* DMAC dyes were expected to show greater electron injection from the excited states to the TiO$_2$ CB owing to their high LHE and low λ_i.

4. Conclusions

Two D–π–A metal-free organic dyes featuring an azobenzene spacer were designed, and their structural, electronic, and optical properties were investigated. Moreover, the effects of the substituted donor groups, including the *trans-cis-trans* conformational change of the azobenzene π-spacer, on the photovoltaic properties were computationally investigated using DFT and TDDFT methods before and after dye adsorption on TiO$_2$ for DSSCs. The adsorption energy, FT-IR spectra, cation-to-TiO$_2$ distance, FMO, orbital energy gaps, UV-Vis absorption spectra, and other electronic and optical properties of the two dyes, such as MEP, CDD, NBO, polarizability, hyperpolarizability, and NTO were investigated. Additionally, the chemical reactivity parameters of the two dyes, including EA, IP, chemical hardness, electrophilicity power, and electron-donating strength were calculated. Moreover, the key parameters that were closely related to the short-circuit current density and open circuit voltage, including LHE, dipole moment, coupling constant, EBE, excited state lifetime, driving force of electron injection, dye regeneration, total reorganization energy, total dipole moment, and CB edge of the semiconductor were elucidated to determine the primary reasons for the difference in the photovoltaic performance of the two dyes.

The following conclusions were drawn from the calculated results: (i) All the dyes adsorbed well on the TiO$_2$ surface, with the DMAC dyes showing a higher electron transfer rate. (ii) The electron-donating strength affected the geometric properties of the dyes, owing to the alteration of the bond lengths. (iii) The DPAC dyes showed a bathochromic shift, compared to the DMAC dyes. (iv) The *cis* dyes accelerated the recombination processes and facilitated electron back-transfer to either the cation or the electrolyte. (v) All the dyes showed ICT, which is essential for charge transfer. (vi) The *cis-trans* conformation did not significantly affect the ICT and the distribution of the FMO electrons, which indicated that azobenzene was a good π-spacer for ICT under illumination. (vii) The dye/TiO$_2$ complexes exhibited an indirect injection route because no new absorption bands appeared in the absorption spectra. (viii) After binding onto the TiO$_2$ surface, the dyes showed a lower HOMO-LUMO energy gap. (ix) NBO analysis revealed that the *trans* dyes showed a greater charge difference between the donor and acceptor moieties. (x) The lower chemical hardness and IP and the higher electrophilicity power and EA of the E-DMAC dye led to a higher J_{SC}, resulting in excellent PCE. (xi) Because of higher ΔG_{inject}, ΔG_{regen}, τ, μ_{normal}, eV_{OC}, and ΔE_{CB}, and smaller EBE and λ_I, the E-DMAC dye exhibited higher J_{SC} and V_{OC}. Thus, the DMAC dye was an outstanding candidate for DSSCs. It is expected that molecules with structures similar to that of the DMAC dye can retain photoelectric properties by molecular regulation. However, other properties like the stability (mechanical and thermal) and operability of the dye in actual environments, amount of dye adsorbed on the TiO$_2$ surface, and dye aggregation effects, which are not accounted for in this study, must be considered for better understanding of the photoelectrical properties and photovoltaic performance. These findings offer a new approach for the molecular design of dyes with desired absorption colors and will, thus, contribute to the development of novel dyes while providing crucial insights for elucidating the experimental data of DSSCs.

Supplementary Materials: The following are available online at http://www.mdpi.com/2079-4991/10/5/914/s1, Figure S1: FT-IR spectra of *trans* and *cis* isomers of (a) DMAC and (b) DPAC as isolated dyes and dye/TiO$_2$ complexes, Figure S2: Cation-to-TiO$_2$ surface distance of *trans* and *cis* isomers of (a) DMAC and (b) DPAC dyes, Figure S3: Natural transition orbitals for *trans* and *cis* isomers of DMAC and DPAC as (a) isolated dyes and (b) dye/TiO$_2$ complexes, Figure S4: Total density of states and partial density of states of *trans* and *cis* isomers of DMAC and DPAC as (a) isolated dyes and (b) dye/TiO$_2$ complexes, Figure S5: (a) Polarizability (α_{total}) and (b) hyperpolarizability (β_{total}) of *trans* and *cis* isomers of isolated DMAC and DPAC dyes., Table S1: Conjugative interaction energies ($\Delta E^{(2)}$, in kcal/mol) between the selected π and π^* orbitals, Table S2: Polarizability and hyperpolarizability of DMAC and DPAC dyes, Table S3: Electrophilicity index (ω), electron-accepting power (ω^+), ionization potential (IP), electron affinity (EA), and chemical hardness (η) of DMAC and DPAC as isolated dyes and dye/TiO$_2$ complexes, Table S4: Hole extraction potential, electron extraction potential, hole reorganization energy, electron reorganization energy, and total reorganization energy of DMAC and DPAC dyes.

Author Contributions: Conceptualization, J.H.; formal analysis, M.A.M.R.; writing—original draft preparation, M.A.M.R.; writing—review and editing, M.A.M.R., K.K., and J.H.; visualization, M.A.M.R. and D.H.; supervision, K.K. and J.H.; project administration, J.H.; funding acquisition, J.H. All authors have read and agreed to the published version of the manuscript.

Funding: This research was supported by the Korea Institute of Energy Technology Evaluation and Planning (KETEP) grant funded by the Ministry of Trade, Industry and Energy (MTIE) of Korea (No. 20193010014740), the National Research Foundation of Korea (NRF) Grant funded by the Ministry of Science and ICT (MSIT) for First-Mover Program for Accelerating Disruptive Technology Development (NRF-2018M3C1B9088457), and Basic Science Research Program through the NRF funded by the Ministry of Education (NRF-2019R1F1A1063669).

Acknowledgments: We acknowledge the Chung-Ang University Research Grants in 2019.

Conflicts of Interest: The authors declare no conflict of interest.

References

1. Hardin, B.E.; Snaith, H.J.; McGehee, M.D. The renaissance of dye-sensitized solar cells. *Nat. Photon.* **2012**, *6*, 162–169. [CrossRef]
2. Zhang, S.; Yang, X.; Numata, Y.; Han, L. Highly efficient dye-sensitized solar cells: Progress and future challenges. *Energy Environ. Sci.* **2013**, *6*, 1443–1464. [CrossRef]
3. Fakharuddin, A.; Jose, R.; Brown, T.M.; Fabregat-Santiago, F.; Bisquert, J. A perspective on the production of dye-sensitized solar modules. *Energy Environ. Sci.* **2014**, *7*, 3952–3981. [CrossRef]
4. Liang, M.; Chen, J. Arylamine organic dyes for dye-sensitized solar cells. *Chem. Soc. Rev.* **2013**, *42*, 3453–3488. [CrossRef]
5. Xu, M.; Li, R.; Pootrakulchote, N.; Shi, D.; Guo, J.; Yi, Z.; Zakeeruddin, S.M.; Grätzel, M.; Wang, P. Energy-level and molecular engineering of organic D-π-A sensitizers in dye-sensitized solar cells. *J. Phys. Chem. C* **2008**, *112*, 19770–19776. [CrossRef]
6. Cai, N.; Moon, S.-J.; Cevey-Ha, L.; Moehl, T.; Humphry-Baker, R.; Wang, P.; Zakeeruddin, S.M.; Grätzel, M. An organic D-π-A dye for record efficiency solid-state sensitized heterojunction solar cells. *Nano Lett.* **2011**, *11*, 1452–1456. [CrossRef] [PubMed]
7. Seo, D.; Park, K.W.; Kim, J.; Hong, J.; Kwak, K. DFT computational investigation of tuning the electron-donating ability in metal-free organic dyes featuring a thienylethynyl spacer for dye-sensitized solar cells. *Comput. Theor. Chem.* **2016**, *1081*, 30–37. [CrossRef]
8. Park, K.-W.; Serrano, L.A.; Ahn, S.; Baek, M.H.; Wiles, A.A.; Cooke, G.; Hong, J. An investigation of the role the donor moiety plays in modulating the efficiency of 'donor-π-acceptor-π-acceptor' organic DSSCs. *Tetrahedron* **2017**, *73*, 1098–1104. [CrossRef]
9. Rashid, M.A.M.; Hayati, D.; Kwak, K.; Hong, J. Computational Investigation of Tuning the Electron-Donating Ability in Metal-Free Organic Dyes Featuring an Azobenzene Spacer for Dye-Sensitized Solar Cells. *Nanomaterials* **2019**, *9*, 119. [CrossRef]
10. Merino, E.; Ribagorda, M. Control over molecular motion using the *cis-trans* photoisomerization of the azo group. *Beilstein J. Org. Chem.* **2012**, *8*, 1071–1090. [CrossRef]
11. Matczyszyn, K.; Sworakowski, J. Phase Change in Azobenzene Derivative-Doped Liquid Crystal Controlled by the Photochromic Reaction of the Dye. *J. Phys. Chem. B* **2003**, *107*, 6039–6045. [CrossRef]
12. Yu, X.; Wang, Z.; Buchholz, M.; Füllgrabe, N.; Grosjean, S.; Bebensee, F.; Bräse, S.; Wöll, C.; Heinke, L. Cis-to-*Trans* Isomerization of Azobenzene Investigated by Using Thin Films of Metal–Organic Frameworks. *Phys. Chem. Chem. Phys.* **2015**, *17*, 22721–22725. [CrossRef] [PubMed]
13. Teng, C.; Yang, X.; Yang, C.; Tian, H.; Li, S.; Wang, X.; Hagfeldt, A.; Sun, L. Influence of triple bonds as p-spacer units in metal-free organic dyes for dye-sensitized solar cells. *J. Phys. Chem. C* **2010**, *114*, 11305–11313. [CrossRef]

14. Martisnovich, N.; Troisi, A. Theoretical studies of dye-sensitized solar cells: From electronic structure to elementary processes. *Energy Environ. Sci.* **2011**, *4*, 4473–4495. [CrossRef]
15. Labat, F.; Bahers, T.L.; Ciofini, I.; Adamo, C. First-principles modeling of dye-sensitized solar cells: Challenges and perspectives. *Acc. Chem. Res.* **2012**, *45*, 1268–1277. [CrossRef]
16. Al-Eid, M.; Lim, S.; Park, K.-W.; Fitzpatrick, B.; Han, C.-H.; Kwak, K.; Hong, J.; Cooke, G. Facile synthesis of metal-free organic dyes featuring a thienylethynyl spacer for dye sensitized solar cell. *Dyes Pigment* **2014**, *104*, 197–203. [CrossRef]
17. Yang, J.; Ganesan, P.; Teuscher, J.; Moehl, T.; Kim, Y.J.; Yi, C.; Comte, P.; Pei, K.; Holcombe, T.W.; Nazeeruddin, M.K.; et al. Influence of the donor size in D–pi–A organic dyes for dye-sensitized solar cells. *J. Am. Chem. Soc.* **2014**, *136*, 5722–5730. [CrossRef]
18. Zarate, X.; Verdugo, S.S.; Serrano, A.R.; Schott, E. The Nature of the Donor Motif in Acceptor-Bridge-Donor Dyes as an Influence in the Electron Photo-Injection Mechanism in DSSCs. *J. Phys. Chem. A* **2016**, *120*, 1613–1624. [CrossRef]
19. Novir, S.B.; Hashemianzadeh, S.M. Quantum chemical investigation of structural and electronic properties of *trans*- and *cis*-structures of some azo dyes for dye-sensitized solar cells. *Comput. Theor. Chem.* **2017**, *1102*, 87–97. [CrossRef]
20. Hay, P.J.; Wadt, W.R. Ab initio Effective Core Potentials for Molecular Calculations. Potentials for the Transition Metal Atoms Sc to Hg. *J. Chem. Phys.* **1985**, *82*, 270–283. [CrossRef]
21. Yanai, T.; Tew, D.P.; Handy, N.C. A new hybrid exchange–correlation functional using the Coulomb-attenuating method (CAM-B3LYP). *Chem. Phys. Lett.* **2004**, *393*, 51–57. [CrossRef]
22. Tomasi, J.; Mennucci, B. Quantum Mechanical Continuum Solvation Models. *Chem. Rev.* **2005**, *105*, 2999–3093. [CrossRef] [PubMed]
23. Glendening, E.D.; Badenhoop, J.K.; Reed, A.E.; Carpenter, J.E.; Bohmann, J.A.; Morales, C.M.; Weinhold, F. *NBO 5.0*; Theoretical Chemistry Institute, University of Wisconsin: Madison, WI, USA, 2001.
24. Frisch, M.J.; Trucks, G.W.; Schlegel, H.B.; Scuseria, G.E.; Robb, M.A.; Cheeseman, J.R.; Scalmani, G.; Barone, V.; Petersson, G.A.; Nakatsuji, H.; et al. *Gaussian 16*; Gaussian, Inc.: Wallingford, CT, USA, 2016.
25. Hagfeldt, A.; Boschloo, G.; Sun, L.; Kloo, L.; Pettersson, H. Dye-Sensitized Solar Cells. *Chem. Rev.* **2010**, *110*, 6595–6663. [CrossRef] [PubMed]
26. Sánchez-de-Armas, R.; San Miguel, M.A.; Oviedo, J.; Sanz, J.F. Coumarin Derivatives for Dye Sensitized Solar Cells: A TD-DFT Study. *Phys. Chem. Chem. Phys.* **2012**, *14*, 225–233. [CrossRef]
27. Sánchez-de-Armas, R.; Oviedo López, J.; San-Miguel, M.A.; Sanz, J.F.; Ordejón, P.; Pruneda, M. Real-Time TD-DFT Simulations in Dye Sensitized Solar Cells: The Electronic Absorption Spectrum of Alizarin Supported on TiO_2 Nanoclusters. *J. Chem. Theory Comput.* **2010**, *6*, 2856–2865. [CrossRef]
28. Kreglewski, M. The geometry and inversion-internal rotation potential function of methylamine. *J. Mol. Spectrosc.* **1989**, *113*, 10–21. [CrossRef]
29. Pearson, R.; Lovas, F.J. Microwave spectrum and molecular structure of methylenimine (CH_2NH). *J. Chem. Phys.* **1977**, *66*, 4149–4156. [CrossRef]
30. Bouwstra, J.A.; Schouten, A.; Kroon, J. Structural studies of the system *trans*-azobenzene/*trans*-stilbene. I. A reinvestigation of the disorder in the crystal structure of *trans*-azobenzene, $C_{12}H_{10}N_2$. *Acta Crystallogr. Sect. C* **1983**, *39*, 1121–1123. [CrossRef]
31. Oviedo, M.B.; Zarate, X.; Negre, C.F.A.; Schott, E.; Arratia-Pérez, R.; Sánchez, C.G. Quantum dynamical simulations as a tool for predicting photoinjection mechanisms in dye-sensitized TiO_2 solar cells. *J. Phys. Chem. Lett.* **2012**, *3*, 2548–2555. [CrossRef]
32. Sánchez-de-Armas, R.; Oviedo, J.; San Miguel, M.Á.; Sanz, J.F. Direct vs indirect mechanisms for electron injection in dye-sensitized solar cells. *J. Phys. Chem. C* **2011**, *115*, 11293–11301. [CrossRef]
33. Asbury, J.B.; Wang, Y.-Q.; Hao, E.; Ghosh, H.N.; Lian, T. Evidences of hot excited state electron injection from sensitizer molecules to TiO_2 nanocrystalline thin films. *Res. Chem. Intermed.* **2001**, *27*, 393–406. [CrossRef]
34. Wu, Y.; Marszalek, M.; Zakeeruddin, S.M.; Zhang, Q.; Tian, H.; Grätzel, M.; Zhu, W. High-Conversion Efficiency Organic Dye-Sensitized Solar Cells: Molecular Engineering on D-A-π-A Featured Organic Indoline Dyes. *Energy Environ. Sci.* **2012**, *5*, 8261. [CrossRef]
35. Kleinman, D.A. Nonlinear dielectric polarization in optical media. *Phys. Rev.* **1962**, *126*, 1977–1979. [CrossRef]
36. Hsu, C. The electronic couplings in electron transfer and excitation energy transfer. *Acc. Chem. Res.* **2009**, *42*, 509–518. [CrossRef]

37. Koopmans, T.A. Über die Zuordnung von Wellenfunktionen und Eigenwerten zu den Einzelnen Elektronen Eines Atoms. *Physica* **1934**, *1*, 104–113. [CrossRef]
38. Mozer, A.J.; Wagner, P.; Officer, D.L.; Wallace, G.G.; Campbell, W.M.; Miyashita, M.; Sunahara, K.; Mori, S. The origin of open circuit voltage of porphyrin-sensitised TiO_2 solar cells. *Chem. Commun.* **2008**, *39*, 4741–4743. [CrossRef]
39. Li, M.; Kou, L.; Diao, L.; Zhang, Q.; Li, Z.; Wu, Q.; Lu, W.; Pan, D.; Wei, Z. Theoretical study of WS-9-Based organic sensitizers for unusual vis/NIR absorption and highly efficient dye-sensitized solar cells. *J. Phys. Chem. C* **2015**, *119*, 9782–9790. [CrossRef]
40. Grätzel, M. Recent Advances in sensitized mesoscopic solar cells. *Acc. Chem. Res.* **2009**, *42*, 1788–1798. [CrossRef]
41. Ardo, S.; Meyer, G.J. Photodriven heterogeneous charge transfer with transition-metal compounds anchored to TiO_2 semiconductor surfaces. *Chem. Soc. Rev.* **2009**, *38*, 115–164. [CrossRef]
42. Preat, J.; Michaux, C.; Jacquemin, D.; Perpète, E.A. Enhanced Efficiency of Organic Dye-Sensitized Solar Cells: Triphenylamine Derivatives. *J. Phys. Chem. C* **2009**, *113*, 16821–16833. [CrossRef]
43. Katoh, R.; Furube, A.; Yoshihara, T.; Hara, K.; Fujihashi, G.; Takano, S.; Murata, S.; Arakawa, H.; Tachiya, M. Efficiencies of electron injection from excited N3 dye into nanocrystalline semiconductor (ZrO_2, TiO_2, ZnO, Nb_2O_5, SnO_2, In_2O_3) films. *J. Phys. Chem. B* **2004**, *108*, 4818–4822. [CrossRef]
44. Islam, A.; Sugihara, H.; Arakawa, H. Molecular design of ruthenium(II) polypyridyl photosensitizers for efficient nanocrystalline TiO_2 solar cells. *J. Photochem. Photobiol. A* **2003**, *158*, 131–138. [CrossRef]
45. Daeneke, T.; Mozer, A.J.; Uemura, Y.; Makuta, S.; Fekete, M.; Tachibana, Y.; Koumura, N.; Bach, U.; Spiccia, L. Dye regeneration kinetics in dye-sensitized solar cells. *J. Am. Chem. Soc.* **2012**, *134*, 16925–16928. [CrossRef] [PubMed]
46. Marinado, T.; Nonomura, K.; Nissfolk, J.; Karlsson, M.K.; Hagberg, D.P.; Sun, L.; Mori, S.; Hagfeldt, A. How the nature of triphenylamine-polyene dyes in dye-sensitized solar cells affects the open-circuit voltage and electron lifetimes. *Langmuir* **2010**, *26*, 2592–2598. [CrossRef]
47. Preat, J.; Jacquemin, D.; Perpete, E.A. Towards new efficient dye-sensitised solar cells. *Energy Environ. Sci.* **2010**, *3*, 891–904. [CrossRef]
48. Zhang, C.-R.; Liu, Z.-J.; Chen, Y.-H.; Chen, H.-S.; Wu, Y.-Z.; Feng, W.; Wang, D.-B. DFT and TD-DFT study on structure and properties of organic dye sensitizer TA-St-CA. *Curr. Appl. Phys.* **2010**, *10*, 77–83. [CrossRef]
49. Jung, H.S.; Lee, J.K. Dye-Sensitized Solar Cells for Economically Viable Photovoltaic Systems. *J. Phys. Chem. Lett.* **2013**, *4*, 1682–1693. [CrossRef]
50. Marcus, R.A. Electron transfer reactions in chemistry. Theory and experiment. *Rev. Mod. Phys.* **1993**, *65*, 599. [CrossRef]
51. Zhang, Z.L.; Zou, L.Y.; Ren, A.M.; Liu, Y.F.; Feng, J.K.; Sun, C.C. Theoretical studies on the electronic structures and optical properties of star-shaped triazatruxene/heterofluorene co-polymers. *Dyes Pigment* **2013**, *96*, 349–363. [CrossRef]
52. Hutchison, G.R.; Ratner, M.A.; Marks, T.J. Hopping Transport in Conductive Heterocyclic Oligomers: Reorganization Energies and Substituent Effects. *J. Am. Chem. Soc.* **2005**, *127*, 2339–2350. [CrossRef]

© 2020 by the authors. Licensee MDPI, Basel, Switzerland. This article is an open access article distributed under the terms and conditions of the Creative Commons Attribution (CC BY) license (http://creativecommons.org/licenses/by/4.0/).

Article

Quantum-Mechanical Assessment of the Energetics of Silver Decahedron Nanoparticles

Svatava Polsterová [1], Martin Friák [2,3,*], Monika Všianská [1,2] and Mojmír Šob [1,3,4]

[1] Department of Chemistry, Faculty of Science, Masaryk University, Kotlářská 2, 611 37 Brno, Czech Republic; polsterova@mail.muni.cz (S.P.); 230038@mail.muni.cz (M.V.); mojmir@ipm.cz (M.Š.)
[2] Central European Institute of Technology, CEITEC IPM, Institute of Physics of Materials, Czech Academy of Sciences, Žižkova 22, 616 62 Brno, Czech Republic
[3] Institute of Physics of Materials, Czech Academy of Sciences, Žižkova 22, 616 62 Brno, Czech Republic
[4] Central European Institute of Technology, CEITEC MU, Masaryk University, Kamenice 753/5, 625 00 Brno, Czech Republic
* Correspondence: friak@ipm.cz; Tel.: +420-532-290-400

Received: 14 February 2020; Accepted: 7 April 2020; Published: 16 April 2020

Abstract: We present a quantum-mechanical study of silver decahedral nanoclusters and nanoparticles containing from 1 to 181 atoms in their static atomic configurations corresponding to the minimum of the *ab initio* computed total energies. Our thermodynamic analysis compares T = 0 K excess energies (without any excitations) obtained from a phenomenological approach, which mostly uses bulk-related properties, with excess energies from *ab initio* calculations of actual nanoclusters/nanoparticles. The phenomenological thermodynamic modeling employs (i) the bulk reference energy, (ii) surface energies obtained for infinite planar (bulk-related) surfaces and (iii) the bulk atomic volume. We show that it can predict the excess energy (per atom) of nanoclusters/nanoparticles containing as few as 7 atoms with the error lower than 3%. The only information related to the nanoclusters/nanoparticles of interest, which enters the phenomenological modeling, is the number of atoms in the nanocluster/nanoparticle, the shape and the crystallographic orientation(s) of facets. The agreement between both approaches is conditioned by computing the bulk-related properties with the same computational parameters as in the case of the nanoclusters/nanoparticles but, importantly, the phenomenological approach is much less computationally demanding. Our work thus indicates that it is possible to substantially reduce computational demands when computing excess energies of nanoclusters and nanoparticles by *ab initio* methods.

Keywords: nanoparticles; thermodynamics; silver; decahedron; excess energy; *ab initio* calculations

---

1. Introduction

The silver nanoparticles are widely used as antiviral agents [1,2], sensors [3,4], catalysts [5], as nanoparticle solders [6,7] as well as in numerous others application. Nanoclusters, as extreme cases of nanoparticles, have a yet greater surface/volume ratio and different geometries and electronic structures when compared with their bulk counterparts. Theoretical computations constitute a very advantageous tool when studying nanoclusters as they can accurately determine many of their characteristics, such as their surface type, strain energies [8,9], phase diagrams [10] or information on their catalytic activity. Many studies reported that modifications of the surface energy can change the shape of a (nano-)particle and/or its melting temperature [11–13]. The surface energy of a nanoparticle is often considered as the most important factor in catalysis, crystal growth, sintering and other surface-related processes. The most stable surface geometry for nanoparticles of pure fcc transition metals is the {111} facet [14] but the situation can differ in multi-component cases [15].

Relative to the bulk, the {111} facet exhibits the highest density of atoms and the highest coordination number of surface atoms. The most stable structures of fcc nanoclusters include the icosahedron, cuboctahedron and decahedron [8]. Another energy contribution is that related to strain. The strain energy of the particle can be affected by many factors. As the ratio of surface to volume decreases, the effect of surface stress is more significant and leads to the compression of particles [8].

Our study is focused on decahedral particles which have very interesting plasmonic and optical properties [16] as well as catalytic possibilities due to high strain energy [9]. The decahedron and icosahedron are inherently strained due to twinning and unfilled volume [17]. In particular, the decahedral nanoclusters are balancing the surface stability of five tetrahedrons (see Figure 1), which exhibit the {111} facets, against the strain energy related to an internal unfilled gap of 7.35° and distortion induced by their twinned internal structure [9]. The actual shape of the studied nanoparticles can deviate from a prediction by the Wulff construction due to the influence of the internal strain and strain-associated strain energy (in particular in the case of intermediate states [18,19]).

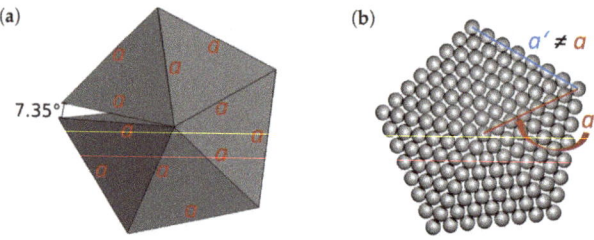

Figure 1. Schematic visualizations of (**a**) five tetrahedrons forming an imperfect decahedron with a gap of 7.35° and (**b**) one of the studied decahedral nanoclusters/nanoparticles (see below) without the gap. A characteristic length a defined here is used below when defining the shape factor.

2. Methods

The energies of studied decahedral nanoclusters and nanoparticles were calculated in two different ways which are both connected with quantum-mechanical Density Functional Theory (DFT) [20,21] calculations. The first method is a phenomenological thermodynamic modeling based on the CALPHAD method when the energy of nanoclusters and nanoparticles is approximated by a sum of relevant energy contributions corresponding (i) to a defect-free bulk material and (ii) surface energies and stresses (related to surfaces of a bulk, not nanoparticles) [22–26]. It is customary now that some or all energy contributions used in CALPHAD approach are computed using quantum-mechanical methods. Let us note that the idea of connecting the CALPHAD method and *ab initio* calculations is not trivial. It was presented first before the end and at the beginning of the new millennium in papers [27–32] and has been used many times since then (see e.g., [33–42]), also in studies of nanoalloys and nanoparticles, as mentioned above.

Our second approach is represented by direct *ab initio* calculation of electronic structure of decahedrons, schematically shown in Figure 2. The quantum-mechanical calculations are very computationally demanding in this case but still feasible for systems of a few hundred of atoms such as the studied decahedral nanoparticles/nanoclusters. Each of the computed nanoclusters/nanoparticles was treated inside a larger computational supercell where it was surrounded by vacuum, but the periodic boundary apply to these supercells. The positions of atoms in the nanoclusters were optimized so as to minimize the total energy which is provided by our *ab initio* software package (see Section 2.2). The total energy includes electronic-structure energy terms such as the Hartree energy, exchange-correlation energy, local ionic pseudopotential energy or kinetic energy as well as Madelung energy of the ions. The total energy is essentially related to T = 0 K without any entropy contributions and its minimum corresponds to the ground state of each nanocluster/nanoparticle.

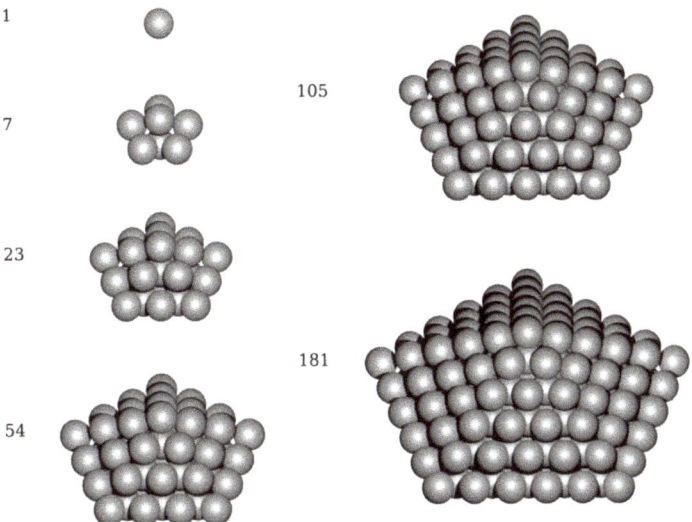

Figure 2. A schematic visualization of studied decahedron nanoclusters and nanoparticles (for higher number of atoms) with each of them accompanied by the number of atoms.

2.1. Phenomenological Thermodynamic Modeling

The phenomenological thermodynamic approach based on the CALPHAD method is very often used for calculations of the total energy of particles as well as for the prediction of phase diagrams [22–26,43]. The computations use an approximation when bulk variables are applied in the case of nanoparticles but not all properties of the (nano-)particles are included (for example, a structural disorder is sometimes omitted). The molar total Gibbs energy G_{tot} is decomposed into a sum of relevant contributions [24]:

$$G_{tot} = G_{ref} + G_{id} + G_E + G_{mag} + G_P + G_{sur}, \qquad (1)$$

where G_{ref} is the molar reference Gibbs energy, G_{id} is the molar energy of ideal mixing of an alloy, G_E is the molar excess Gibbs energy, G_{mag} is the molar contribution related to magnetism (which could be particularly complicated in the case of magnetic nanoparticles, including spin and orbital moment contributions as discussed, e.g., in Ref. [44]), G_P accounts for the influence of pressure P and G_{sur} for the molar contribution of surface energy. When adapting the Equation (1) to the studied case of silver nanoclusters/nanoparticles, i.e., an unary non-magnetic metal at the temperature $T = 0$ K, only the reference Gibbs energy G_{ref} and the surface Gibbs energy G_{sur} remain:

$$G_{tot} = G_{ref} + G_{sur}. \qquad (2)$$

The surface energy contribution is in the case of spherical nanoparticles equal to:

$$G_{sur} = \frac{A_{sph}}{n} \cdot \sigma_{sur} = \frac{A_{sph}}{n} \cdot \frac{V_{sph}}{V_{sph}} \cdot \sigma_{sur} = 3 \cdot \frac{V_m}{r} \cdot \sigma_{sur}, \qquad (3)$$

where σ_{sur} is surface energy, $A_{sph} = 4\pi r^2$ is the surface area of a spherical nanoparticle with the volume $V_{sph} = (4/3)\pi r^3$, n is is the number of moles and the V_m is the molar volume. As the volume of nanoclusters/nanoparticle is ill-defined, we below discuss three different ways of assigning the (molar) volume to the studied nanoclusters/nanoparticles. The radius r is then set equal to the radius of a sphere, which has the volume equal to the product of the number of particles in the nanocluster/nanoparticle and the (specifically assigned) volume (per atom). The surface area (to be multiplied by σ_{sur}) is then put equal to the surface of a sphere with the radius equal to the above discussed radius r. The fact, that the shape of studied nanoclusters/nanoparticle is non-spherical, decahedral, is taken into account by a shape factor C is introduced [25,43,45] into the Equation (3):

$$G_{sur} = 3 \cdot C \cdot \frac{V_m}{r} \cdot \sigma_s, \quad C = A_{shape}/A_{sph}, \qquad (4)$$

where the shape factor C is defined as the ratio of the surface area of the calculated nanoparticle A_{shape} to the surface area of a spherical particle with the same volume. For the decahedron we use

$$A_{dec} = a^2 \cdot 5 \cdot \sqrt{3} \cdot \sin 36° \cdot \sqrt{1 - 0.75 \cdot (\sin 36°)^2} \quad \text{and} \quad V_{dec} = a^3 \cdot (5/4) \cdot \sin 36° \cdot \cos 36° \qquad (5)$$

with the length parameter a defined in Figure 1 (also equal to the height of the decahedron, a pentagonal dipyramid). The values for a few commonly occurring shapes of (nano-)particles are listed in Table 1. Importantly, following the procedure described above we do not need a Tolman length to define the surface of particle [46] as it is defined by molar volume and radius of a spherical particle.

Table 1. Shape factors for different shapes of (nano-)particles as collected from selected literature sources. By a liquid spherical shape we mean an ideal sphere (without any atomic structure manifesting itself) while solid spherical shape represents a spherical nanoparticle with its atomic structure which is making its surface not ideally spherical.

Shape of Particle	Shapefactor	References
spherical - liquid	1.00	[12,22,47,48]
spherical - solid	1.05	[22,47,48]
regular icosahedron	1.06	[12,49]
regular dodecahedron	1.10	[49]
regular octahedron	1.18	[12,49]
cube	1.24	[12,49]
decahedron	1.28	this work
regular tetrahedron	1.49	[12,43]

Analogous to the Equations (2) and (4) is the total Gibbs energy of cluster, g_{tot} equal to:

$$g_{tot} = \frac{G_{tot}}{N_A} \cdot N = g_{ref} + g_{sur} \quad \text{where} \quad g_{ref} = N \cdot g_{Ag}^\phi \qquad (6)$$

where N is the number of atoms in a studied nanocluster/nanoparticle, N_A is the Avogadro constant and the g_{Ag}^ϕ is the atomic Gibbs energy of pure constituent Ag. As far as the Gibbs energy g_{Ag}^ϕ is concerned, we use the Gibbs energy of the bulk fcc Ag per atom E_{bulk}, calculated by DFT.

The equation of surface contribution for one nanocluster/nanoparticle is then changed to:

$$g_{sur} = N \cdot 3 \cdot C \cdot \frac{V_{at}}{r} \cdot \sigma_{sur} = N \cdot 3 \cdot C \cdot \frac{V_m}{r \cdot N_A} \cdot \sigma_{sur}. \qquad (7)$$

where V_{at} corresponds to the volume of one atom. One of the consequences of a high surface Gibbs energy is a surface strain [17]. It is caused by the minimization of the surface energy of the studied (nano-)particles and it leads to the reduction of the mean molar volume of the nanoparticle. As an extreme case, a particle without any surface energy exhibits zero surface strain and its volume is equal to that of the bulk.

Our phenomenological approach of calculating thermodynamic properties of silver nanoclusters/nanoparticles consisting of N atoms at temperature $T = 0\,K$ is based on the following procedure. First, the Gibbs energy of the studied nanoparticle has the two contributions mentioned in Equation (6), i.e., $g_{tot} = g_{ref} + g_{sur}$. Importantly, when evaluating the reference energy $g_{ref} = N \cdot g_{Ag}^\phi$ of the Ag nanoparticles we put the g_{Ag}^ϕ equal to the reference Gibbs energy of the bulk E_{bulk} (per atom), here fcc Ag, which we get from *ab initio* calculations.

All the changes, which are related to the fact, that we assess nanoparticles (and not the bulk), are included in the molar volume of the studied nanoparticles V_m and the surface stress P_{sur}, i.e., in the surface energy term G_{sur}:

$$G_{sur} = P_{sur} \cdot V_m \quad \text{where} \quad P_{sur} = \frac{G_{sur}}{V_m} = \frac{3 \cdot C \cdot \sigma_{sur}}{r}. \tag{8}$$

Two further approximations are made. First, the surface stress P_{sur} is evaluated for each relevant surface orientation, i.e., those existing on the facets of the studied nanoparticles, from the DFT calculations of infinite planar surfaces of the bulk system—see its schematic visualization in Figure 3. It means that we use a bulk-related property, the surface stress, instead of the surface stress (or surface energy) which would be related to any actual nanoparticles (there the surfaces contain edges and vertices where individual facets meet). In our particular case of decahedral nanoparticles, which have only {111} facets, our DFT calculations were performed for the (111) surface of fcc Ag.

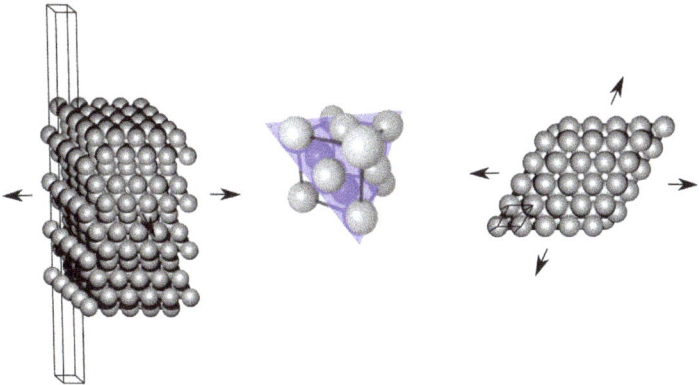

Figure 3. A schematic visualization of a computational cell, so-called slab, used for calculations of surface energy and surface stress in fcc-structure Ag (the visualization shows a 5 × 5 multiple of the studied primitive cell within the surface plane, a side view on the left, and a top view on the right). The surface is formed by the (111) crystallographic plane (see it in the middle also visualized inside a bulk fcc structure elementary cell).

The second important approximative step is related to the evaluation of the molar volume of the studied nanocluster/nanoparticle. In order to assign it to a particular decahedral nanoparticle, we put the surface stress P_{sur} equal to a fictitious hydrostatic pressure p which would be acting on every atom of the studied nanoclusters/nanoparticles. We thus do not take into account any elastic strains and stresses which are inside of decahedral nanoparticles due to the unfilled gap of 7.35°. Instead, we apply the surface stress to all particles as if it were a hydrostatic pressure p acting upon all atoms in the nanoparticle. For calculations of the molar volume of the particle (from the known molar volume of a bulk system) we apply the following three methods.

First, we use the Murnaghan equation [50,51] applied to the bulk:

$$p(V_m) = \frac{B_0}{B_0'} \cdot \left[\left(\frac{V_{m,0}}{V_m}\right)^{B_0'} - 1\right] \quad \text{and we put} \quad p(V_m) = P_{sur}. \tag{9}$$

The Murnaghan equation of state above contains the following (bulk-related) quatities: B_0 is the bulk modulus, B_0' is its pressure derivative and $V_{m,0}$ is the molar volume of the bulk material. The values of B_0, B_0' and $V_{m,0}$ are obtained from energy-volume curve of *ab initio* calculation of the bulk fcc Ag. The equation allows to assign the molar volume V_m to the studied nanocluster/nanoparticle.

The second way of assigning a volume to the studied nanoclusters/nanoparticles is based on the definition of the bulk modulus B_0 [52] where we use finite differences instead of derivatives:

$$B_0 = -V_m \cdot \frac{dp}{dV_m} \approx -V_m \cdot \frac{\Delta p}{\Delta V_m} = -V_m \cdot \frac{0 - p(V_m)}{V_{m,0} - V_m}, \quad \text{and where we put} \quad p(V_m) = P_{sur} \tag{10}$$

and there is one state of the bulk for which $p(V_m)$ is equal to 0, V_m is the value of bulk molar volume and the second state is such that there is a non-zero pressure $p(V_m)$ applied on the particle. Instead of extrapolating from $V_{m,0}$ using B' from the Murnaghan equation as in Equation (9), we put $p(V_m)$ equal to P_{sur} and determine the volume directly from Equation (10). Again, the bulk-related quantities, such as the bulk modulus B_0 and the molar volume of the bulk material $V_{m,0}$ are determined from quantum-mechanical calculations of bulk fcc Ag.

Third, in the following section we also consider the case when the volume of the atoms in the studies nanoclusters/nanoparticles is simply set to be equal to the volume of atoms in the bulk fcc Ag, i.e., the molar volume is not affected by the fact that we study a nanocluster/nanoparticle system.

The above described series of approximative steps, which we apply as a part of our phenomenological thermodynamic approach to nanoparticles, is tested by direct quantum-mechanical calculations of energies of static (T = 0 K) atomic configurations of a series of nanoclusters/nanoparticles visualized in Figure 2. The energy of these nanoclusters/nanoparticles is computed directly and compared to the results of the phenomenological thermodynamic approach described above (including three different ways of assigning the molar volume to the studied nanoparticle).

2.2. Parameters of Our DFT Calculations

All our DFT calculations were performed using the Vienna Ab-initio Simulation Package (VASP) [53,54]. The exchange and correlation energy was treated in the generalized gradient approximation (GGA) as parametrized by Perdew, Burke and Ernzerhof (PBE-96) [55]. The used Ag pseudopotential contains 1 s electron and 10 d electrons. We prefer the GGA-PBE exchange-correlation approximation over the the local density approximation (LDA) [21] because the former gives the value of the bulk modulus of silver closer to the experimental value (see the discussion below). Consequently, we assumed a better description of strained/stressed states. The cut-off plane-wave energy was equal to 550 eV and the employed spacing between k-points amounted to 0.11 Å^{-1}. When minimizing the total energy, the forces acting upon atoms of the surface slabs were reduced under 0.01 meV/Å while those acting upon atoms of the nanoparticles were minimized under 0.1 meV/Å.

The surface energies and stresses were determined from DFT calculations employing computational supercells with so-called slab geometry, see an example for the (111) surface [56] in Figure 3. The surface energy of an infinite slab σ_{hkl} (where {hkl} are mainly {111}, {100} and {110} for fcc-structure faces) was calculated as a difference of the relaxed surface energy $E_{sur}(N)$ and the relaxed bulk $E_{bulk}(N)$ per surface area S:

$$\sigma_{hkl} = E_{sur}(N) - E_{bulk}(N)/2 \cdot S, \tag{11}$$

with both energies being related to systems with the same number of atoms N.

Due to various shape of nanoparticles, the mean surface energy of nanoclusters/nanoparticles is computed according to the approach suggested by Guisbiers and Abudukelimu in [57]

$$\sigma_{sur} = \left(\sum A_{hkl} \cdot \sigma_{hkl} \right) / \sum A_{hkl}, \tag{12}$$

where A_{hkl} are areas of facets with different {hkl} crystallographic orientation on the surface of a nanocluster/nanoparticle. In the following, we put the energy $\sigma_{\{111\}}$ equal to the surface energy σ_{sur} of the whole decahedron particle as its surface contains only the {111} facets.

Next to the energy we also make an attempt to determine the molar volume of the studied nanoparticles from our quantum-mechanical calculations by the following steps. We first compute the mean radius as a half of the average inter-atomic distance between all the atoms and all their nearest neighbors—see Figure 4. Second, we put this mean radius equal to a radius of equally sized touching spheres in a fcc bulk crystal (as when computing the atomic packing factor, for fcc equal to 0.74). Third, we assign the atomic volume in such a fcc bulk crystal to each atom in our nanocluster.

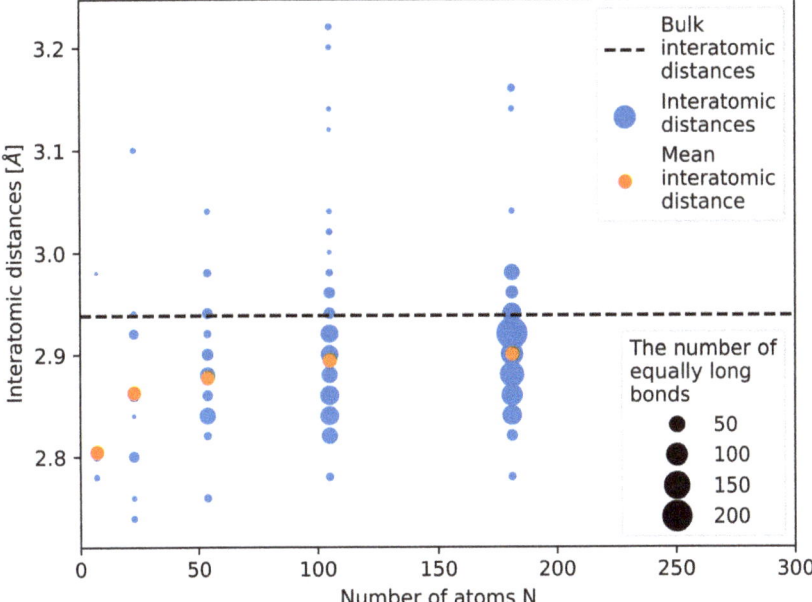

Figure 4. Computed bond lengths between pairs of nearest neighbors in the studied nanoclusters/nanoparticles. Blue symbols represent all inter-atomic distances in calculated nanoparticles, orange color marks the mean inter-atomic distances and the black dashed line shows the inter-atomic distance of bulk fcc structure of Ag.

The figure neatly shows that the studied nanoclusters/nanoparticles are highly strained. The majority of bond lengths (interatomic distances) is well below the bulk value of fcc Ag (see the horizontal black dashed line in Figure 4). In particular, this is true for the two nanoclusters with the number of atoms equal to 7 and 23. The mean (average) interatomic distance (see the orange data points) clearly demonstrates this reduction of the interatomic distances. It is worth noting that internal elastic strains (and the corresponding energies) are not included in our approximative phenomenological thermodynamic description of nanoclusters/nanoparticles (as described in the subsection above) but all particles are subject to a fictitious hydrostatic pressure (which we put equal to the surface stress value).

3. Results and Discussion

The molar volumes determined from direct quantum-mechanical calculations using the procedure of averaging the interatomic distances (see Figure 4) are compared with those determined from the Murhaghan equation and the definition of the bulk modulus within our thermormodynamic approach in Figure 5a. The volumes obtained from the direct calculations of the electronic structure of nanoclusters/nanoparticles are represented by the DFT black data points, and the molar volumes from our phenomenological thermodynamic approach are continuous blue and red lines for the volumes based on the Murnaghan equation and the definition of the bulk modulus, respectively. It is evident that the volumes from direct DFT calculations of nanoclusters/nanoparticles agree very well with those based on the definition of the bulk modulus (Equation (10)).

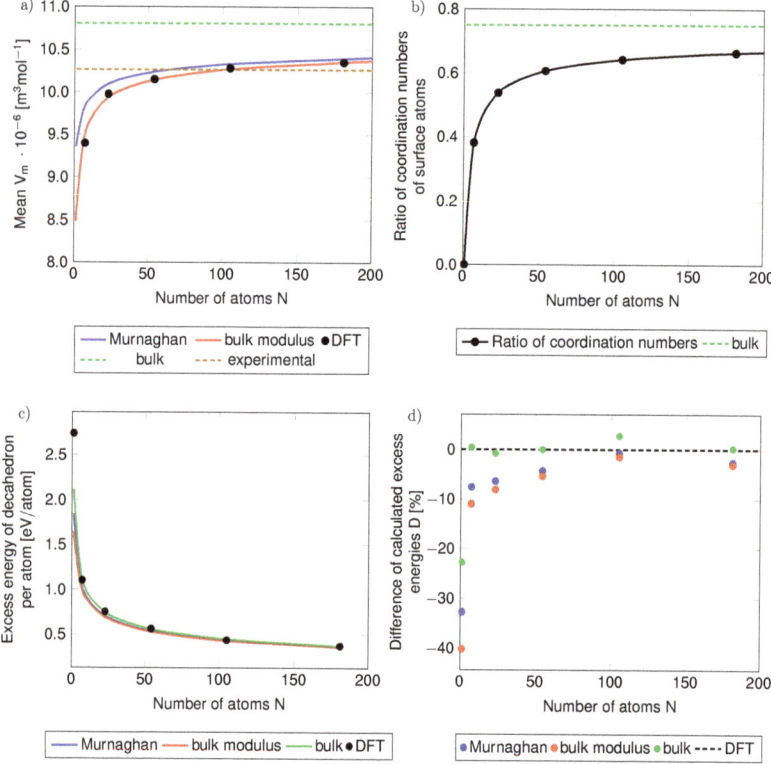

Figure 5. The dependencies of (**a**) the molar volume and (**b**) the ratio of coordination numbers of nanocluster surface atoms with respect to that of {111} surface of the bulk (which is equal to 9) as functions of the number of atoms in the studied nanoclusters/nanoparticles. The green dashed lines in parts (**a**,**b**) are the values corresponding to the bulk (or its {111} surface), blue and red lines represent volumes assigned to nanoparticles using Murnaghan equation and the definition of the bulk modulus, respectively. The full black circles represent results of direct DFT calculations of nanoparticles. The part (**c**) shows absolute (and part (**d**) also relative) differences of excess energies per atom determined by our phenomenological thermodynamic approach w.r.t. to the excess energies of the direct DFT calculations of nanoparticles when the volume of nanoparticles in the phenomenological thermodynamic modeling is determined from the Murnaghan equation (blue curves and blue data points in part (**d**)), from the definition of the bulk modulus (red curves and red data points) and from determining the volume of the nanoparticles from the volume of bulk fcc Ag (green lines and green data points). Also added is the experimental molar volume (the horizontal brown dashed line in (**a**)).

In order to determine the molar volumes from the Murnaghan equation of state and the definition of the bulk modulus (shown in Figure 5a) we used the hydrostatic pressure equal to the surface stress $p = P_{sur}$ which was found from the calculations of the surface energy of Ag for T = 0 K for the {111} and {100} terminations of the bulk fcc Ag. The obtained value of the surface energy for the {111} facet is equal to 0.80 Jm^{-2} and for the {100} surface orientation to 1.14 Jm^{-2}. Our values agree quite well with the experimental mean surface energy of 1.1–1.3 Jm^{-2} (see Table 2), reported for much higher temperature of 1073 K in Ref. [58], or with the theoretical values obtained using the LDA approximation in Ref. [59].

Table 2. Our computed surface energies for Ag surfaces with different crystallographic orientations in comparison with available experimental data [58].

	eV/atom	J/m^2
(111)	0.409	0.881
(100)	0.646	1.206
(110)	0.801	1.057
exp. [58] (1073 K)		1.1–1.3

Our calculations also reproduce fairly well the lattice constant and the bulk modulus of the bulk fcc Ag. Our theoretical lattice constant of fcc Ag is equal to 4.1555 Å in an acceptable agreement with the experimental value of 4.0853 Å. Our computed bulk modulus of 90 GPa lies between the experimental values of 84 GPa and 118 GPa [60].

In a similar way we analyze also the ratio of the coordination number of surface atoms of the studied nanoparticles, the coordination number of an fcc bulk lattice is 12. The coordination number of the surface atoms is lower. For an infinite surface of a bulk fcc (see the schematics in Figure 3) it is equal to 9 and so the ratio of the coordination numbers of surface atoms of the bulk with respect to the coordination number of atoms in the fcc bulk is 0.75 (see this value as the horizontal green dashed line Figure 5b). The coordination numbers of surface atoms at the {111} facets of the studied nanoclusters/nanoparticles apparently converge to the coordination number of surface atoms at the {111} surface of the bulk only very slowly as a function of the number of atoms in the nanoparticle.

Using our computational approaches it is now possible to evaluate an energy contribution related to the fact that the studied systems are nanoclusters/nanoparticles (with respect to the energy of the bulk). As this energy has a character of an excess energy E_{ex} (the total energy of nanoclusters/nanoparticles without the cohesion energy of the bulk):

$$E_{ex} = \frac{E_{tot} - N \cdot E_{bulk}}{N} \quad (13)$$

we show it (per atom) in Figure 5c as a function of the number of atoms in the studied nanoclusters/nanoparticles. The excess energy per atom decreases with increasing radius of nanoparticles. Let us note that this excess energy is different from the excess Gibbs energy G_E employed in Equation 1, similarly as in other papers dealing with nanoparticles, e.g., Ref. [61].

While Figure 5c clearly shows that the absolute values of the excess energies (per atom) as determined using (i) our phenomenological thermodynamic approach based on bulk-related properties (obtained by DFT calculations) very well match (ii) those from direct DFT calculations of actual nanoclusters/nanoparticles E_{ex}^{DFT}, it is important to evaluate the differences more precisely. Therefore, we analyze the excess energy differences as relative values:

$$D = \frac{E_{ex} - E_{ex}^{DFT}}{E_{ex}^{DFT}} \cdot 100\%. \quad (14)$$

The results are presented in Figure 5d again for differently defined volumes of nanoclusters/nanoparticles. Importantly, the Figure 5d clearly demonstrates that when using the phenomenological thermodynamic approach based on (i) the bulk reference energy, (ii) the bulk surface stress (slab calculations in Figure 3) and (iii) the bulk atomic volume, then the relative differences with respect to the energies obtained from direct DFT calculations of nanoclusters are only a few %, see the green data points in Figure 5d. The only exception from this nice agreement is the limiting case of a single silver atom (the relative error of the excess energy is over -22%). The actual values of the relative differences of the excess energy are summarized in Table 3.

The agreement can be interpreted so that that the surface-related energy of the phenomenological thermodynamic model of static configuration of atoms in a nanocluster/nanoparticle at $T = 0$ K (blue, red and green data points in Figure 5d) covers a vast majority of the excess energy which is determined by the DFT calculations of the actual nanoclusters (black horizontal dashed line in Figure 5d). We thus demonstrate that, in the case of the total energy of static atomic configurations of nanoparticles, the top-down phenomenological approach can be extended from the bulk down to nanoclusters containing essentially only a few atoms. The only necessary information related to the nanoparticle of interest is then (i) the number of atoms, (ii) the type of surface facets (their crystallographic orientation) and (iii) the shape of the nanocluster/nanoparticle.

The last aspect enters our phenomenological approach via the shape factor C (see Equation (4) and Table 1). Its importance is demonstrated in Figure 6 where the predictions of the phenomenological thermodynamic modeling are visualized for the same set of DFT values related to the bulk but for different values of the shape factor C corresponding to differently shaped nanoparticles. For a spherical nanoparticle, when the surface is not formed by planar {111} facets as in the case of decahedron, the surface energy was put equal to the average of surfaces energies obtained by DFT calculations of {100} and {111} surfaces.

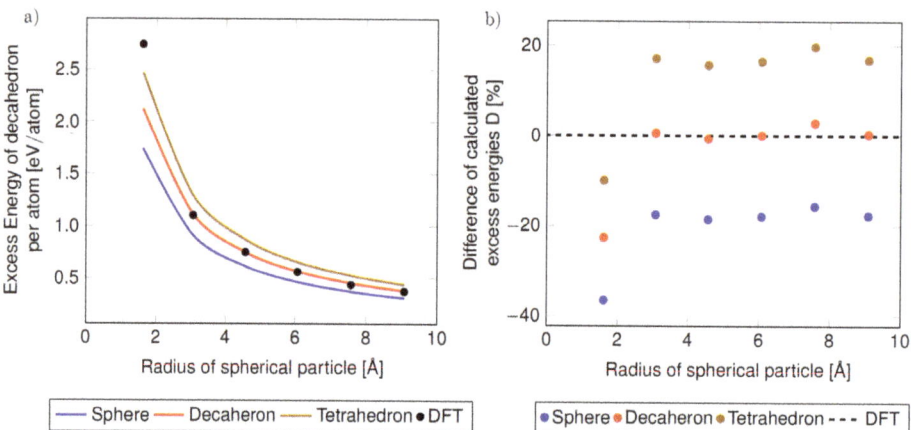

Figure 6. The computed excess energies of the studied nanoclusters/nanoparticles (per atom and as functions of the nanoparticle radius) as obtained from the phenomenological thermodynamic modeling when considering different shapes of the nanoparticles with the same number of atoms. The absolute excess energies are shown in part (**a**) and compared with the excess energies from the direct DFT calculations of the actual decahedral nanoparticles (shown as black symbols in part (**a**)). The differences of the excess energies are presented also relatively in part (**b**) with respect to the DFT values (horizontal black dashed line). The blue curve in part (**a**) and blue data points in part (**b**) correspond to the spherical shape, red to the decahedral shape and brown to the tetrahedral shape.

Table 3. Relative differences (Equation (14)) of the excess energy per atom (Equation (13)) shown in Figures 5 and 6.

D [%] of the Excess Energy E_{ex} Per Atom	Number of Atoms in the Nanocluster/Nanoparticle				
	1	7	23	54	105	181
from the Murnaghan in Figure 5d	−32.7	−7.5	−6.3	−4.3	−0.7	−2.5
from the bulk modulus in Figure 5d	−40.1	−10.9	−8.0	−5.3	−1.4	−3.0
from the bulk value in Figure 5d	−22.7	0.5	−0.8	0.0	2.7	0.2
spherical shape in Figure 6d	−36.6	−17.6	−18.6	−18.0	−15.7	−17.8
decahedral shape in Figure 6d	−22.7	0.5	−0.8	0.0	2.7	0.2
tetrahedral shape in Figure 6d	−10.0	16.9	15.5	16.4	16.6	16.7

Regarding the tetrahedron, the {111} surface energy was used similarly as in the case of the decahedron. As seen in Figure 6a,b, the best agreement between the total energies of nanoparticles determined from phenomenological thermodynamic modeling and those obtained from direct DFT calculations of nanoparticles is found when the actual (decahedral) shape of nanoparticles is considered. Our calculations also confirm the previous findings that the surface energy and strain energy change rapidly with the change of structure (accompanying the change of temperature [8]). In agreement with results published by Vollath et al. in Ref. [62] our analysis also demonstrates that changes of surface energy are not noticeable for nanocluster/nanoparticles sizes less than 4 Å.

Finally, it is worth mentioning that our conclusion, that the energy of Ag nanoclusters and nanoparticles can be quite reliably assessed using the volume of the bulk fcc Ag, agrees well with the concept of so-called surface area correction [61,63] which is related to the expansion of the electronic cloud around the nanoclusters/nanoparticles. This phenomenon is specifically important for small nanoparticles and the resulting volume is, in the case of nano-sized systems, put equal to that of the bulk material. Our recommended choice of the volume in the case of nanoclusters/nanoparticles (to be set equal to the volume of the same number of atoms in the bulk) is thus neatly justified also by the electronic structure of the discussed nanoparticles.

4. Conclusions

With the help of first-principles calculations, we investigated properties of silver decahedral nanoclusters/nanoparticles containing 1–181 atoms in their static atomic configurations corresponding to the minimum of the quantum-mechanically computed total energies. Our T = 0 K thermodynamic analysis compares excess energies (per atom) obtained from a phenomenological approach, which is mostly based on bulk-related properties, with excess energies of direct quantum-mechanical DFT calculations of actual nanoclusters/nanoparticles. We show that the phenomenological thermodynamic modeling, which uses (i) the bulk reference energy, (ii) surface energies obtained for infinite planar (bulk-related) surfaces and (iii) the bulk atomic volume can predict the excess energy per atom of the studied nanoclusters/nanoparticles with the error lower than 3% with the only exception being the limiting case of a single silver atom. This agreement is achieved when the bulk-related properties (the bulk reference energy, the atomic volume and surface energy) are determined by the *ab initio* calculations performed as much as possible on equal footing with direct quantum-mechanical calculations of the studied nanoclusters/nanoparticles, i.e., with the same computational parameters (the same exchange-correlation functional, energy cut-off, k-point density, ...). The only necessary information related to the nanoclusters/nanoparticles of interest, which enters the phenomenological thermodynamic modeling, is the number of atoms in the nanocluster/nanoparticle, their shape and the crystallographic orientations of facets. Importantly, the quantum-mechanical calculations of bulk-related properties are much less computationally demanding and we demonstrate that a top-down phenomenological approach can be extended from the bulk down to nanoclusters containing only a few atoms.

Our work thus indicates that it is possible to substantially reduce computational demands when assessing thermodynamic properties of nanoclusters and nanoparticles by quantum-mechanical methods. We would also like to emphasize that, importantly, the agreement between (i) our phenomenological modelling and (ii) the DFT energies for the actual nanoclusters has not been found sensitive to minor deviations of the shape of the studied nanoclusters from a geometrically ideal decahedral case (due to atomic relaxations in our DFT calculations). On the other hand, it should be noted that (i) our study does not cover any excitations, such as phonons, and (ii) whenever the absolute value of the excess energy, i.e., not per atom, is needed when thermodynamically assessing the stability of nanoclusters/nanoparticles, the deviation of the absolute excess energies as obtained from our method may change with the number of atoms (with respect to absolute excess energies from direct *ab initio* calculations of the studied nanoclusters/nanoparticles).

Author Contributions: Conceptualization, S.P. and M.F.; methodology, S.P., M.F. and M.V.; formal analysis, M.V.; resources, M.Š.; writing—original draft preparation, S.P.; writing—review and editing, M.F., M.V. and M.Š.; visualization, S.P. and M.V.; supervision, M.F.; project administration, M.Š. and M.F. All authors have read and agreed to the published version of the manuscript.

Funding: This research was funded by the Ministry of Education, Youth and Sports of the Czech Republic under the Project CEITEC 2020 (Project No. LQ1601) and by the Czech Science Foundation under the Projects "Stability and phase equilibria of bimetallic nanoparticles" (Project No. GA14-12653S) and "Structure and properties of selected nanocomposites" (Project No. GA 16-24711S).

Acknowledgments: We are very grateful to Jana Pavlů from Masaryk University in Brno, Czech Republic, for many fruitful discussions related to the CALPHAD modeling. Computational resources were provided by the Ministry of Education, Youth and Sports of the Czech Republic under the Projects CESNET (Project No. LM2015042), the Project CERIT Scientific Cloud (Project No. LM2015085) and by IT4Innovations National Supercomputer Center (Project No. LM2015070) within the program Projects of Large Research, Development and Innovations Infrastructures. Figures 2 and 3 were visualized using the VESTA package [64–66].

Conflicts of Interest: The authors declare no conflict of interest.The funders had no role in the design of the study; in the collection, analyses, or interpretation of data; in the writing of the manuscript, or in the decision to publish the results.

References

1. Daniel, S.C.G.K.; Tharmaraj, V.; Sironmani, T.A.; Pitchumani, K. Toxicity and Immunological Activity of Silver Nanoparticles. *Appl. Clay Sci.* **2010**, *48*, 547–551. [CrossRef]
2. Galdiero, S.; Falanga, A.; Vitiello, M.; Cantisani, M.; Marra, V.; Galdiero, M. Silver Nanoparticles as Potential Antiviral Agents. *Molecules* **2011**, *16*, 8894–8918. [CrossRef] [PubMed]
3. Bindhu, M.R.; Umadevi, M. Silver and Gold Nanoparticles for Sensor and Antibacterial Applications. *Spectrochim. Acta Part A Mol. Biomol. Spectrosc.* **2014**, *128*, 37–45. [CrossRef] [PubMed]
4. Chapman, R.; Mulvaney, P. Electro-Optical Shifts in Silver Nanoparticle Films. *Chem. Phys. Lett.* **2001**, *349*, 358–362. [CrossRef]
5. Grouchko, M.; Kamyshny, A.; Ben-Ami, K.; Magdassi, S. Synthesis of Copper Nanoparticles Catalyzed by Pre-Formed Silver Nanoparticles. *J. Nanoparticle Res.* **2009**, *11*, 713–716. [CrossRef]
6. Sopoušek, J.; Buršík, J.; Zálešák, J.; Buršíková, V.; Brož, P. Interaction of Silver Nanopowder with Copper Substrate. *Sci. Sinter.* **2011**, *43*, 33–38. [CrossRef]
7. Sopoušek, J.; Buršík, J.; Zálešák, J.; Pešina, Z. Silver Nanoparticles Sintering at Low Temperature on a Copper Substrate: In Situ Characterisation under Inert Atmosphere and Air. *J. Min. Metall. Sect. B Metall.* **2012**, *48*, 63–71. [CrossRef]
8. Ali, S.; Myasnichenko, V.S.; Neyts, E.C. Size-Dependent Strain and Surface Energies of Gold Nanoclusters. *Phys. Chem. Chem. Phys.* **2016**, *18*, 792–800. [CrossRef]
9. Patala, S.; Marks, L.D.; De La Cruz, M.O. Elastic Strain Energy Effects in Faceted Decahedral Nanoparticles. *J. Phys. Chem. C* **2013**, *117*, 1485–1494. [CrossRef]
10. Wang, J.; Lu, X.G.; Sundman, B.; Su, X. Thermodynamic Assessment of the Au-Ni System. *Calphad Comput. Coupling Phase Diagrams Thermochem.* **2005**, *29*, 263–268. [CrossRef]

11. Lu, H.M.; Li, P.Y.; Cao, Z.H.; Meng, X.K. Size-, Shape-, and Dimensionality-Dependent Melting Temperatures Of. *J. Phys. Chem. C* **2009**, *113*, 7598–7602. [CrossRef]
12. Qi, W.H.; Wang, M.P. Size and Shape Dependent Melting Temperature of Metallic Nanoparticles. *Mater. Chem. Phys.* **2004**, *88*, 280–284. [CrossRef]
13. Barnard, A.S. Using Theory and Modelling to Investigate Shape at the Nanoscale. *J. Mater. Chem.* **2006**, *16*, 813–815. [CrossRef]
14. Zhang, J.M.; Ma, F.; Xu, K.W. Calculation of the surface energy of FCC metals with modified embedded-atom method. *Appl. Surf. Sci.* **2004**, *229*, 34–42. [CrossRef]
15. Quesne, M.G.; Roldan, A.; de Leeuw, N.H.; Catlow, C.R.A. Bulk and surface properties of metal carbides: Implications for catalysis. *Phys. Chem. Chem. Phys.* **2018**, *20*, 6905–6916. [CrossRef]
16. Gao, Y.; Jiang, P.; Song, L.; Wang, J.X.; Liu, L.F.; Liu, D.F.; Xiang, Y.J.; Zhang, Z.X.; Zhao, X.W.; Dou, X.Y.; et al. Studies on Silver Nanodecahedrons Synthesized by PVP-Assisted N,N-Dimethylformamide (DMF) Reduction. *J. Cryst. Growth* **2006**, *289*, 376–380. [CrossRef]
17. Sneed, B.T.; Young, A.P.; Tsung, C.K. Building up Strain in Colloidal Metal Nanoparticle Catalysts. *Nanoscale* **2015**, *7*, 12248–12265. [CrossRef]
18. Pietrobon, B.; Kitaev, V. Photochemical Synthesis of Monodisperse Size-Controlled Silver Decahedral Nanoparticles and Their Remarkable Optical Properties. *Chem. Mater.* **2008**, *20*, 5186–5190. [CrossRef]
19. Zhao, H.; Qi, W.; Ji, W.; Wang, T.; Peng, H.; Wang, Q.; Jia, Y.; He, J. Large Marks-decahedral Pd nanoparticles synthesized by a modified hydrothermal method using a homogeneous reactor. *J. Nanoparticle Res.* **2017**, *19*, 162. [CrossRef]
20. Hohenberg, P.; Kohn, W. Inhomogeneous Electron Gas. *Phys. Rev.* **1964**, *136*, B864–B871. [CrossRef]
21. Kohn, W.; Sham, L.J. Self-Consistent Equations Including Exchange and Correlation Effects. *Phys. Rev.* **1965**, *140*, A1133–A1138. [CrossRef]
22. Sopoušek, J.; Vřešťál, J.; Pinkas, J.; Brož, P.; Buršík, J.; Styskalik, A.; Skoda, D.; Zobač, O.; Lee, J. Cu–Ni Nanoalloy Phase Diagram – Prediction and Experiment. *Calphad* **2014**, *45*, 33–39. [CrossRef]
23. Sopoušek, J.; Pinkas, J.; Brož, P.; Buršík, J.; Vykoukal, V.; Škoda, D.; Stýskalík, A.; Zobač, O.; Vřešťál, J.; Hrdlička, A.; et al. Ag-Cu Colloid Synthesis: Bimetallic Nanoparticle Characterisation and Thermal Treatment. *J. Nanomater.* **2014**, *2014*, 638964. [CrossRef]
24. Kroupa, A.; Káňa, T.; Buršík, J.; Zemanová, A.; Šob, M. Modelling of Phase Diagrams of Nanoalloys with Complex Metallic Phases: Application to Ni–Sn. *Phys. Chem. Chem. Phys.* **2015**, *17*, 28200–28210. [CrossRef] [PubMed]
25. Kroupa, A.; Vykoukal, V.; Káňa, T.; Zemanová, A.; Pinkas, J.; Šob, M. The Theoretical and Experimental Study of the Sb-Sn Nano-Alloys. *Calphad* **2019**, *64*, 90–96. [CrossRef]
26. Vykoukal, V.; Zelenka, F.; Bursik, J.; Kana, T.; Kroupa, A.; Pinkas, J. Thermal properties of Ag@Ni core-shell nanoparticles. *Calphad* **2020**, *69*, 101741. [CrossRef]
27. Wang, L.; Šob, M.; Havránková, J.; Vřešťál, J. First-principles Calculations of Formation Energy in Cr-based σ-phases. In Proceedings of the CALPHAD XXVII, Beijing, China, 17–22 May 1998; Abstract Book; p. 14.
28. Vřešťál, J.; Houserová, J.; Šob, M.; Friák, M. Calculation of Phase Equilibria with σ-phase in Some Cr-based Systems Using First-principles Calculation Results. In Proceedings of the 16th Discussion Meeting on Thermodynamics of Alloys (TOFA), Stockholm, Sweden, 8–11 May 2000; Abstract Book; p. 33.
29. Friák, M.; Šob, M.; Houserová, J.; Vřešťál, J. Modeling the σ-phase Based on First-principles Calculations Results. In Proceedings of the CALPHAD XXIX, Cambridge, MA, USA, 18–23 June 2000; Abstract Book; p. 4.
30. Vřešťál, J. Recent progress in modelling of sigma-phase. *Arch. Metall.* **2001**, *46*, 239–247.
31. Havránková, J.; Vřešťál, J.; Wang, L.G.; Šob, M. Ab initio analysis of energetics of σ-phase formation in Cr-based systems. *Phys. Rev. B* **2001**, *63*, 174104. [CrossRef]
32. Burton, B.; Dupin, N.; Fries, S.; Grimvall, G.; Guillermet, A.; Miodownik, P.; Oates, W.; Vinograd, V. Using *ab initio* calculations in the CALPHAD environment. *Z. Met.* **2001**, *92*, 514–525.
33. Kaufman, L.; Turchi, P.; Huang, W.; Liu, Z.K. Thermodynamics of the Cr-Ta-W system by combining the Ab Initio and CALPHAD methods. *Calphad* **2001**, *25*, 419–433. [CrossRef]
34. Houserová, J.; Vřešťál, J.; Šob, M. Phase diagram calculations in the Co–Mo and Fe–Mo systems using first-principles results for the sigma phase. *Calphad* **2005**, *29*, 133–139. [CrossRef]
35. Turchi, P.E.A.; Abrikosov, I.A.; Burton, B.; Fries, S.G.; Grimvall, G.; Kaufman, L.; Korzhavyi, P.; Manga, V.R.; Ohno, M.; Pisch, A.; et al. Interface between quantum-mechanical-based approaches, experiments,

and CALPHAD methodology. *Calphad-Comput. Coupling Phase Diagrams Thermochem.* **2007**, *31*, 4–27. [CrossRef]
36. Joubert, J.M. Crystal chemistry and Calphad modeling of the sigma phase. *Prog. Mater. Sci.* **2008**, *53*, 528–583. [CrossRef]
37. Liu, Z.K. First-Principles Calculations and CALPHAD Modeling of Thermodynamics. *J. Phase Equilibria Diffus.* **2009**, *30*, 517–534. [CrossRef]
38. Cacciamani, G.; Dinsdale, A.; Palumbo, M.; Pasturel, A. The Fe-Ni system: Thermodynamic modelling assisted by atomistic calculations. *Intermetallics* **2010**, *18*, 1148–1162. [CrossRef]
39. Schmetterer, C.; Khvan, A.; Jacob, A.; Hallstedt, B.; Markus, T. A New Theoretical Study of the Cr-Nb System. *J. Phase Equilibria Diffus.* **2014**, *35*, 434–444. [CrossRef]
40. Jacob, A.; Schmetterer, C.; Singheiser, L.; Gray-Weale, A.; Hallstedt, B.; Watson, A. Modeling of Fe-W phase diagram using first principles and phonons calculations. *CALPHAD-Comput. Coupling Phase Diagrams Thermochem.* **2015**, *50*, 92–104. [CrossRef]
41. Bigdeli, S.; Ehtesami, H.; Chen, Q.; Mao, H.; Korzhavy, P.; Selleby, M. New description of metastable hcp phase for unaries Fe and Mn: Coupling between first-principles calculations and CALPHAD modeling. *Phys. Status Solidi Basic Solid State Phys.* **2016**, *253*, 1830–1836. [CrossRef]
42. Wang, W.; Chen, H.L.; Larsson, H.; Mao, H. Thermodynamic constitution of the Al–Cu–Ni system modeled by CALPHAD and *ab initio* methodology for designing high entropy alloys. *Calphad* **2019**, *65*, 346–369. [CrossRef]
43. Leitner, J.; Sedmidubský, D. Thermodynamic Equilibria in Systems with Nanoparticles. In *Thermal Physics and Thermal Analysis: From Macro to Micro, Highlighting Thermodynamics, Kinetics and Nanomaterials*; Šesták, J., Hubík, P., Mareš, J.J., Eds.; Hot Topics in Thermal Analysis and Calorimetry; Springer International Publishing: Cham, Switzerlands, 2017; pp. 385–402. [CrossRef]
44. Hucht, A.; Sahoo, S.; Sil, S.; Entel, P. Effect of anisotropy on small magnetic clusters. *Phys. Rev. B* **2011**, *84*, 104438. [CrossRef]
45. Kaptay, G. The Gibbs Equation versus the Kelvin and the Gibbs-Thomson Equations to Describe Nucleation and Equilibrium of Nano-Materials. *J. Nanosci. Nanotechnol.* **2012**, *12*, 2625–2633. [CrossRef] [PubMed]
46. Molleman, B.; Hiemstra, T. Size and Shape Dependency of the Surface Energy of Metallic Nanoparticles: Unifying the Atomic and Thermodynamic Approaches. *Phys. Chem. Chem. Phys.* **2018**, *20*, 20575–20587. [CrossRef] [PubMed]
47. Lee, J.; Tanaka, T.; Lee, J.; Mori, H. Effect of Substrates on the Melting Temperature of Gold Nanoparticles. *Calphad Comput. Coupling Phase Diagrams Thermochem.* **2007**, *31*, 105–111. [CrossRef]
48. Sopoušek, J.; Vřešťál, J.; Zemanová, A.; Buršík, J. Phase Diagram Prediction and Particle Characterization of Sn-Ag Nano Alloy for Low Melting Point Lead-Free Solders. *J. Min. Metall. Sect. B Metall.* **2012**, *48*, 419–425. [CrossRef]
49. Yang, X.; Lu, T.; Kim, T. Effective Thermal Conductivity Modelling for Closed-Cell Porous Media with Analytical Shape Factors. *Transp. Porous. Med.* **2013**, *100*, 211–244. [CrossRef]
50. Tyuterev, V.; Vast, N. Murnaghan's Equation of State for the Electronic Ground State Energy. *Comput. Mater. Sci.* **2006**, *38*, 350–353. [CrossRef]
51. Murnaghan, F.D. The Compressibility of Media under Extreme Pressures. *Proc. Natl. Acad. Sci. USA* **1944**, *30*, 244–247. [CrossRef]
52. Timoshenko, S.; Goodier, J.N. (Eds.) *Theory of Elasticity*; McGraw-Hill: New York, NY, USA, 1951.
53. Kresse, G.; Hafner, J. Ab Initio Molecular Dynamics for Liquid Metals. *Phys. Rev. B* **1993**, *47*, 558–561. [CrossRef]
54. Kresse, G.; Furthmüller, J. Efficient Iterative Schemes for ab Initio Total-Energy Calculations Using a Plane-Wave Basis Set. *Phys. Rev. B* **1996**, *54*, 11169–11186. [CrossRef]
55. Perdew, J.P.; Burke, K.; Ernzerhof, M. Generalized Gradient Approximation Made Simple. *Phys. Rev. Lett.* **1996**, *77*, 3865–3868. [CrossRef] [PubMed]
56. Hjorth Larsen, A.; Jørgen Mortensen, J.; Blomqvist, J.; Castelli, I.E.; Christensen, R.; Dułak, M.; Friis, J.; Groves, M.N.; Hammer, B.; Hargus, C.; et al. *The Atomic Simulation Environment—A Python Library for Working with Atoms*; IOP Publishing: Bristol, UK, 2017; Volume 29. [CrossRef]
57. Guisbiers, G.; Abudukelimu, G. Influence of Nanomorphology on the Melting and Catalytic Properties of Convex Polyhedral Nanoparticles. *J. Nanoparticle Res.* **2013**, *15*, 1431. [CrossRef]

58. He, L.B.; Zhang, L.; Tan, X.D.; Tang, L.P.; Xu, T.; Zhou, Y.L.; Ren, Z.Y.; Wang, Y.; Teng, C.Y.; Sun, L.T.; et al. Surface Energy and Surface Stability of Ag Nanocrystals at Elevated Temperatures and Their Dominance in Sublimation-Induced Shape Evolution. *Small* **2017**, *13*, 1700743. [CrossRef] [PubMed]
59. Vitos, L.; Ruban, A.V.; Skriver, H.L.; Kollár, J. The Surface Energy of Metals. *Surf. Sci.* **1998**, *411*, 186–202. [CrossRef]
60. Properties: Silver—Applications and Properties of Silver. Available online: https://www.azom.com/properties.aspx?ArticleID=600 (accessed on 12 April 2020).
61. Holec, D.; Dumitraschkewitz, P.; Vollath, D.; Fischer, F.D. Surface Energy of Au Nanoparticles Depending on Their Size and Shape. *Nanomaterials* **2020**, *10*, 484. [CrossRef] [PubMed]
62. Vollath, D.; Fischer, F.D.; Holec, D. Surface Energy of Nanoparticles—Influence of Particle Size and Structure. *Beilstein J. Nanotechnol.* **2018**, *9*, 2265–2276. [CrossRef]
63. Holec, D.; Fischer, F.D.; Vollath, D. Structure and surface energy of Au$_{55}$ nanoparticles: An ab initio study. *Comput. Mater. Sci.* **2017**, *134*, 137–144. [CrossRef]
64. Momma, K.; Izumi, F. An integrated three-dimensional visualization system VESTA using wxWidgets. *Comm. Crystallogr. Comput. Iucr Newslett.* **2006**, *7*, 106. [CrossRef]
65. Momma, K.; Izumi, F. VESTA: a three-dimensional visualization system for electronic and structural analysis. *J. Appl. Crystallogr.* **2008**, *41*, 653–658. [CrossRef]
66. Momma, K.; Izumi, F. VESTA 3 for three-dimensional visualization of crystal, volumetric and morphology data. *J. Appl. Crystallogr.* **2011**, *44*, 1272–1276. [CrossRef]

© 2020 by the authors. Licensee MDPI, Basel, Switzerland. This article is an open access article distributed under the terms and conditions of the Creative Commons Attribution (CC BY) license (http://creativecommons.org/licenses/by/4.0/).

Article

Surface Energy of Au Nanoparticles Depending on Their Size and Shape

David Holec [1,*], Phillip Dumitraschkewitz [2], Dieter Vollath [3] and Franz Dieter Fischer [4]

1. Department of Materials Science, Montanuniversität Leoben, Franz Josef Straße 18, A-8700 Leoben, Austria
2. Chair of Nonferrous Metallurgy, Department of Metallurgy, Montanuniversität Leoben, Franz-Josef-Straße 18, A-8700 Leoben, Austria; phillip.dumitraschkewitz@unileoben.ac.at
3. NanoConsulting, Primelweg 3, D-76297 Stutensee, Germany; dieter.vollath@nanoconsulting.de
4. Institute of Mechanics, Montanuniversität Leoben, Franz Josef Straße 18, A-8700 Leoben, Austria; mechanik@unileoben.ac.at
* Correspondence: david.holec@unileoben.ac.at

Received: 14 February 2020; Accepted: 5 March 2020; Published: 8 March 2020

Abstract: Motivated by often contradictory literature reports on the dependence of the surface energy of gold nanoparticles on the variety of its size and shape, we performed an atomistic study combining molecular mechanics and ab initio calculations. We show that, in the case of Au nanocubes, their surface energy converges to the value for $(0\,0\,1)$ facets of bulk crystals. A fast convergence to a single valued surface energy is predicted also for nanospheres. However, the value of the surface energy is larger in this case than that of any low-index surface facet of bulk Au crystal. This fact can be explained by the complex structure of the surface with an extensive number of broken bonds due to edge and corner atoms. A similar trend was obtained also for the case of cuboctahedrons. Since the exact surface area of the nanoparticles is an ill-defined quantity, we have introduced the surface-induced excess energy and discuss this quantity as a function of (i) number of atoms forming the nano-object or (ii) characteristic size of the nano-object. In case (i), a universal power-law behaviour was obtained independent of the nanoparticle shape. Importantly, we show that the size-dependence of the surface energy is hugely reduced, if the surface area correction is considered due to its expansion by the electronic cloud, a phenomenon specifically important for small nanoparticles.

Keywords: surface energy; nanoparticles; gold; ab initio; molecular mechanics

---

1. Introduction

Surface energy is an important thermodynamic quantity. Particularly in cases where the volume-to-surface ratio becomes small, as is the case of nanoparticles, its relevance must not be underestimated [1,2].

There has been a vivid discussion concerning the qualitative trend of the surface energy as a function of the nanoparticle size. On the one hand, in many cases one finds reports on decreasing surface energy with decreasing particle size, e.g., in a study by Vollath and Fischer [3] or earlier studies [4,5]. This trend has been conventionally explained with an increasing tendency to form a liquid-like structure at the surface of the particles [6]. On the other hand, there exists a number of primarily theoretical papers finding a significant increase of the surface energy with decreasing particle size, see, e.g., Refs. [7–9]. Furthermore, there are also some heavily disputed experimental results indicating an increasing surface stress (and hence, due to a conventional assumption, also surface energy) with decreasing particle size [10,11]. Nanda et al. [11] pointed out that the difference between various reported trends stems from the nanoparticle nature. The surface energy is expected to increase for free nanoparticles with decreasing particle size, while the opposite trend is obtained for nanoparticles embedded in a matrix.

Wei and Chen [12] pointed out that, from the theoretical point of view, the trend could be qualitatively altered by changing the definition of a nanoparticle surface area. Unlike the energy change related to forming the free surface of a nanoparticle, the area is not well defined. Consequently, small changes of the radius/size yield large changes of the surface area, especially for nanometre-sized particles [12]. The rather geometrical argumentation of Ref. [12] was later linked to a physical quantity, a spatial expansion of the electronic cloud [13]. Using a refined, physically-based surface for small nanoparticles consequently leads to a weak-to-no size dependence of surface energy [14,15]. The latter reference also provided a thermodynamical-based model with predictive capabilities, hence seemingly resolving the enigma regarding the size dependence of the surface energy.

Nanoparticles, and particularly gold nanoparticles, nonetheless present a rich area of application as well as curiosity-driven research. Their applications span from biomimetic materials, over printed electronics to electrochemical biosensors [16,17]. Quite counterintuitively, the most preferable structure of a 55 Au atoms cluster was shown to be an amorphous structure even at 0 K [18], being a consequence of the small nanoparticle size. This prediction, however, was experimentally corroborated [18]. Ali et al. [9] predicted a rapid increase of the surface energy upon the nanoparticle melting. In agreement with earlier work of Shim et al. [19], they also predicted the decrease of melting temperature with decreasing nanoparticle size. Spontaneous segregation to some facets has been reported for Au-Ni nanoparticles, leading to an overall isotropic elastic response [20]. Another interesting effect is the shape variety of nanoparticles, accessible via solution synthesis modifying the surface energy in its very essence [17,21].

In the present study we, therefore, employ atomistic simulations to study the impact of nanoparticle shape on the resulting surface energy estimation. We focus on shapes ranging from rather artificial but geometrically simple nanocubes, over cuboctahedrons (special members of the truncated octahedrons, which have been reported as equilibrium shapes of Au nanoparticle), to nanospheres. In the final section we discuss how is the shape and size dependence of the surface-induced excess energy (i.e., the total nanoparticle surface energy) are related to number of broken bonds due to the creation of the free surface.

2. Methodology

Molecular mechanics (MM) simulations were performed using the LAMMPS package [22] together with an interatomic potential describing the gold interatomic interaction within the embedded atom method (EAM) as parametrised by Grochola et al. [23]. The individual idealised nanoparticles with well-defined shapes were cut out from bulk fcc structure with lattice constants of 4.0694 Å. This was obtained from fitting calculated total energies corresponding to different bulk volumes with Birch-Murnaghan equation of state [24], and agrees well with the values 4.0701 Å obtained by Grochola et al. [23]. All models were structurally relaxed using conjugate-gradient energy minimisation scheme at 0 K with force-stopping convergence criterion set to 10^{-12} eV/Å.

Additionally, a few ab initio runs were performed to benchmark our MM calculations. We used Vienna Ab initio Simulation Package (VASP) [25,26] implementation of Density Functional Theory (DFT) [27,28]. Two common approximations of the electronic exchange and correlation effects were considered: local density approximation (LDA) [28] and the Perdew–Wang parametrisation of the generalised gradient approximation (GGA) [29]. The contribution of ions and core electrons were described by projector augmented wave (PAW) pseudopotentials [30]. The plane wave cut-off energy was set to 400 eV, and the reciprocal space sampling was equivalent to $10 \times 10 \times 10$ k-mesh for the fcc-conventional cell. In directions, where periodicity should be avoided (e.g., the direction of the slab, all 3 directions in the case of nanoparticles), only a single k-point was used. In other directions, the number of k-points was scaled so that the k-point spacing in the reciprocal space was kept constant, i.e., $\approx \pi/(10 \cdot 4.069)$ Å^{-1} = 0.077 Å^{-1}, where 4.069 Å is the lattice parameter of fcc-Au.

Due to the employed periodic boundary conditions, we used a simulation box ≈20 Å larger than the actual (unrelaxed) nanoparticle to avoid any undesired interactions through the vacuum separating neighbouring nanoparticles. Similarly, ≈15 Å vacuum in the direction perpendicular to a free surface was used to separate slabs for calculating the surface energies of bulk Au. The electron charge was considered converged when the total energy of two subsequent self-consistency cycles differed by less than 10^{-4} eV, whereas structural optimisations were stopped when the total energy of two subsequent configurations differed by less than 10^{-3} eV. These criteria provide a total energy accuracy in the order of 1 meV/at. or better.

Finally, the qhull program [31] was used to calculate an area of a convex hull of ionic positions for each nanoparticle, to be used as an estimate of the surface area.

3. Results

3.1. Low-Index Facets of Bulk Au

The results presented in this chapter serve the subsequent discussion of the MM results, and their accuracy with respect to first principles calculations. Surface energy, γ, of a surface facet (hkl) can be calculated as

$$\gamma = \frac{1}{2A}(E_{slab} - NE_{bulk}), \tag{1}$$

where E_{slab} is energy of a slab composed of N layers. E_{bulk} is the energy of the bulk material per one layer of cross-section A. The factor 2 results from the fact that the slab has two surfaces. A layer is understood as a surface primitive cell, i.e., when the desired facet (hkl) is perpendicular to one of the lattice vectors (for a detailed description of the surface primitive cells, see e.g., Ref. [32]). Due to the interaction of the two free surfaces, either through the vacuum (i.e., not well separated slabs in the case of periodic boundary conditions) or the bulk of the slab (i.e., too thin slab), the value γ has to be converged with respect to both of these. In the case of MM simulations, only the latter convergence needs to be tested if the simulation is run in a box without periodic boundary conditions in the direction perpendicular to the free surface.

Test calculations revealed that vacuum of 10 Å is sufficient to get surface energy results converged to well below 1 meV/Å^2. Similarly, a slab thickness of about 40 Å is needed in order to avoid interactions of the free surfaces through the gold layer. The obtained values from the DFT benchmarks and MD simulations are summarised in Table 1. The here obtained DFT values are comparable with data from the literature. They exhibit the same ordering ($\gamma_{(110)} > \gamma_{(100)} > \gamma_{(111)}$) as reported earlier [33]. In a simplified picture, the surface energy expresses energy penalty related to the areal density of broken bonds [32,34]. This is $8/a_0^2$ for the (100) surface, $7.07/a_0^2$ for (110), and $4.33/a_0^2$ for the (111) surface (a_0 being the fcc lattice constant). The density of broken bonds is similar for the (100) and (110) surfaces, while it is significantly lower for the (111) orientated facet, hence providing a qualitative explanation for the surface energy ordering.

Table 1. Calculated surface energies for three low-index facets, including data from the literature for comparison.

	(100) [meV/Å²]	(100) [J/m²]	(110) [meV/Å²]	(110) [J/m²]	(111) [meV/Å²]	(111) [J/m²]
DFT-GGA (this work)	54.5	0.87	57.0	0.91	45.2	0.72
DFT-GGA (Ref. [35])					50	0.80
FCD-GGA [†] (Ref. [33])	101.5	1.63	106.1	1.70	80	1.28
MM (this work)	80.9	1.30			72.5	1.16
DFT-LDA (this work)	83.5	1.34	89.2	1.43	78.4	1.26
DFT-LDA (Ref. [35])					80	1.28
experiment (Ref. [36])					93.6	1.50
experiment (Ref. [37])					94.0	1.51

[†] FCD = full charge density.

The DFT and MM values exhibit an almost constant difference between the corresponding surface energies. Moreover, the MM values are very close to the DFT-LDA results. This is a somewhat surprising result since the EAM potential has been fitted to the DFT-GGA data using the same parametrisation by Perdew and Wang [29] as used here. We speculate that this is caused by fixing 4.07 Å as the lattice constant during the EAM potential fitting [23], as our LDA and GGA calculations yielded 4.061 and 4.176 Å, respectively. Nevertheless, since LDA and GGA are known to overestimate and underestimate, respectively, binding [38], and since the MM values are in between the two DFT-based estimations, we conclude that the interatomic potential used here is suitable for studying trends in surface energies. Moreover, the resulting values are expected to be very close to DFT-LDA calculations.

3.2. Impact of Shape and Size on the Nanoparticles Surface Energy

The surface energy of a gold nanoparticle consisting of N atoms is defined as an excess energy with respect to the energy of N atoms of bulk fcc gold, normalised to the nanoparticle surface area, A:

$$\gamma = \frac{E_{\text{nanoparticle}} - N E_{\text{bulk}}}{A}. \quad (2)$$

In the above, $E_{\text{nanoparticle}}$ is the total energy of the nanoparticle, while E_{bulk} is energy per atom of bulk fcc Au. Unlike the total energies, the surface area A is not a well defined quantity. In the following sections, an area of a convex hull of the relaxed ionic positions is consistently used as an estimate for A.

3.2.1. Nanocubes

In order to calculate the total energy of $\{100\}$-faceted nanocubes, structural models with a side length up to 20 nm were fully structurally relaxed. As a consequence of the surface tension, the apexes "popped in" as is apparent from the snapshot of relaxed atomic positions shown in Figure 1.

Supercells up to $3 \times 3 \times 3$ conventional fcc cell (172 atoms) were treated using the DFT, while nanocubes up to $50 \times 50 \times 50$ (515 151 atoms) were calculated using MM. A nanocube formed from $n \times n \times n$ conventional cubic fcc cells (4 atoms per cell) contains $N = 4n^3 + 6n^2 + 3n + 1$ of atoms. The calculated surface energy values shown in Figure 2 were fitted with an exponential relationship

$$\gamma = \gamma_0 \exp\left(\frac{\mathcal{L}}{a}\right), \quad (3)$$

where $a = n \cdot a_0$ is the side length of a cube formed by $n \times n \times n$ conventional fcc cells with the lattice parameter a_0. The quantities γ_0 and $\mathcal{L}$ are used as two fitting parameters. The thus obtained values of the pre-exponential parameter, $\gamma_0^{\text{GGA}} = 57.4\,\text{meV}/\text{Å}^2$, $\gamma_0^{\text{LDA}} = 89.8\,\text{meV}/\text{Å}^2$, and $\gamma_0^{\text{MM}} = 81.4\,\text{meV}/\text{Å}^2$ agree well with the bulk surface energies for the (100) facets ($\gamma_{(100)}^{\text{GGA}} = 54.5\,\text{meV}/\text{Å}^2$,

$\gamma_{(100)}^{LDA} = 83.5\,\text{meV}/\text{Å}^2$, and $\gamma_{(100)}^{MM} = 80.9\,\text{meV}/\text{Å}^2$). This is an expected result as the bulk values are limits for infinitely large cubes. It is, however, surprising that such a good agreement is obtained for the DFT data where only three data points are available for the fitting procedure. The same fitting procedure yielded for the parameter $\mathcal{L}$ (Equation (3)) values of 0.397 nm, 0.392 nm, and 0.661 nm for DFT-GGA, DFT-LDA, and MM data sets, respectively.

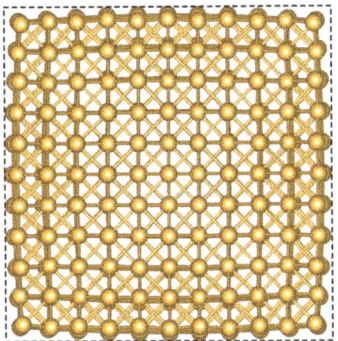

Figure 1. Relaxed structure of a nanocube with side $a = 2.035$ nm (666 atoms). The dashed line is a guide for the eye showing an ideal square shape.

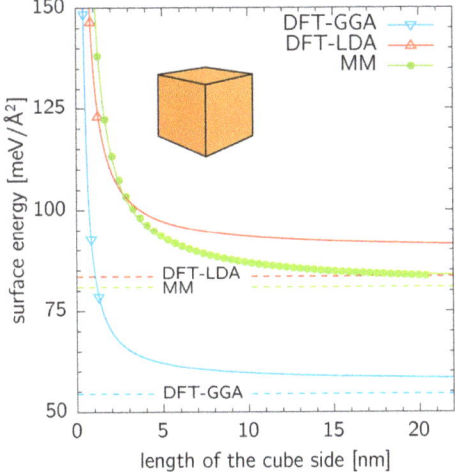

Figure 2. Surface energy of nanocubes calculated by DFT and MM. The calculated datapoints were fitted with Equation (3). The dashed lines are (100) surface energies as listed in Table 1.

3.2.2. Nanospheres

Nanospheres with all possible facet orientations were considered as an opposite extreme to the nanocubes with only a single orientation of their facets. They were constructed by cutting material contained in an ideal sphere of a given radius out of an infinitely large fcc Au crystal. The DFT calculations were performed up to $r = 0.9$ nm (152 atoms), while the MM calculations allowed easily for spheres up to $r = 20.3$ nm (2 094 177 atoms) (Figure 3). In comparison to the case of nanocubes, the surface energy of the nanospheres converges faster to a constant value of ≈ 94 meV/Å^2. This is a slightly higher value than γ of any low-index facet (cnf. Table 1) reflecting the fact that a spherical surface composes (from the atomistic point of view) of a large number differently orientated facets.

Places where these facets meet (i.e., edges) are composed of atoms with the same or higher number of broken bonds than atoms in the surrounding planar facets, thus, further increasing the surface energy.

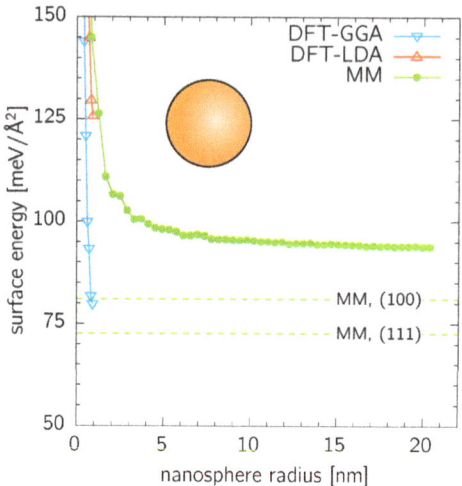

Figure 3. Surface energy of nanospheres calculated by DFT and MM. The dashed lines are the MM values for single-orientated (1 0 0) and (1 1 1) surfaces as listed in Table 1.

3.2.3. Cuboctahedrons

The last class of objects studied in this work are cuboctahedrons, i.e., (1 0 0)-faceted cubes with all apexes cut by (1 1 1) planes (see inset in Figure 4b). Cuboctahedrons are a special subset of truncated octahedrons with the all sites equally long. Figure 4a shows the total energy per atom plotted against the nanoparticle size in terms of the number of forming atoms for cuboctahedrons together with more general truncated octahedrons. The latter were generated with a build-in function of Atomic Simulation Environment toolkit [39] for various sizes of the truncated octahedron apexes. Obviously, the cuboctahedrons are not always the most convenient shape for a given number of atoms.

The surface energy of cuboctahedrons (Figure 4b) oscillates between two values, ≈78 and ≈90 meV/Å^2. This behaviour is caused by the changing ratio of surface atoms forming the (1 0 0) and (1 1 1) facets and the edges and corners, which directly corresponds with the atomistic nature of the nanoparticle. A detailed analysis of the coordination of the surface atoms reveals that the number of 9-coordinated surface atoms, corresponding to ideal (1 1 1) facets, is in anti-phase with the surface energy as shown in Figure 4b. The 8-coordinated (1 0 0) surface atoms also show small steps hence causing a non-monotonous increase of their number as a function of the cuboctahedron size. At the same time, the numbers of 10-, 7-, 6-, and 5-coordinated surface atoms forming edges and corners (i.e., atoms with even smaller coordination and, consequently, more broken bonds than those on ideal (1 0 0) and (1 1 1) facets, and hence increasing the overall surface energy), exhibit the same "oscillations" concerning the cuboctahedron size as the surface energy itself. Therefore, the oscillations are expected to decrease with increasing cuboctahedron size. It is interesting to note that the two limit values for the surface energies, ≈90 and ≈80 meV/Å^2, represent approximately the same range as the two values, 80.9 and 72.5 meV/Å^2 for pure (1 0 0) and (1 1 1) facets, respectively. Similarly to the case of nanospheres, the values are somewhat higher than the ideal single-orientated facets due to the presence of the edges and corners.

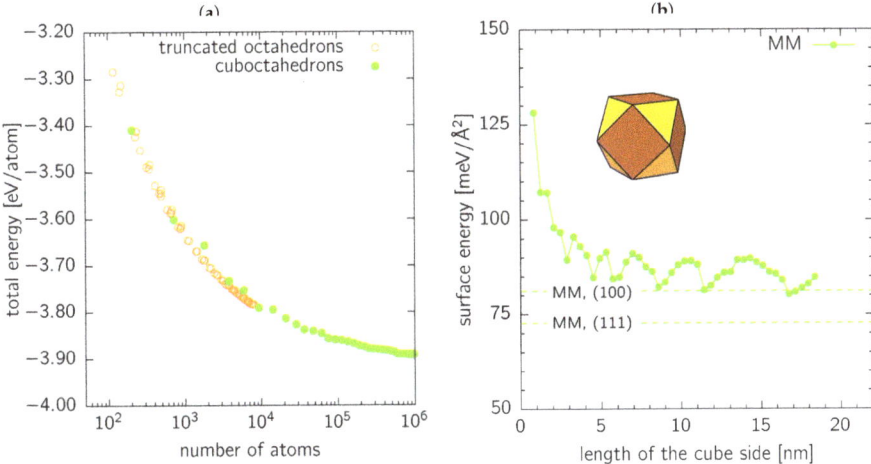

Figure 4. (a) Total energy per atom as a function of the nanoparticle size (in terms of number of forming atoms) for cuboctahedrons (full circles) and general truncated octahedrons (open circles). (b) Surface energy of cuboctahedrons calculated by MM and showed as a function of the size of "parent" cube. The dashed lines are the MM values for single-orientated (1 0 0) and (1 1 1) surfaces as listed in Table 1.

4. Discussion

4.1. Correction of the Surface Area for Electronic Cloud

The surface areas calculated in the previous parts corresponds to the convex hull of ionic positions. In our recent paper [13] dealing with predicting surface energy of Au_{55} cluster, we have discussed the error made by neglecting extend of the electronic cloud. There, a radius correction of 1.3–1.4 Å has been proposed under the assumption that the mass density of the nanocluster is the same as that of bulk fcc-Au. Note that, radius corrections of 0.5–0.8 Å have been proposed by de Heer [40].

In order to see how neglecting the electronic cloud layer actually influences the predicted surface energies, we re-evaluate the surface areas. Let $\{\vec{R}_i\}$ be a set of the atomic (ionic) positions defined with respect to the nanoparticle centre of mass, i.e.,

$$\sum_i \vec{R}_i = \vec{0}, \qquad (4)$$

where the sum is performed over all atoms in the nanoparticle. Subsequently, a new set of coordinates, $\{\vec{\tilde{R}}_i\}$, is defined as

$$\vec{\tilde{R}}_i = \left(|\vec{R}_i| + \Delta\right) \vec{R}_i^0 \qquad (5)$$

where $\vec{R}_i^0 = \vec{R}_i / |\vec{R}_i|$ is a unit vector along the direction of $\vec{R}_i$. This means that all atoms, and in particular those on the convex hull envelope, are shifted by Δ away from the nanoparticle centre of mass. A new surface area is calculated as a convex hull of $\{\vec{\tilde{R}}_i\}$ positions for several representative values of Δ.

The results are summarised in Figure 5 for all three nanoparticle geometries considered in the present work. In all cases, the surface energy decreases with increasing values of Δ, which is a simple consequence of the surface energy definition in Equation (2). It is, however, remarkable to notice that even for the largest nanoparticle sizes the surface energy reduction is still larger than 1% for the DFT-based electron cloud thickness. We, therefore, conclude that, especially for nanoparticles with specific sizes below 5 nm, the correction of the surface area due to the electronic cloud is essential. Moreover, it is likely that for the small nanoparticle sizes, the surface energies calculated here are overestimated due to that fact that even a lower energy can be obtained for a different atomic ordering than fcc (e.g., Mackay icosahedrons as in the case of Au_{55}) or even amorphous liquid-like structures [13]. Finally, it is worth noting that the problem of electronic cloud is not an issue in standard calculations of single orientated flat single crystal facets since it does not influence the actual surface area.

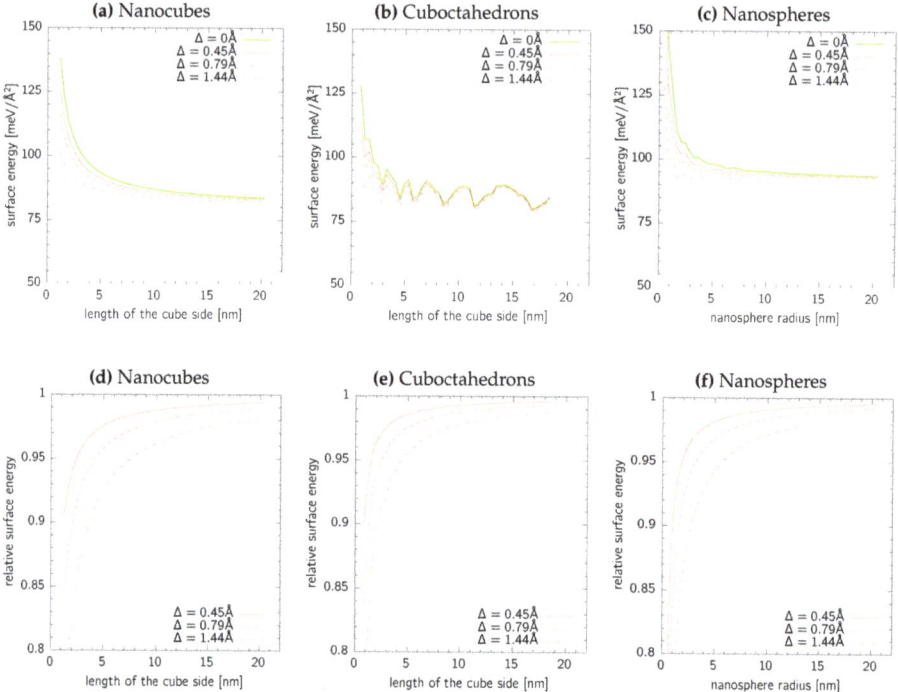

Figure 5. Corrected absolute (upper row) and relative values (lower row) of the surface energies for (a,d) nanocubes, (b,e) cuboctahedrons, and (c,f) nanosheres. The relative surface energies are calculated with respect to the values without correction for the electronic cloud thickness ($\Delta = 0$).

4.2. Surface Induced Excess Energy

As mentioned above and discussed in the literature, the surface area of nanoparticles is an ill-defined quantity. In order to eliminate this problem, we introduce a new quantity, E_{excess}, expressing the surface-induced excess energy with respect to the bulk energy corresponding to the same number, N, of atoms as in the nanoparticle, normalised to 1 atom, as

$$E_{excess} = \frac{E_{nanoparticle} - N E_{fcc\text{-}Au}}{N}. \tag{6}$$

A similar concept has been previously demonstrated to work also for energetics of carbon fullerenes [41], or even for elasticity of nanoporous gold [42]. If the excess energy, E_{excess}, is evaluated for nanocubes, nanospheres, and cuboctahedrons, a linear relationship between $\log E_{excess}$ and $\log N$ is obtained independent of the nanoparticle shape (Figure 6a). This suggests that the excess energy is a power law function of the total number of atoms (nanoparticle size). This fit (the dashed line in Figure 6a) gives

$$E_{excess} = 3523.3 \, \text{meV/atom} \times N^{-0.346}. \tag{7}$$

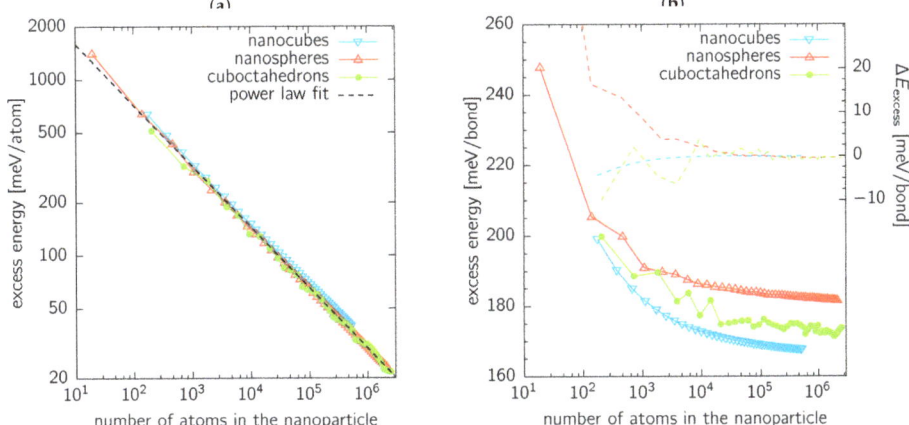

Figure 6. Excess energy, E_{excess}, of nanoparticles with respect to the bulk fcc Au as a function of the number, N, of atoms forming the nanoobject. E_{excess} is normalised to (**a**) number of the atoms forming the nanoparticle, and (**b**) to the number of broken bonds. The dashed lines in (**b**) show the difference between the actual value of E_{excess} as calculated by MM, and a fitted value using Equation (10).

Recalling the idea that the surface energy is genuinely connected with the broken bonds (bb), we now establish the energy needed to "break" a bond. Let us consider an $n \times n \times n$ nanocube containing atoms with 4 different nearest neighbour coordinations: 8 atoms with 9 bb forming corners (i.e., 3-coordinated atoms), $(12n - 12)$ atoms with 7 bb forming the edges (i.e., 5-coordinated atoms), $(12n^2 - 12n + 6)$ atoms with 4 bb forming the surface facets (i.e., 8-coordinated atoms), and $(4n^3 - 6n^2 + 3n - 1)$ bulk atoms with no bb (i.e., fully 12-coordinated atoms). If we simply assume that all bonds "cost" the the same energy E_{bond} to break them, the excess energy, E_{excess}, i.e., the sum of the contributions described above, follows as

$$E_{excess} = \left[9 \times 8 + 7 \times (12n - 12) + 4 \times (12n^2 - 12n + 6)\right] E_{bond}, \tag{8}$$

yielding $E_{bond} = 168.1$ meV/bond from fitting the nanocubes data.

However, the red triangles in Figure 6b, showing the nanocubes excess energy normalised to the number of broken bonds, clearly exhibit a non-constant value for E_{bond}. Consequently, we propose a slightly modified description in which the energy needed to break a bond is a (non-linear) function of the coordination. Hence, it costs different energy to create, e.g., a corner atom (9 broken bonds) than a facet atom (4 broken bonds). Thus the excess energy becomes

$$E_{excess} = 72 E_{corner} + 84(n-1) E_{edge} + 24(2n^2 - 2n + 1) E_{facet}. \tag{9}$$

Fitting yields $E_{corner} = 272.1\,meV/bond$, $E_{edge} = 215.2\,meV/bond$, and $E_{facet} = 166.0\,meV/bond$. It turns out that for nanocubes with side $\gtrsim 5\,nm$, Equation (9) provides predictions with an accuracy better than ≈ 1 meV/bond. Energy of a broken bond, corresponding to an infinitely large (100) facet, can be estimated from the surface energies as given in Table 1. This value is 167.5 meV/bond, which is close to $E_{bond} = 168.1\,meV/bond$ (Equation (8)) as well as $E_{facet} = 166.0\,meV/bond$ (Equation (9)).

The complex shapes of cuboctahedrons and nanospheres somewhat restrict the intuitive analysis of the excess energy above presented. When the excess energy is fitted with a single valued energy per broken bond (equivalent to Equation (8)), values of 172.8 meV/bond and 181.9 meV/bond are obtained for cuboctahedrons and nanospheres, respectively. These values represent an excellent estimation of the excess energies in the limit of large nanoparticles, as shown in Figure 6b. Moreover, the excess energy value for cuboctahedrons lies between the values estimated for (100) ($E_{(100)} = 167.5\,meV/bond$) and (111) ($E_{(111)} = 173.3\,meV/bond$) facets. This fact further illustrates that the surface energy values, as presented in Section 3.2, are remarkably influenced by the evaluation of the actual surface area (which is, from the atomistic point of view, ill-defined). Consequently, the mean value of the surface energy of cuboctahedrons as shown in Figure 4b lies outside the range bounded by $\gamma_{(100)}$ and $\gamma_{(111)}$ values.

Finally, in order to obtain a non-constant behaviour, we fit the excess energy with

$$E_{excess} = \sum_{i=1}^{11}(12-i)N(i)E(i) \qquad (10)$$

where $N(i)$ is the number of i-coordinated atoms (i.e., those having $(12-i)$ broken bonds) and $E(i)$ is the corresponding excess energy contribution. Equation (10) is a generalised formulation of Equation (9) reflecting that all possible coordinations may occur due to the shape of nanoparticles. We note that the smallest coordination obtained was 3 and 4 for the case of cuboctahedrons and nanospheres, respectively. The fitted values of $E(i)$ are given in Table 2, and the difference between the actual E_{excess} from MM and values predicted using Equation (10) is shown in Figure 6b with dashed lines. Obviously, the fit provides excellent agreement for nanoparticles containing $\approx 10^4$ atoms and more.

Table 2. Fitted coefficients $E(i)$ for the excess energy expression according to Equation (10). The index i expresses the coordination of atoms (i.e., $(12-i)$ is the number of broken bonds, bb).

	Nanocubes	Cuboctahedrons	Nanospheres
$E(3)$ [meV/bond]	272.1	287.3	0
$E(4)$ [meV/bond]	0	161.1	426.3
$E(5)$ [meV/bond]	215.2	243.4	258.3
$E(6)$ [meV/bond]	0	163.1	232.0
$E(7)$ [meV/bond]	0	239.5	212.2
$E(8)$ [meV/bond]	166.0	170.3	181.1
$E(9)$ [meV/bond]	0	162.2	159.2
$E(10)$ [meV/bond]	0	93.6	100.7
$E(11)$ [meV/bond]	0	16.9	46.0

Our analysis provides an insight into the here predicted trends. Regardless of the nanoparticle shape, the surface energy decreases with the increasing particle size. The reason is that the smaller is the nanoparticle, the larger is the fraction of the surface atoms with small coordination, i.e., those with lots of broken bonds. Moreover, the energy to break a bond increases (generally) with the decreasing atom coordination.

4.3. Contribution of Surface Stress State

As it has been recently stressed out [43], the excess energy due to a free surface has two contributions: the surface energy contribution related to the energy penalty of broken bond and the contribution due to the elastic strain energy generated by the surface stress state. The latter depends on the surface curvature. As an illustrative example let us assume a spherical body and a homogeneous surface stress state with the value σ leading to a pressure with value $2\sigma/R$ in the whole spherical body. From this description it becomes clear that the energetic surface stress contribution is zero for the slab approach. Similarly, the energetic surface stress contribution will be negligible for rather large nanocubes with only a marginal amount of corner and edge atoms (see discussion in the Section 4.2).

We now try to estimate the energetic surface stress contribution to the excess energy for the case of a spherical nanoparticle using classical continuum mechanics. Let us denote R the nanosphere's radius, and γ its surface energy. Furthermore, let us keep to the reasonable assumption that the value of σ and γ are of the same order of magnitude. The corresponding total surface energy is then

$$\mathcal{E}_\gamma = 4\pi R^2 \gamma. \tag{11}$$

For sake of simplicity, we further assume isotropic elastic properties of the nanoparticle, with ν and E being its Poisson's ratio and Young's modulus, respectively. The elastic strain energy caused by the surface stress σ, activating an internal pressure $2\sigma/R$, is

$$\mathcal{E}_\sigma = \frac{4}{3}\pi R^3 \frac{6(1-2\nu)}{E} \frac{\sigma^2}{R^2}, \tag{12}$$

for details, see, e.g., Ref. [1], Appendix 3. The ratio of the energetic surface stress contribution to the surface energy follows with $\sigma = \gamma$ as

$$\frac{\mathcal{E}_\sigma}{\mathcal{E}_\gamma} = \frac{2(1-2\nu)}{E} \frac{\gamma}{R}. \tag{13}$$

Taking a representative values for gold, $\gamma = 1\,\text{J/m}^2$, $E = 78\,\text{GPa}$, $\nu = 0.44$, and $R = 1\,\text{nm}$, Equation (13) yields 0.359×10^{-2}, i.e., the energetic surface stress contribution to the total excess energy is less than 1% of the surface induced excess energy. This ratio becomes even smaller (negligible) for larger nanospheres.

To corroborate this rather simplistic estimation, we plot the excess energy distribution over a cross section including the centre for a nanosphere (Figure 7a) and a nanocube (Figure 7b) as obtained from the MM simulations. Several observations can be made. Firstly, the excess energy is concentrated at the nanoparticle surface irrespective of its shape. The surface stress (and hence the corresponding elastic strain energy) could be only of relevance for a nanosphere. However, we can conclude that this contribution is effectively zero (or negligible). A similar situation can be expected for a nanocube, where the excess energy is concentrated to the nanocube edges (corner of the cross section in Figure 7b). This fact nicely agrees with the fitted values of $E_{\text{edge}} = 215.2\,\text{meV/bond}$ being larger than $E_{\text{facet}} = 166.0\,\text{meV/bond}$, estimated in Section 4.2.

Even though the term surface energy was used in a slightly imprecise way throughout the Section 3.2 (more accurate would be to talk about surface induced excess energy), we conclude that the energy contribution of surface stress can be neglected and the two quantities, surface energy and surface induced excess energy, are equivalent (or at least of the same order of magnitude) for practical cases with nanoparticles larger than ≈ 1 nm.

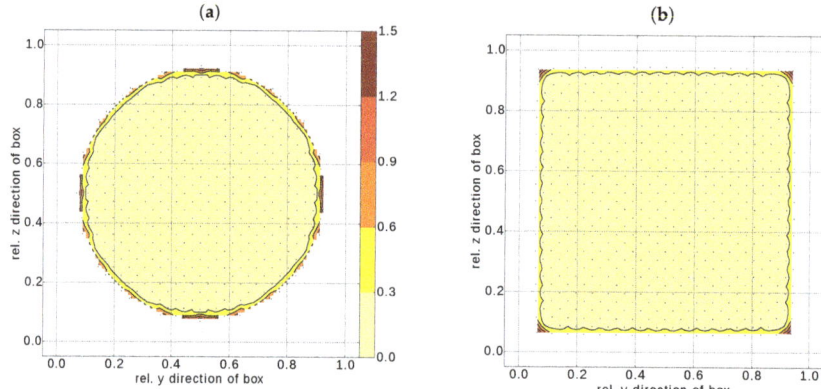

Figure 7. Contour plots of the distribution of the surface stress induced excess energy (in eV/at.) contribution for a cross section of (**a**) a nanosphere ($R = 3.25$ nm) and (**b**) a nanocube ($a = 6.92$ nm). Both cross sections include the nanoparticle centre. The dots represent actual atoms in the cross section, e.g., real locations, where the excess energy is stored. For sake of clear demonstration, the discrete data were interpolated over the whole cross sectional area.

5. Conclusions

A molecular mechanics study, complemented by first principles Density Functional Theory calculations, was performed to obtain surface energy of small gold nanoclusters of various sizes and (geometrically well defined) shapes. The employed interatomic pair potential was shown to give structural parameters and surface energies comparable with DFT-LDA calculations. The surface energy of nanocubes and nanospheres has been shown to converge to a constant value. The convergence was faster in the case of nanospheres compared with nanocubes. The surface energy, γ, is practically constant for any particles with radius larger than ≈ 3 nm. Truncated cubes (cuboctahedrons) did not achieve a single value for the surface energy within the studied range of nanoparticle sizes but, instead, an oscillating behaviour between two values. The range of these oscillations equals to the difference between γ of $(1\,0\,0)$ and $(1\,1\,1)$ facets. Finally, the surface-induced excess energy obviously follows a universal power-law dependence on the number of atoms forming the nanoparticle and is, to a large extent, related to the number of broken bonds (reduced coordination of the surface atoms). Importantly, the size-dependence of surface energy becomes significantly reduced when the actual surface area is corrected by the thickness of the electronic cloud, leading to almost constant values particularly for nanocube and nanosphere sizes of about 5 nm and more.

As outlined above, this study has found an increase of the surface energy with decreasing particle size (which is in agreement with other theoretical studies). Two remarks may be useful in this regard. Firstly, this fact should not be confused with experimental works on liquid solution–solid nanoparticle interface energies of gold nanoparticles, moreover often having irregular shapes or even liquid-like surface layer. Secondly, we note that small nanoparticles, specifically the Au_{55}, were shown to be amorphous rather than crystalline. Hence the values predicted here for the smallest particle sizes of a few nanometers are not relevant for amorphous or glassy particles.

In conclusion, this work contributes to understanding of surface energy (solid phase–vacuum interface) of crystalline nanoparticles and its relation to the their structure.

Author Contributions: Conceptualization, D.H., D.V., and F.D.F.; methodology, D.H. and P.D.; software, D.H.; investigation, D.H. and P.D.; resources, D.H.; writing—original draft preparation, D.H.; writing—review and editing, D.H., D.V. and F.D.F.; visualization, D.H. and P.D.; supervision, D.H.; funding acquisition, D.H. and F.D.F. All authors have read and agreed to the published version of the manuscript.

Funding: F.D.F. and D.H. appreciate financial support by the Austrian Federal government (in particular from the Bundesministerium für Verkehr, Innovation and Technologie and the Bundesministerium für Wirtschaft und Arbeit) and the Styrian Provincial Government, represented by Österreichische Forschungsförderungsgesellschaft mbH and by Steirische Wirtschaftsförderungsgesellschaft mbH, within the research activities of the K2 Competence Centre on "Integrated Research in Materials, Processing and Product Engineering", operated by the Materials Center Leoben Forschung GmbH in the framework of the Austrian COMET Competence Centre Programme, Projects A1.23 and A2.32.

Acknowledgments: The computational results presented have been achieved in part using the Vienna Scientific Cluster (VSC).

Conflicts of Interest: The authors declare no conflict of interest.

References

1. Fischer, F.D.; Waitz, T.; Vollath, D.; Simha, N.K. On the role of surface energy and surface stress in phase-transforming nanoparticles. *Prog. Mater Sci.* **2008**, *53*, 481–527. [CrossRef]
2. Guisbiers, G. Advances in thermodynamic modelling of nanoparticles. *Adv. Phys. X* **2019**, *4*, 1668299. [CrossRef]
3. Vollath, D.; Fischer, F.D. Estimation of thermodynamic data of metallic nanoparticles based on bulk values. In *Metal Nanopowders: Production, Characterization, and Energetic Applications*; Gromov, A.A., Teipel, U., Eds.; Wiley: Hoboken, NJ, USA, 2014; pp. 1–24.
4. Safaei, A.; Attarian Shandiz, M. Melting entropy of nanocrystals: an approach from statistical physics. *Phys. Chem. Chem. Phys.* **2010**, *12*, 15372–15381. [CrossRef] [PubMed]
5. Attarian Shandiz, M.; Safaei, A. Melting entropy and enthalpy of metallic nanoparticles. *Mater. Lett.* **2008**, *62*, 3954–3956. [CrossRef]
6. Chang, J.; Johnson, E. Surface and bulk melting of small metal clusters. *Philos. Mag.* **2005**, *85*, 3617–3627. [CrossRef]
7. Medasani, B.; Park, Y.H.; Vasiliev, I. Theoretical study of the surface energy, stress, and lattice contraction of silver nanoparticles. *Phys. Rev. B Condens. Matter* **2007**, *75*, 235436. [CrossRef]
8. Medasani, B.; Vasiliev, I. Computational study of the surface properties of aluminum nanoparticles. *Surf. Sci.* **2009**, *603*, 2042–2046. [CrossRef]
9. Ali, S.; Myasnichenko, V.S.; Neyts, E.C. Size-dependent strain and surface energies of gold nanoclusters. *Phys. Chem. Chem. Phys.* **2015**, *18*, 792–800. [CrossRef]
10. Nanda, K.K.; Maisels, A.; Kruis, F.E. Surface tension and sintering of free gold nanoparticles. *J. Phys. Chem. C* **2008**, *112*, 13488–13491. [CrossRef]
11. Nanda, K.K.; Maisels, A.; Kruis, F.E.; Fissan, H.; Stappert, S. Higher surface energy of free nanoparticles. *Phys. Rev. Lett.* **2003**, *91*, 106102. [CrossRef]
12. Wei, Y.; Chen, S. Size-dependent surface energy density of spherical face-centered-cubic metallic nanoparticles. *J. Nanosci. Nanotechnol.* **2015**, *15*, 9457–9463. [CrossRef] [PubMed]
13. Holec, D.; Fischer, F.D.; Vollath, D. Structure and surface energy of Au_{55} nanoparticles: An ab initio study. *Comput. Mater. Sci.* **2017**, *134*, 137–144. [CrossRef]
14. Vollath, D.; Fischer, F.D.; Holec, D. Surface energy of nanoparticles – influence of particle size and structure. *Beilstein J. Nanotechnol.* **2018**, *9*, 2265–2276. [CrossRef] [PubMed]
15. Molleman, B.; Hiemstra, T. Size and shape dependency of the surface energy of metallic nanoparticles: Unifying the atomic and thermodynamic approaches. *Phys. Chem. Chem. Phys.* **2018**, *20*, 20575–20587. [CrossRef]
16. Yáñez-Sedeño, P.; Pingarrón, J.M. Gold nanoparticle-based electrochemical biosensors. *Anal. Bioanal. Chem.* **2005**, *382*, 884–886. [CrossRef]
17. Grzelczak, M.; Pérez-Juste, J.; Mulvaney, P.; Liz-Marzán, L.M. Shape control in gold nanoparticle synthesis. *Chem. Soc. Rev.* **2008**, *37*, 1783–1791. [CrossRef]

18. Vollath, D.; Holec, D.; Fischer, F.D. Au$_{55}$, a stable glassy cluster: results of ab initio calculations. *Beilstein J. Nanotechnol.* **2017**, *8*, 2221–2229. [CrossRef]
19. Shim, J.H.; Lee, B.J.; Cho, Y.W. Thermal stability of unsupported gold nanoparticle: a molecular dynamics study. *Surf. Sci.* **2002**, *512*, 262–268. [CrossRef]
20. Herz, A.; Friák, M.; Rossberg, D.; Hentschel, M.; Theska, F.; Wang, D.; Holec, D.; Šob, M.; Schneeweiss, O.; Schaaf, P. Facet-controlled phase separation in supersaturated Au-Ni nanoparticles upon shape equilibration. *Appl. Phys. Lett.* **2015**, *107*, 073109. [CrossRef]
21. Chen, Y.; Gu, X.; Nie, C.G.; Jiang, Z.Y.; Xie, Z.X.; Lin, C.J. Shape controlled growth of gold nanoparticles by a solution synthesis. *Chem. Commun.* **2005**, 4181–4183. [CrossRef]
22. Plimpton, S. Fast parallel algorithms for short-range molecular dynamics. *J. Comput. Phys.* **1995**, *117*, 1–19. [CrossRef]
23. Grochola, G.; Russo, S.P.; Snook, I.K. On fitting a gold embedded atom method potential using the force matching method. *J. Chem. Phys.* **2005**, *123*, 204719. [CrossRef] [PubMed]
24. Birch, F. Finite elastic strain of cubic crystals. *Phys. Rev.* **1947**, *71*, 809–824. [CrossRef]
25. Kresse, G.; Furthmüller, J. Efficient iterative schemes for ab initio total-energy calculations using a plane-wave basis set. *Phys. Rev. B* **1996**, *54*, 11169–11186. [CrossRef]
26. Kresse, G.; Furthmüller, J. Efficiency of ab-initio total energy calculations for metals and semiconductors using a plane-wave basis set. *Comput. Mater. Sci.* **1996**, *6*, 15–50. [CrossRef]
27. Hohenberg, P.; Kohn, W. Inhomogeneous electron gas. *Phys. Rev.* **1964**, *136*, B864–B871. [CrossRef]
28. Kohn, W.; Sham, L.J. Self-consistent equations including exchange and correlation effects. *Phys. Rev.* **1965**, *140*, A1133–A1138. [CrossRef]
29. Wang, Y.; Perdew, J.P. Correlation hole of the spin-polarized electron gas, with exact small-wave-vector and high-density scaling. *Phys. Rev. B* **1991**, *44*, 13298–13307. [CrossRef]
30. Kresse, G.; Joubert, D. From ultrasoft pseudopotentials to the projector augmented-wave method. *Phys. Rev. B* **1999**, *59*, 1758–1775. [CrossRef]
31. Barber, C.B.; Dobkin, D.P.; Huhdanpaa, H. The Quickhull Algorithm for Convex Hulls. *ACM Trans. Math. Softw.* **1996**, *22*, 469–483. [CrossRef]
32. Holec, D.; Mayrhofer, P.H. Surface energies of AlN allotropes from first principles. *Scripta Mater.* **2012**, *67*, 760–762. [CrossRef] [PubMed]
33. Vitos, L.; Ruban, A.V.; Skriver, H.L.; Kollár, J. The surface energy of metals. *Surf. Sci.* **1998**, *411*, 186–202. [CrossRef]
34. Kozeschnik, E.; Holzer, I.; Sonderegger, B. On the potential for improving equilibrium thermodynamic databases with kinetic simulations. *J. Phase Equilib. Diffus.* **2007**, *28*, 64–71. [CrossRef]
35. Crljen, Ž.; Lazić, P.; Šokčević, D.; Brako, R. Relaxation and reconstruction on (111) surfaces of Au, Pt, and Cu. *Phys. Rev. B* **2003**, *68*, 1–8. [CrossRef]
36. Galanakis, I.; Papanikolaou, N.; Dederichs, P.H. Applicability of the broken-bond rule to the surface energy of the fcc metals. *Surf. Sci.* **2002**, *511*, 1–12. [CrossRef]
37. Tyson, W.; Miller, W. Surface free energies of solid metals: Estimation from liquid surface tension measurements. *Surf. Sci.* **1977**, *62*, 267–276. [CrossRef]
38. Haas, P.; Tran, F.; Blaha, P. Calculation of the lattice constant of solids with semilocal functionals. *Phys. Rev. B* **2009**, *79*, 085104. [CrossRef]
39. Larsen, A.H.; Mortensen, J.J.; Blomqvist, J.; Castelli, I.E.; Christensen, R.; Dułak, M.; Friis, J.; Groves, M.N.; Hammer, B.; Hargus, C.; et al. The atomic simulation environment—A Python library for working with atoms. *J. Phys. Condens. Matter* **2017**, *29*, 273002. [CrossRef]
40. de Heer, W.A. The physics of simple metal clusters: experimental aspects and simple models. *Rev. Mod. Phys.* **1993**, *65*, 611–676. [CrossRef]
41. Holec, D.; Hartmann, M.A.; Fischer, F.D.; Rammerstorfer, F.G.; Mayrhofer, P.H.; Paris, O. Curvature-induced excess surface energy of fullerenes: Density functional theory and Monte Carlo simulations. *Phys. Rev. B* **2010**, *81*, 235403. [CrossRef]

42. Mameka, N.; Markmann, J.; Jin, H.J.; Weissmüller, J. Electrical stiffness modulation-confirming the impact of surface excess elasticity on the mechanics of nanomaterials. *Acta Mater.* **2014**, *76*, 272–280. [CrossRef]
43. Müller, P.; Saùl, A.; Leroy, F. Simple views on surface stress and surface energy concepts. *Adv. Nat. Sci. Nanosci. Nanotechnol.* **2014**, *5*, 013002. [CrossRef]

© 2020 by the authors. Licensee MDPI, Basel, Switzerland. This article is an open access article distributed under the terms and conditions of the Creative Commons Attribution (CC BY) license (http://creativecommons.org/licenses/by/4.0/).

Article

Ab Initio Study of Ferroelectric Critical Size of SnTe Low-Dimensional Nanostructures

Takahiro Shimada [1,*], Koichiro Minaguro [1], Tao Xu [2], Jie Wang [3] and Takayuki Kitamura [1]

1. Department of Mechanical Engineering and Science, Kyoto University, Nishikyo-ku, Kyoto 615-8540, Japan; minaguro.koichiro.77w@st.kyoto-u.ac.jp (K.M.); takayuki.kitamura.kyoto@gmail.com (T.K.)
2. Materials Genome Institute, Shanghai University, Shanghai Materials Genome Institute, Shanghai 200444, China; xutao6313@shu.edu.cn
3. Department of Engineering Mechanics & Key Laboratory of Soft Machines and Smart Devices of Zhejiang Province, School of Aeronautics and Astronautics, Zhejiang University, Hangzhou 310027, China; jw@zju.edu.cn
* Correspondence: shimada@me.kyoto-u.ac.jp; Tel.: +81-75-383-3633

Received: 12 February 2020; Accepted: 8 April 2020; Published: 11 April 2020

Abstract: Beyond a ferroelectric critical thickness of several nanometers existed in conventional ferroelectric perovskite oxides, ferroelectricity in ultimately thin dimensions was recently discovered in SnTe monolayers. This discovery suggests the possibility that SnTe can sustain ferroelectricity during further low-dimensional miniaturization. Here, we investigate a ferroelectric critical size of low-dimensional SnTe nanostructures such as nanoribbons (1D) and nanoflakes (0D) using first-principle density-functional theory calculations. We demonstrate that the smallest (one-unit-cell width) SnTe nanoribbon can sustain ferroelectricity and there is no ferroelectric critical size in the SnTe nanoribbons. On the other hand, the SnTe nanoflakes form a vortex of polarization and lose their toroidal ferroelectricity below the surface area of 4 × 4 unit cells (about 25 Å on one side). We also reveal the atomic and electronic mechanism of the absence or presence of critical size in SnTe low-dimensional nanostructures. Our result provides an insight into intrinsic ferroelectric critical size for low-dimensional chalcogenide layered materials.

Keywords: ferroelectricity; SnTe; nanoribbon; nanoflakes; critical size; density-functional theory

1. Introduction

Ferroelectrics exhibit spontaneous polarization that can be reversed by an external electric field, due to their noncentrosymmetric crystal structure having a relative displacement of cations and anions in a ferroelectric (FE) phase. Ferroelectric properties have attracted attention due to their technological applications such as ferroelectric memory (FeRAM), sensors, MEMS/NEMS, and actuators [1–3]. To enhance the performance of these devices, it is necessary to reduce the size of ferroelectrics and integrate them at a high density. In recent years, with the progress of manufacturing technology, nanoscale ferroelectric materials with low-dimensional structures such as nano-thin films [4,5] (two-dimensional; 2D), nanowires [6,7], nanotubes [8,9] (one-dimensional; 1D), and nanodots [10,11] (zero-dimensional; 0D) have been synthesized for the high-performance, high-integration of nano-devices.

However, ferroelectricity disappears when the size of the ferroelectric material becomes nanoscale (ferroelectric critical size): in perovskite oxide $PbTiO_3$ and $BaTiO_3$ nanofilms, ferroelectricity disappears when the thickness of the films becomes 2 nm or less [12–15]. The appearance of ferroelectric critical size was explained by two aspects: (I) effect of electrostatic (depolarization) field and (II) the reconstruction and rearrangement of atomic and electronic structure at surfaces or edges. At the surface of ferroelectric materials, the surface polarization charge is formed due to the termination of polarization, and the depolarization field formed in the opposite direction of spontaneous polarization suppresses the ferroelectricity [16]. In particular, when the material dimensions become nanoscale, the ratio of the surface or edge to the entire volume increases, the influence of the depolarization field formed by the surface charge becomes dominant, and the ferroelectricity of the entire material disappears. This is factor (I). In general, ferroelectricity originates from a delicate balance between long-range interaction due to Coulomb force, which is the driving force for the relative displacement (ferroelectric displacement) of cations and anions in the crystal, and short-range interaction due to covalent bonds that stabilize the centrosymmetric structure [17]. In nanoscale materials, the long-range interaction is reduced due to the absence of atoms outside of material surfaces, and thereby the balance between long-range and short-range interaction is broken. In particular, such interactions are also changed due to the reconstruction and rearrangement of atomic and electronic structures at surfaces or edges. This is factor (II). For these reasons, the ferroelectric critical size appears. This physical limitation prevents the miniaturization of ferroelectric materials beyond the critical size.

In recent years, however, ferroelectricity was discovered in the monolayer structure of chalcogenide SnTe in the in-plane direction [18]. This indicates that ferroelectricity can exist in a structure with an atomic thickness. Obviously, this discovery is beyond the long-believed ferroelectric critical thickness of several nanometers. Since the nanostructure is commonly utilized in a low-dimensional form, it is scientifically interesting and technologically important to investigate whether ferroelectricity is also sustained in ultimate SnTe nanoribbons (1D) and SnTe nanoflakes (0D) in addition to the discovered monolayer (2D) form. However, the ferroelectric critical size for SnTe nanoribbons and nanoflakes has not yet been reported.

In this study, we investigate whether a ferroelectric critical size exists in low-dimensional structure of SnTe, the nanoribbons (1D) and nanoflakes (0D) using first-principle, density-functional theory (DFT) calculations.

2. Materials and Methods

We focus on SnTe nanoribbons and nanoflakes with an edge structure. SnTe has two types of edges formed along the [110] and [100] directions. Henceforth, these edge structures are called [110] edge and [100] edge, respectively. Table 1 shows the preliminarily calculated formation energies of the [110] and [100] edges. Here, the edge formation energy is calculated by $E_{edge} = (E_{nanoribbon} - E_{monolayer})/2l$, where $E_{nanoribbon}$ and $E_{monolayer}$ are the total energies of SnTe nanoribbons and SnTe monolayer, respectively, and l is the length of edge in the nanoribbon model, shown later. The formation energy of the [110] edge is lower, and thus more stable, than that of the [100] edge. In addition, the [110] edge structure was experimentally observed at the edge of the SnTe monolayers [18–20]. Following these experimental and theoretical results, we thus analyze the nanoribbons and nanoflakes consisting of the [110] edges, as shown in Figure 1. Figure 1 shows the paraelectric phase of SnTe monolayer with a space group of $Fm3m$. In the ferroelectric phase, Sn and Te atoms are spontaneously displaced along the [110] direction. The space group of the ferroelectric SnTe monolayer is $Pmn2_1$. The following SnTe nanoribbons and nanoflakes are in the ferroelectric phase, and thus modeled with a small initial displacement along [110]. Note that the electronic origin of ferroelectricity and alternating short and long bonds has already been discussed by Liu et al. [21], and they revealed that the stabilization of the ferroelectric phase and large distortion originates from an interplay between hybridization interactions of Sn-Te, which act as a driving force for the ferroelectricity, and Pauli repulsions, which tend to suppress the ferroelectricity.

Table 1. Calculated edge formation energy E_{edge} of [110] and [100] edge in layered SnTe.

Edge Direction	[110]	[100]
E_{edge} (eV/Å)	0.097	0.127

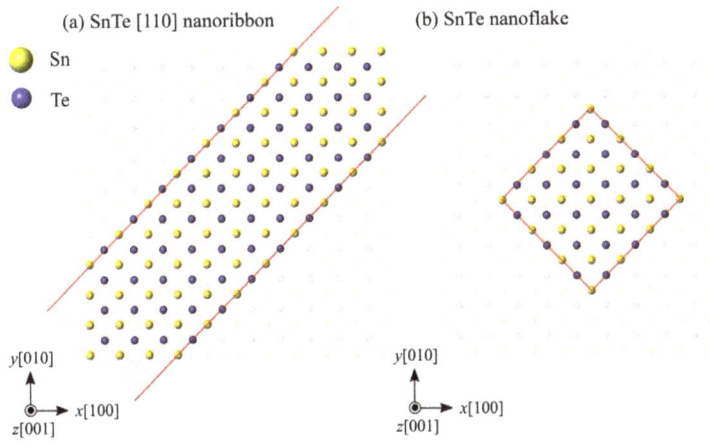

Figure 1. Schematic illustration of (**a**) SnTe nanoribbon and (**b**) SnTe nanoflake.

Figure 2 shows the simulation model of the SnTe nanoribbons. Here, m_{110} denotes the number of unit cells constituting the nanoribbon width. To explore the critical ferroelectric size, we calculate several SnTe nanoribbons with different widths of $m_{110} = 1$ to 10 (6 to 65 Å). The model of a nanoribbon width 3-unit-cells width ($m_{110} = 3$) is shown in Figure 2 as an example. In this simulation model, there are $4m_{110} + 2$ Sn and Te atoms each, for a total of $8m_{110} + 4$ atoms. The three-dimensional periodic boundary condition is applied to the simulation cell. To prevent undesirable interactions between the nanoribbons in the neighboring image cells, a vacuum region of $l_v = 20$ Å in the y- and z directions. a_1, a_2, and a_3 in the figure are the simulation cell vectors, and are represented by

$$a_1 = (a, 0, 0), \tag{1}$$

$$a_2 = (0, m_{110}b + l_v, 0), \tag{2}$$

$$a_3 = (0, 0, c + l_v), \tag{3}$$

where a, b, and c are the equilibrium lattice constants of the SnTe monolayer, $a = 6.520$ Å, $b = 6.479$ Å, and $c = 3.240$ Å.

Figure 3 shows the simulation model of SnTe nanoflakes. Here, m_f denotes the number of unit cells constituting each side of the SnTe nanoflake. To explore the critical ferroelectric size, we calculate several SnTe nanoflakes with $m_f = 1$ to 7 (6 to 45 Å) on one side. The model of a SnTe nanoflake with 5 × 5 unit-cells surface area ($m_f = 5$) (hereinafter referred to as a 5 × 5 nanoflake) as an example. In this simulation model, there are $(2m_f + 1)^2$ Sn and Te atoms each, for a total of $2 \times (2m_f + 1)^2$ atoms. The three-dimensional periodic boundary condition is applied to the simulation cell. To avoid undesirable interactions from neighboring nanoribbons in image cells, a vacuum region of $l_v = 20$ Å is introduced to the x, y, and z directions of the simulation cell. a_1, a_2, and a_3 in the figure are the simulation cell vectors, and are represented by

$$a_1 = (m_f a + l_v, 0, 0), \tag{4}$$

$$a_2 = (0, m_f b + l_v, 0), \tag{5}$$

$$a_3 = (0, 0, c + l_v),\qquad(6)$$

where, a, b, and c are the equilibrium lattice constants of the SnTe monolayer.

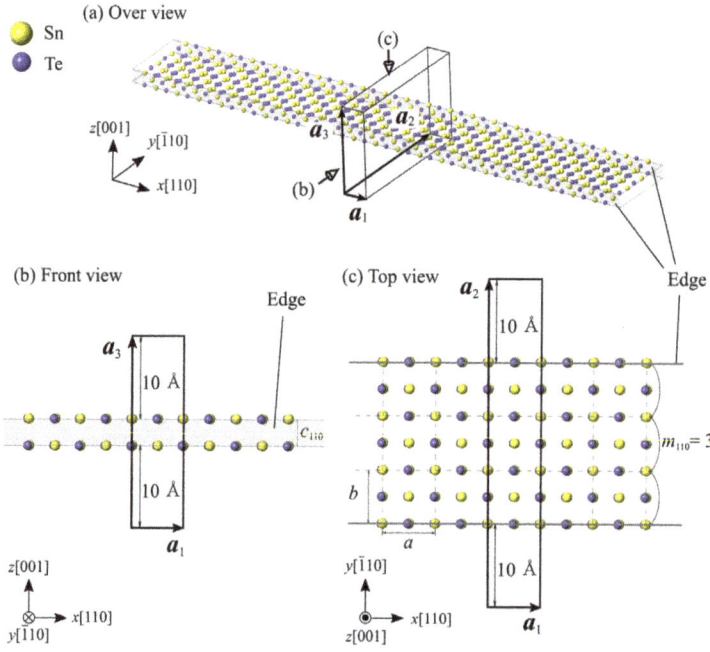

Figure 2. Simulation models of SnTe nanoribbon with 3-unit-cells width. The gray area and solid lines indicate edges of SnTe nanoribbon. The solid boxes represent the simulation cells. a_1, a_2 and a_3 indicate the simulation cell vectors.

We perform first-principle, density-functional theory (DFT) calculations [22,23]. The effects of nuclei and inner shells are expressed by the project-augmented wave (PAW) method [24,25], and the Sn $4d$, $5s$, $5p$ orbitals and the Te $5s$, $5p$ orbitals are explicitly treated as valence electrons. The electronic wave function is expanded in plane-waves, and the cutoff energy of the plane waves is set to 450 eV. The Brillouin zone integration is performed using a $10 \times 1 \times 1$ Monkhorst-Pack k-point mesh for the nanoribbon models and a $1 \times 1 \times 1$ k-point mesh for the nanoflake models [26]. The PBE-D3 functional is used for the evaluation of the exchange correlation term [27]. The stable structure is determined by relaxing atomic positions using the conjugate gradient method until the force acting on the atoms became 1.0×10^{-3} eV/Å or less. All the first-principles calculations are performed using the Vienna Ab-initio Simulation Package (VASP) code [28,29]. The present calculation condition was confirmed to reproduce the electronic (band structure) and ferroelectric properties of SnTe monolayer via the comparison of experimental data [18].

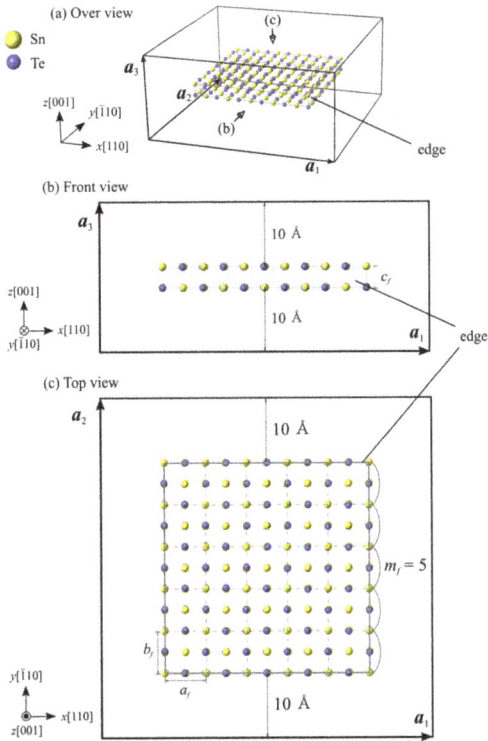

Figure 3. Simulation models of the SnTe nanoflake with a surface area of 5 × 5 unit cells. The gray area and solid lines indicate edges of the SnTe nanoflake. The solid boxes represent the simulation cells. a_1, a_2 and a_3 indicate the simulation cell vectors.

3. Results and Discussion

3.1. Ferroelectric Critical Size of SnTe Nanoribbons

Figure 4 shows the change in the spontaneous polarization P of the nanoribbon with respect to the width of the SnTe nanoribbon. The polarization of SnTe nanoribbons are aligned along the longitudinal direction ([110] direction). The black dashed line indicates the polarization value of the SnTe monolayer. Here, the ferroelectric polarization is calculated by the Berry phase approach [30]. Note that, in this study, the indeterminacy of spontaneous polarization via the Berry phase calculations is treated by using the paraelectric phase SnTe structure as the reference (zero-polarization) state. The SnTe nanoribbons with a 10-unit-cells width exhibits the polarization of $P = 26.6$ μC/cm^2, which is almost the same magnitude as the SnTe monolayer, $P = 26.5$ μC/cm^2. This indicates that there is no size effect in the nanoribbons with a 10-unit-cell width (65 Å). Even when the nanoribbon width is reduced, the ferroelectric polarization is almost constant at $P = 26.6$ to 27.9 μC/cm^2, and all simulated nanoribbons show ferroelectric polarization comparable to that of the SnTe monolayer. This means that the SnTe nanoribbon does not exhibit any size-dependence, unlike conventional three-dimensional ferroelectrics such as BaTiO$_3$ and PbTiO$_3$. In addition, the SnTe nanoribbon with the minimum one-unit-cell width exhibits non-zero ferroelectric polarization. Our result indicates that ferroelectricity does not disappear even in the smallest nanoribbon, and there is, therefore, no critical dimension in which the ferroelectricity disappears in the SnTe nanoribbons.

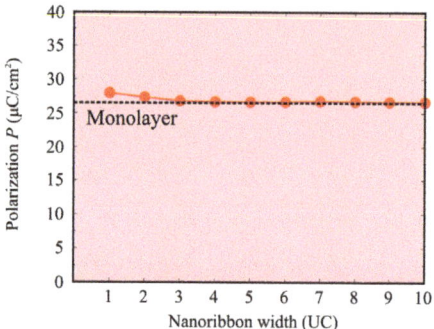

Figure 4. Polarization P in SnTe nanoribbon as a function of the width. The red dotted line shows the ferroelectric polarization of 2D SnTe monolayer. UC denotes the unit cells.

Conventional three-dimensional crystalline ferroelectrics, such as the perovskite oxides $BaTiO_3$ and $PbTiO_3$, exhibit remarkable size-dependence, and the ferroelectricity disappears when the material size reaches several nanometers [31–33]: for example, the critical ferroelectric size of $BaTiO_3$ nano-films was reported to be 2-nm thickness [12,13], that of $PbTiO_3$ nano-films was 1.2-nm thickness [14,34], and $Pb(Zr,Ti)O_3$ nanodot lost its ferroelectricity when it became 3.2 nm in diameter [10]. Regardless of the material and shape, the ferroelectricity is reduced and finally disappears as the material size decreases. As was explained in the introduction, such size effects and critical dimensions on ferroelectricity originate from two factors: (I) electrostatic effects due to the formation of a depolarization field [16,35–38], and (II) the reconstruction and rearrangement of atomic and electronic structures due to the low coordination number at the surface (or edge) [36–39].

Considering the above discussion on the conventional ferroelectrics, here we investigate the absence of a critical ferroelectric size of SnTe nanoribbons in terms of factors (I) and (II). Considering factor (I), the ferroelectric polarization direction of the SnTe nanoribbon is almost parallel to the edge line, and thereby, no surface polarization charge is induced and no depolarization field is generated. Therefore, the electrostatic factor (I) due to the depolarization field does not occur in SnTe nanoribbons. Next, the dangling bond formation at the edge of the SnTe nanoribbon is examined to consider factor (II). Here, we first refer the bonding structure of the SnTe monolayer (i.e., without dangling bonds) as a reference, as shown in Figure 5. In the SnTe monolayer, spontaneous polarization P appears in the [110] direction due to the relative displacement of Sn^{2+} ions in the [110] direction with respect to Te^{2-} ions in the ferroelectric phase (see Figure 5a). Due to the ionic displacement, the SnTe monolayer forms an Sn-Te bond along the [110] direction, which is the same as the direction of spontaneous polarization. This means that the relative displacement and bonding of the Sn^{2+} and Te^{2-} ions in the [110] direction in the SnTe monolayer corresponds to the spontaneous polarization and, thereby, is a characteristic of the ferroelectric manifestation of the SnTe monolayer. Figure 6 compares the bonding situation between in the SnTe monolayer and the SnTe nanoribbon. The white lines in the figure indicate Sn-Te bonds, while the white dashed circles and lines in the SnTe nanoribbon indicate the imaginary Sn or Te position and bond, respectively, which were formed in the SnTe monolayer. As described above, the SnTe monolayer forms an armchair-shaped Sn-Te bond along the [110] direction (see Figure 5b-2), and this situation can be seen alternately appearing on the Sn-Te bond from the top view of the monolayer, as shown in Figure 6a. In general, near the surface or edge, the rearrangement of electrons occurs due to the presence of dangling bonds, which affects ferroelectricity [36–38]. On the other hand, focusing on the electron density distribution at the edge of the SnTe nanoribbon in Figure 6b, Sn-Te is also found along the [110] direction, which is almost the same as the electron density distribution of the SnTe monolayer; i.e., the absence of a dangling bond at the edge of SnTe monolayer. This is because the bonding sequence in the SnTe monolayer is mainly along the polar direction of [110], and thereby, the formation of the [110] edge does not introduce any dangling bond. The absence of a dangling bond at the [110] edge in the SnTe nanoribbons makes the ferroelectricity same as that in the SnTe monolayer. Therefore, the absence of factors (I) and (II) leads to the absence of critical ferroelectric size in SnTe nanoribbons.

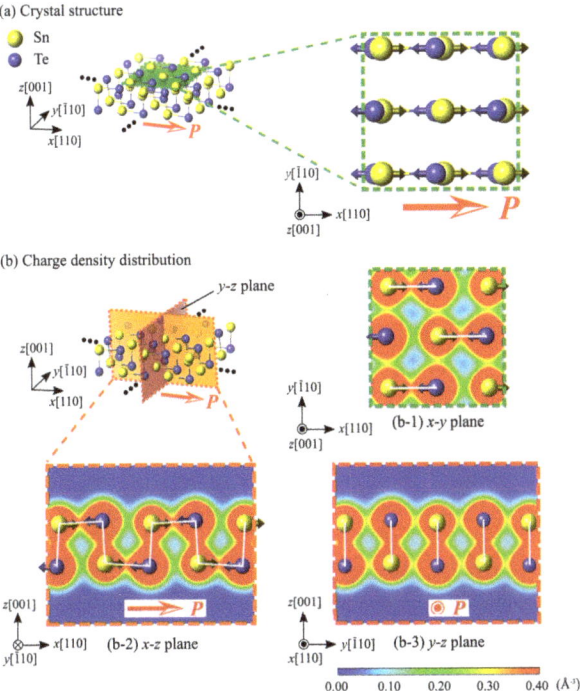

Figure 5. (a) Crystal structure of SnTe monolayer and (b) charge density distributions in SnTe monolayer. The covalent Sn-Te bonds are shown by white lines. Red arrows P indicate the direction of spontaneous polarization in the SnTe monolayer. Yellow and blue arrows indicate the displacement of Sn and Te atoms, respectively.

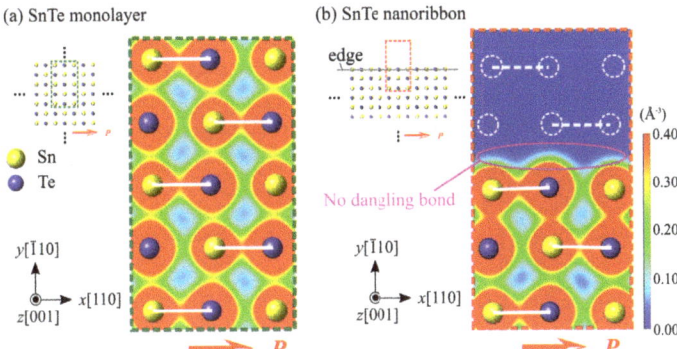

Figure 6. Charge density distributions (a) in the SnTe monolayer and (b) in the SnTe nanoribbon with 6-unit-cells width. Red arrows P indicate the spontaneous polarization. The covalent Sn-Te bonds are shown by white lines. White dotted circles and lines indicate the imaginary Sn or Te atom positions and Sn-Te bonds formed in the SnTe monolayer.

3.2. Ferroelectric Critical Size of SnTe Nanoflakes

Figure 7 shows the local polarization distribution in the 5 × 5 nanoflakes. Spontaneous polarization exists and forms a vortex polarization order in the counterclockwise direction. Similar vortex polarization distributions are also observed in the other 6 × 6 and 7 × 7 nanoflakes. Such a polar vortex is characteristic of the polarization order in ferroelectric nanostructures [11,40] because the surface component of polarization is aligned along a surface or edge to prevent the formation of the depolarization field and minimize the electrostatic energy efficiently. Since SnTe nanoflakes are surrounded by edges on all sides, a surface polarization charge is induced at the edges, and a depolarization field is generated inside the SnTe nanoflakes. Since the parallel polarization to the edge of the nanoflake does not produce any surface polarization charge or depolarization field, the formation of vortex polarization is more energetically stable than the original straight form of ferroelectric polarization.

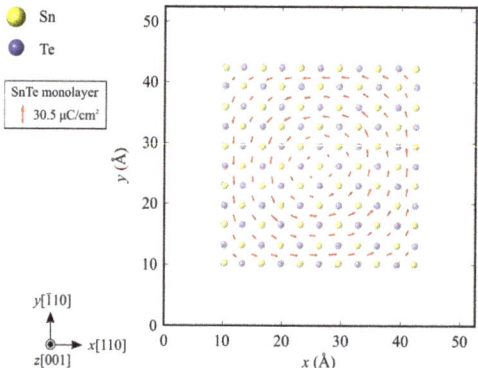

Figure 7. Vector-field representation of local polarization distribution in the 5 × 5 nanoflake. Red arrows indicate the spontaneous polarization.

To evaluate the critical dimension of the ferroelectricity in SnTe nanoflakes with vortex polarization, here we consider the toroidal moment G. The toroidal moment G is used as a physical quantity that characterizes vortex polarization appearing in nanoscale ferroelectric materials [10,41,42], and is given by the following equation [42]

$$G = \frac{1}{2N} \sum_k r_k \times P_k \qquad (7)$$

where r_k is the position vector of the k-th local unit cell, P_k is the local spontaneous polarization at position r_k, and N is the number of local unit cells included in the simulation cell. The sum is taken of all unit cells in the simulation cell. The local polarization is evaluated using the Born effective charge tensors [30]. The site-by-site local polarization can be calculated by

$$P_i = \frac{e}{\Omega_c} w_j Z_j u_j \qquad (8)$$

where Ω_c is the volume of the local unit cell i; e and u_j denote the electron charge and the atomic displacement vector relative to the ideal lattice site (paraelectric lattice site) of atom j, respectively. The index j covers all atoms in the local unit cell i. Z_j is the Born effective charge tensor of atom j and w_j is a weight factor.

Figure 8 shows the toroidal moment G_z and the average polarization P for each SnTe nanoflake size. Note that we show the z component of toroidal moment G because all of the vortex polarization appears on the xy plane. The toroidal moment G_z decreases as the size of the SnTe nanoflakes decreases.

When the size of the SnTe nanoflakes become 4 × 4 unit-cell size or less, the toroidal moment G_z becomes zero. Here, we also show the averaged polarization in panel (b), as defined by

$$\bar{P} = \frac{1}{N}\sum_k |P_k| \qquad (9)$$

Such size-dependent behavior is also seen in the averaged polarization. The average polarization value decreases as the size of the SnTe nanoflakes decreases, and the polarization becomes 0 when one side is less than four unit cells (25 Å). These results indicate that, in contrast to the SnTe monolayer (2D) and nanoribbons (1D), the SnTe nanoflakes (0D) exhibit remarkable size-dependence and a critical dimension at which ferroelectricity disappears. The critical dimension is evaluated to be four unit cells on one side (about 25 Å). This suggests that structural low-dimensionality can affect the ferroelectricity of SnTe system and lead to the appearance of a ferroelectric critical size.

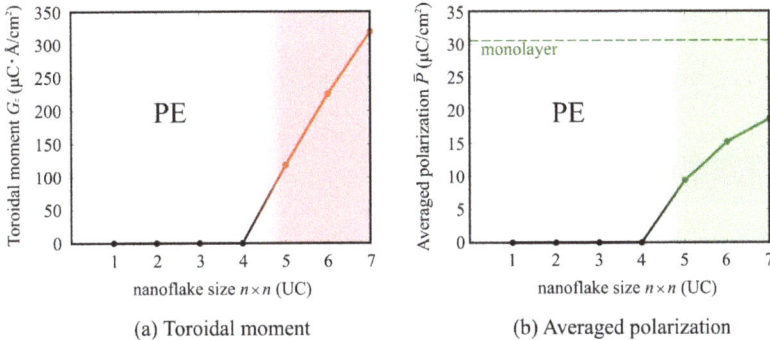

Figure 8. (a) Toroidal moment, G, and (b) averaged polarization, P, in SnTe nanoflake as a function of SnTe nanoflake size. The green dotted line indicates the polarization of the SnTe monolayer. PE indicates the paraelectric phase.

In the following, we discuss the appearance of critical dimension of the vortex polarization in the SnTe nanoflakes. As discussed in Section 3.1, the factors that cause the size effect and critical dimension appear are: (i) the electrostatic effect, due to the formation of a depolarization field [16,35], and (ii) the reconstruction and rearrangement of the atomic and electronic structure due to a lower coordination number at a surface or edge [36–38]. In order to examine the effect of these factors, we calculate an imaginary model of an edge-free SnTe monolayer with a vortex polarization that is the same as the nanoflakes, as shown in Figure 9, and compare the results. This model consists of periodically arranged clockwise and counterclockwise polarization vortices in a SnTe monolayer, and each polarization vortex mimics a nanoflake with a vortex polarization but without any edge structures. Since this imaginary edge-free SnTe monolayer model is free from the edge and the resulting (coinciding) electrostatic depolarization field and dangling bonds, through comparison between the SnTe nanoflakes and this imaginary edge-free model, one can extract how the existence of edge and electrostatic field and/or dangling bonds affect the ferroelectricity of the SnTe nanoflakes. Figure 10 shows the calculated local polarization field of the imaginary edge-free SnTe models with polarization-vortex periodicity of 5 × 5 and 4 × 4 unit cells. The edge-free SnTe model with a 5 × 5 unit cell size exhibits a quasi-stable vortex polarization, as shown in Figure 10a, which is almost same as that observed in the 5 × 5 SnTe nanoflake, as shown in Figure 7. On the other hand, no spontaneous polarization is observed in the edge-free SnTe model with a 4 × 4 unit cell or less (Figure 10b). This is also consistent with the absence of polarization and paraelectric nature of the 4 × 4 SnTe nanoflake. These results indicate that the presence or absence of edges does not affect the appearance of a ferroelectric critical size of SnTe nanoflakes. Therefore, the effects of factors (i) the depolarization field and (ii) the dangling formation at the edges are not the primary causes of the disappearance of the vortex polarization in the SnTe nanoflakes. The situation of the SnTe nanoflakes is clearly different from that of the conventional ferroelectrics, where the critical size appears due to these two factors.

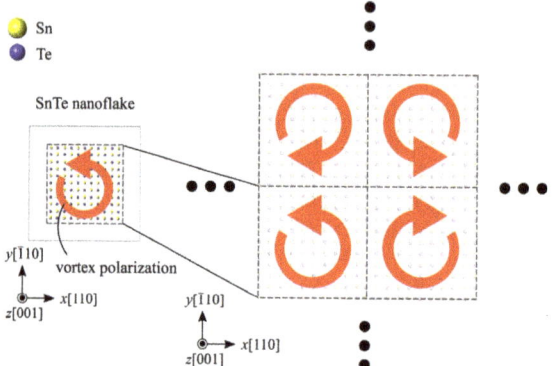

Figure 9. Simulation model of the edge-free SnTe with periodic polarization vortices. Red arrows indicate the direction of vortex polarization.

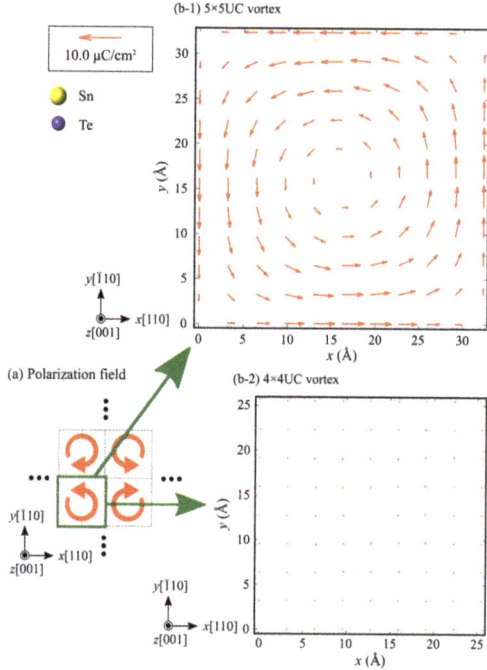

Figure 10. Vector-field representation of the spontaneous polarization of the imaginary edge-free SnTe models with a polarization-vortex periodicity of (a) 5 × 5 and (b) 4 × 4 unit cells. Red arrows indicate spontaneous polarization.

From the above discussion, there is the possibility that the critical size of SnTe nanoflakes is not due to the presence of edges, but the intrinsic size dependence of the vortex polarization itself. To confirm this possibility, we investigate the energetics of the edge-free SnTe models with different vortex sizes. Again, this model is free from the edge and can only extract the effect of the size of vortex polarization. Figure 11 shows the total energy difference of the polar vortex phase from the total energy of the paraelectric (PE) phase ΔE_{vortex} as a function of periodic vortex size. For comparison, the ferroelectric (FE) phase with straight polarization (single domain) is also shown in Figure 11. Here, the total energy difference ΔE_{vortex} is normalized by dividing by the number of unit cells in each system. ΔE_{vortex} is negative at 7 × 7 vortex size, and the vortex size is more stable than the paraelectric phase. ΔE_{vortex} increases with decreasing vortex polarization size, and, finally, ΔE_{vortex} may reach the energy of the PE phase at the 4 × 4 vortex size or less. As shown in Figure 10b-2, the 4 × 4 vortex model becomes paraelectric, and thus the total energy of 4 × 4 size or less is same as that of the PE phase. This indicates that the vortex polarization increases its total energy as the vortex size decreases, and finally the vortex polarization with a smaller than 5 × 5 size becomes more energetically unstable than the PE phase and the vortex polarization cannot be formed. We thus confirm that the vortex form of polarization intrinsically exhibits the size dependence, and there exists a critical size of vortex polarization itself. Therefore, the ferroelectric critical size observed in the SnTe nanoflakes originates from the intrinsic size limit of polarization vortices.

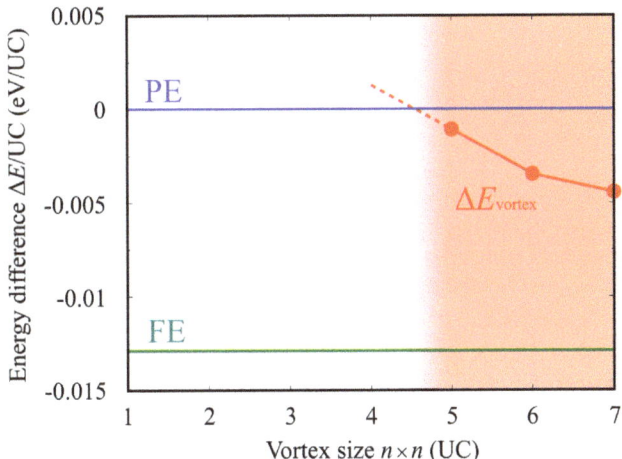

Figure 11. Total energy difference of the vortex polarization ΔE_{vortex} and the ferroelectric phase of the SnTe monolayer on the basis of the paraelectric phase as a function of the size of vortex polarization.

The increase in the total energy of vortex polarization due to the decrease in the size of vortices is due to the increase in domain wall densities. As shown in Figure 12, the vortex structure has four domains (white area), and they are separated by four 90° domain walls (green area). As the size of the vortex polarization decreases, the ratio of the domain wall per unit surface area increases, and the total energy increases. Such a high density of domain walls in the smaller vortex polarization is the primary cause of the loss of polarization in the smaller SnTe nanoflakes.

With the recent advance in manufacturing technology for two-dimensional materials, such as graphene, there are numerous experimental studies which reported the fabrication of various nanostructures of 2D materials, including graphene nanoribbons, nanoflakes, nanotubes and nanohorns. Using the fabrication techniques of graphene and its nanostructures, the fabrication of SnTe nanoribbons and nanoflakes presented in this study should be experimentally feasible. Thus, our results may stimulate an experimental study to fabricate and characterize the unique ferroelectric properties of SnTe monolayer and nanostructures.

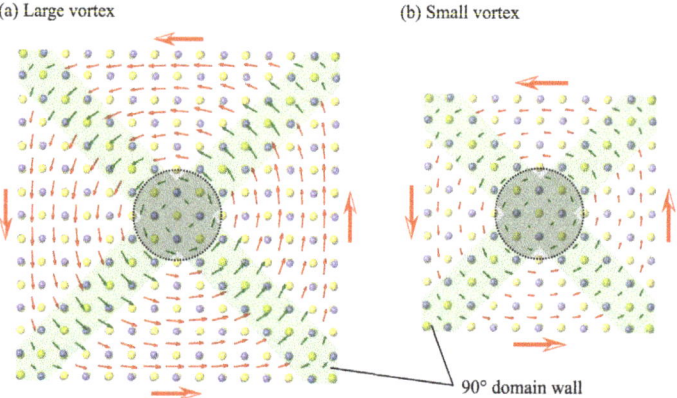

Figure 12. Schematic illustration of domain structure in the SnTe nanoflake consisting of 90° domain walls. Red arrows indicate spontaneous polarization and green areas indicate domain walls.

4. Conclusions

In this study, we investigated a ferroelectric critical size of low-dimensional SnTe nanostructures such as nanoribbons (1D) and nanoflakes (0D) using first-principle density-functional theory calculations. We demonstrated that the smallest (one-unit-cell width) SnTe nanoribbon could sustain ferroelectricity and there was no ferroelectric critical size in the SnTe nanoribbons. On the other hand, the SnTe nanoflakes formed a vortex of polarization and lost its toroidal ferroelectricity below the surface area of 4 × 4 unit cells (about 25 Å on one side). We also revealed the atomic and electronic mechanism of the absence or presence of critical size in SnTe low-dimensional nanostructures. Our result provides an insight into intrinsic ferroelectric critical sizes for low-dimensional chalcogenide layered materials.

Author Contributions: Conceptualization, T.S. and T.K.; methodology, T.S. and K.M.; formal analysis, T.S. and K.M.; investigation, T.S. and K.M.; writing—original draft preparation, T.S. and K.M.; writing—review and editing, T.S., T.X. and J.W.; visualization, T.S. and K.M.; supervision, T.K. All authors have read and agreed to the published version of the manuscript.

Funding: This research was funded by JSPS KAKENHI, grant number 17H03145, 18H05241, 18K18806, and 19K21918.

Conflicts of Interest: The authors declare no conflict of interest.

References

1. Ahn, C.H.; Rabe, K.M.; Triscone, J.M. Ferroelectricity at the nanoscale: Local polarization in oxide thin films and heterostructures. *Science* **2004**, *303*, 488–491. [CrossRef]
2. Gruverman, A.; Kholkin, A. Nanoscale ferroelectrics: Processing, characterization and future trends. *Rep. Prog. Phys.* **2005**, *69*, 2443–2474. [CrossRef]
3. Scott, J. Applications of modern ferroelectrics. *Science* **2007**, *315*, 954–959. [CrossRef]
4. Wasa, K.; Haneda, Y.; Sato, T.; Adachi, H.; Setsune, K. Crystal growth of epitaxially grown PbTiO$_3$ thin films on miscut SrTiO$_3$ substrate. *Vacuum* **1998**, *51*, 591–594. [CrossRef]
5. Fujisawa, H.; Shimizu, M.; Niu, H.; Nonomura, H.; Honda, K. Ferroelectricity and local currents in epitaxial 5-and 9-nm-thick Pb(Zr, Ti)O$_3$ ultrathin films by scanning probe microscopy. *Appl. Phys. Lett.* **2005**, *86*, 012903. [CrossRef]
6. Gu, H.; Hu, Y.; You, J.; Hu, Z.; Yuan, Y.; Zhang, T. Characterization of single-crystalline PbTiO$_3$ nanowire growth via surfactant-free hydrothermal method. *J. Appl. Phys.* **2007**, *101*, 024319. [CrossRef]
7. Yun, W.S.; Urban, J.J.; Gu, Q.; Park, H. Ferroelectric properties of individual barium titanate nanowires investigated by scanned probe microscopy. *Nano Lett.* **2002**, *2*, 447–450. [CrossRef]

8. Wang, W.; Varghese, O.K.; Paulose, M.; Grimes, C.A.; Wang, Q.; Dickey, E.C. A study on the growth and structure of titania nanotubes. *J. Mater. Res.* **2004**, *19*, 417–422. [CrossRef]
9. Morrison, F.D.; Ramsay, L.; Scott, J.F. High aspect ratio piezoelectric strontium-bismuth-tantalate nanotubes. *J. Phys. Condens. Matter* **2003**, *15*, L527. [CrossRef]
10. Naumov, I.I.; Bellaiche, L.; Fu, H. Unusual phase transitions in ferroelectric nanodisks and nanorods. *Nature* **2004**, *432*, 737. [CrossRef]
11. Schilling, A.; Byrne, D.; Catalan, G.; Webber, K.; Genenko, Y.; Wu, G.; Scott, J.; Gregg, J. Domains in ferroelectric nanodots. *Nano Lett.* **2009**, *9*, 3359–3364. [CrossRef] [PubMed]
12. Fong, D.D.; Stephenson, G.B.; Streiffer, S.K.; Eastman, J.A.; Auciello, O.; Fuoss, P.H.; Thompson, C. Ferroelectricity in ultrathin perovskite films. *Science* **2004**, *304*, 1650–1653. [CrossRef] [PubMed]
13. Junquera, J.; Ghosez, P. Critical thickness for ferroelectricity in perovskite ultrathin films. *Nature* **2003**, *422*, 506. [CrossRef]
14. Despont, L.; Koitzsch, C.; Clerc, F.; Garnier, M.; Aebi, P.; Lichtensteiger, C.; Triscone, J.M.; de Abajo, F.G.; Bousquet, E.; Ghosez, P. Direct evidence for ferroelectric polar distortion in ultrathin lead titanate perovskite films. *Phys. Rev. B* **2006**, *73*, 094110. [CrossRef]
15. Shimada, T.; Wang, X.; Kondo, Y.; Kitamura, T. Absence of ferroelectric critical size in ultrathin $PbTiO_3$ nanotubes: A density-functional theory study. *Phys. Rev. Lett.* **2012**, *108*, 067601. [CrossRef] [PubMed]
16. Mehta, R.; Silverman, B.; Jacobs, J. Depolarization fields in thin ferroelectric films. *J. Appl. Phys.* **1973**, *44*, 3379–3385. [CrossRef]
17. Cochran, W. Crystal stability and the theory of ferroelectricity. *Adv. Phys.* **1960**, *9*, 387–423. [CrossRef]
18. Chang, K.; Liu, J.; Lin, H.; Wang, N.; Zhao, K.; Zhang, A.; Jin, F.; Zhong, Y.; Hu, X.; Duan, W.; et al. Discovery of robust in-plane ferroelectricbloity in atomic-thick SnTe. *Science* **2016**, *353*, 274–278. [CrossRef]
19. Chang, K.; Kaloni, T.P.; Lin, H.; Bedoya-Pinto, A.; Pandeya, A.K.; Kostanovskiy, I.; Zhao, K.; Zhong, Y.; Hu, X.; Xue, Q.-K.; et al. Enhanced spontaneous polarization in ultrathin SnTe films with layered antipolar structure. *Adv. Mater.* **2019**, *31*, 1804428. [CrossRef]
20. Chang, K.; Parkin, S.S. The growth and phase distribution of ultrathin SnTe on graphene. *APL Mater.* **2019**, *7*, 041102. [CrossRef]
21. Liu, K.; Lu, J.; Picozzi, S.; Bellaiche, L.; Xiang, H. Intrinsic origin of enhancement of ferroelectricity in SnTe ultrathin films. *Phys. Rev. B* **2018**, *121*, 027601.
22. Hohenberg, P.; Kohn, W. Inhomogeneous electron gas. *Phys. Rev.* **1964**, *136*, B864. [CrossRef]
23. Kohn, W.; Sham, L.J. Self-consistent equations including exchange and correlation effects. *Phys. Rev.* **1965**, *140*, A1133. [CrossRef]
24. Blöchl, P.E. Projector augmented-wave method. *Phys. Rev. B* **1994**, *50*, 17953. [CrossRef] [PubMed]
25. Kresse, G.; Joubert, D. From ultrasoft pseudopotentials to the projector augmented-wave method. *Phys. Rev. B* **1999**, *59*, 1758. [CrossRef]
26. Monkhorst, H.J.; Pack, J.D. Special points for Brillouin-zone integrations. *Phys. Rev. B* **1976**, *13*, 5188. [CrossRef]
27. Grimme, S.; Antony, J.; Ehrlich, S.; Krieg, H. A consistent and accurate ab initio parametrization of density functional dispersion correction DFT-D for the 94 elements H-Pu. *J. Chem. Phys.* **2010**, *132*, 154104. [CrossRef]
28. Kresse, G.; Hafner, J. Ab initio molecular dynamics for liquid metals. *Phys. Rev. B* **1993**, *47*, 558. [CrossRef]
29. Kresse, G.; Furthmüller, J. Efficient iterative schemes for ab initio total-energy calculations using a plane-wave basis set. *Phys. Rev. B* **1996**, *54*, 11169. [CrossRef]
30. Meyer, B.; Vanderbilt, D. Ab initio study of ferroelectric domain walls in $PbTiO_3$. *Phys. Rev. B* **2002**, *65*, 104111. [CrossRef]
31. Zhang, Y.; Li, G.-P.; Shimada, T.; Wang, J.; Kitamura, T. Disappearance of ferroelectric critical thickness in epitaxial ultrathin $BaZrO_3$ films. *Phys. Rev. B* **2014**, *90*, 184107. [CrossRef]
32. Hong, J.; Fang, D. Size-dependent ferroelectric behaviors of $BaTiO_3$ nanowires. *Appl. Phys. Lett.* **2008**, *92*, 012906. [CrossRef]
33. Polking, M.J.; Han, M.-G.; Yourdkhani, A.; Petkov, V.; Kisielowski, C.F.; Volkov, V.V.; Zhu, Y.; Caruntu, G.; Alivisatos, A.P.; Ramesh, R. Ferroelectric order in individual nanometre-scale crystals. *Nat. Mater.* **2012**, *11*, 700. [CrossRef] [PubMed]

34. Béa, H.; Fusil, S.; Bouzehouane, K.; Bibes, M.; Sirena, M.; Herranz, G.; Jacquet, E.; Contour, J.-P.; Barthélémy, A. Ferroelectricity down to at least 2 nm in multiferroic BiFeO$_3$ epitaxial thin films. *Jpn. J. Appl. Phys.* **2006**, *45*, L187. [CrossRef]
35. Batra, I.P.; Silverman, B. Thermodynamic stability of thin ferroelectric films. *Solid State Commun.* **1972**, *11*, 291–294. [CrossRef]
36. Meyer, B.; Padilla, J.; Vanderbilt, D. Theory of PbTiO$_3$, BaTiO$_3$, and SrTiO$_3$ surfaces. *Faraday Discuss.* **1999**, *114*, 395. [CrossRef]
37. Bungaro, C.; Rabe, K. Coexistence of antiferrodistortive and ferroelectric distortions at the PbTiO$_3$ (001) surface. *Phys. Rev. B* **2005**, *71*, 035420. [CrossRef]
38. Umeno, Y.; Shimada, T.; Kitamura, T.; Elsässer, C. Ab initio density functional theory study of strain effects on ferroelectricity at PbTiO$_3$ surfaces. *Phys. Rev. B* **2006**, *74*, 174111. [CrossRef]
39. Shimada, T.; Tomoda, S.; Kitamura, T. Ab initio study of ferroelectricity in edged PbTiO$_3$ nanowires under axial tension. *Phys. Rev. B* **2009**, *79*, 024102. [CrossRef]
40. Wang, X.; Yan, Y.; Shimada, T.; Wang, J.; Kitamura, T. Ferroelectric critical size and vortex domain structures of PbTiO$_3$ nanodots: A density functional theory study. *J. Appl. Phys.* **2018**, *123*, 114101. [CrossRef]
41. Shimada, T.; Wang, X.; Tomoda, S.; Marton, P.; Elsässer, C.; Kitamura, T. Coexistence of rectilinear and vortex polarizations at twist boundaries in ferroelectric PbTiO$_3$ from first principles. *Phys. Rev. B* **2011**, *83*, 094121. [CrossRef]
42. Pilania, G.; Alpay, S.; Ramprasad, R. Ab initio study of ferroelectricity in BaTiO$_3$ nanowires. *Phys. Rev. B* **2009**, *80*, 014113. [CrossRef]

© 2020 by the authors. Licensee MDPI, Basel, Switzerland. This article is an open access article distributed under the terms and conditions of the Creative Commons Attribution (CC BY) license (http://creativecommons.org/licenses/by/4.0/).

Article

The Effect of Vacancies on Grain Boundary Segregation in Ferromagnetic *fcc* Ni

Martina Mazalová [1], Monika Všianská [1,2], Jana Pavlů [1,2] and Mojmír Šob [1,2,3,*]

[1] Department of Chemistry, Faculty of Science, Masaryk University, Kotlářská 2, CZ-611 37 Brno, Czech Republic; 394206@mail.muni.cz (M.M.); 230038@mail.muni.cz (M.V.); houserova@chemi.muni.cz (J.P.)
[2] Institute of Physics of Materials, Academy of Sciences of the Czech Republic, Žižkova 22, CZ-616 62 Brno, Czech Republic
[3] Central European Institute of Technology, CEITEC MU, Masaryk University, Kamenice 753/5, CZ-625 00 Brno, Czech Republic
* Correspondence: sob@mail.muni.cz or mojmir@ipm.cz; Tel.: +420-549-497-450

Received: 26 February 2020; Accepted: 31 March 2020; Published: 6 April 2020

Abstract: This work presents a comprehensive and detailed ab initio study of interactions between the tilt Σ5(210) grain boundary (GB), impurities X (X = Al, Si) and vacancies (Va) in ferromagnetic *fcc* nickel. To obtain reliable results, two methods of structure relaxation were employed: the automatic full relaxation and the finding of the minimum energy with respect to the lattice dimensions perpendicular to the GB plane and positions of atoms. Both methods provide comparable results. The analyses of the following phenomena are provided: the influence of the lattice defects on structural properties of material such as lattice parameters, the volume per atom, interlayer distances and atomic positions; the energies of formation of particular structures with respect to the standard element reference states; the stabilization/destabilization effects of impurities (in substitutional (s) as well as in tetragonal (iT) and octahedral (iO) interstitial positions) and of vacancies in both the bulk material and material with GBs; a possibility of recombination of $Si^{(i)}$+Va defect to $Si^{(s)}$ one with respect to the Va position; the total energy of formation of GB and Va; the binding energies between the lattice defects and their combinations; impurity segregation energies and the effect of Va on them; magnetic characteristics in the presence of impurities, vacancies and GBs. As there is very little experimental information on the interaction between impurities, vacancies and GBs in *fcc* nickel, most of the present results are theoretical predictions, which may motivate future experimental work.

Keywords: *fcc* Ni; tilt Σ5(210) grain boundary; vacancy; Si and Al impurity; grain boundary energy; segregation energy; defects binding energies; magnetism

1. Introduction

Various crystal defects such as impurities, vacancies (Va) and grain boundaries (GB) significantly affect material properties and are objects of both theoretical and applied research. Recent investigations deal with topics such as energetics of GB formation and its sensitivity to segregated impurities playing the role of either embrittlers or cohesion enhancers [1–3]. These issues have far-reaching practical implications manifested in the mechanical and magnetic properties. As an example, let us mention the strengthening/embrittling energy of segregated sp-elements from the 3rd, 4th and 5th period at the Σ5(210) grain boundary in ferromagnetic *fcc* nickel and cobalt [2–8]. Similar topics are also studied experimentally when, for example, the diffusion and segregation of silver in copper Σ5(310) grain boundary were investigated [9] or the influence of boron (segregated at the GB) on fracture resistance of Ni$_3$Al [10] was analyzed. For some impurities segregated in Ni, e.g. for S, semiempirical interatomic potentials have been constructed [11]. Recent trends and open problems in grain boundary segregation are discussed in the reviews [3,12].

In nickel-based alloys, aluminum is added to improve high-temperature strength and precipitation hardening [13]. For example, an increased concentration of aluminum in the 718Plus alloy enhances its tensile strength, but has a negative impact on its ductility [14]. The experimental results also reveal that the content of aluminum has a significant effect on the solvus temperature in nickel-based alloys. In IN738 superalloy, silicon segregates mainly in inter-dendritic regions and promotes the segregation of other elements [15].

In this work, we present an ab initio study of the tilt Σ5(210) grain boundary in ferromagnetic *fcc* nickel both in the clean state and with segregated Al and Si impurities accompanied by vacancies. We look for the preferred positions of impurities at the GB and analyze their effect on atomic arrangement and magnetism. Though the physical mechanisms behind GB embrittlement and strengthening have been studied in detail for some materials [16–19], the specific interactions between GBs, impurities and vacancies are usually not taken into account. In our approach to segregation, not only the final equilibrium state of the studied system is characterized, but we also explore the way how it was achieved. We provide a detailed discussion of two methods of equilibration of structural arrangement and show that some of the initial configurations may end up in equilibrium state far from the initial structure (migration and vanishing of vacancies).

2. Materials and Methods

To investigate the influence of the impurities, vacancies and grain boundaries on properties of ferromagnetic *fcc* nickel, the characteristics of bulk material were investigated first. In case of the bulk material (both without and with impurities), the *fcc* supercells with 60 (Figure 1a) or 120 atoms were used. The basic planes of the *fcc* supercells are constituted by (210) planes of standard *fcc* unit cell with 4 Ni atoms. Hence, a and b lattice parameters of the *fcc* supercells are $\sqrt{5}a$ and the c lattice parameter is equal to an integer multiple of a ($3a$ for Ni$_{60}$ and $6a$ for Ni$_{120}$), where a stands for the lattice parameter of the *fcc* Ni unit cell with 4 atoms. The cell with lattice parameters $\sqrt{5}a$, $\sqrt{5}a$, a is called a Coincident Site Lattice (CSL) cell. To get the *fcc* supercells with reasonable size of 60 or 120 atoms, the size of the CSL cell was increased 3 or 6 times in the direction of the c lattice parameter. Thus, the dimensions of the *fcc* supercell are $\sqrt{5}a$, $\sqrt{5}a$, $3a$ for a cell of the bulk material with 60 atoms (consisting of 3 CSL cells, Figure 1a) and $\sqrt{5}a$, $\sqrt{5}a$, $6a$ for a later employed cell of the bulk material with 120 atoms (consisting of 6 CSL cells).

The interstitial impurities were studied with respect to both octahedral (iO) and tetrahedral (iT) positions. In general, an atom in the octahedral position in the *fcc* unit cell is situated in octahedral site between six Ni atoms. One of four octahedral positions in *fcc* unit cell with 4 atoms, where our impurity was placed, is positioned between six Ni atoms at the centers of faces of the *fcc* cell. The fractional coordinates of such a position are $x = y = z = \frac{1}{2}$. An impurity atom in the tetrahedral position in the *fcc* unit cell is situated in the tetrahedral site between four atoms. There are eight tetrahedral positions in the *fcc* unit cell with 4 atoms. One of these positions exhibits the fractional coordinates $x = y = z = \frac{1}{4}$. Supposing the touching spheres, the ratio between the radius of interstitial position (r) and the radius of the constituent atoms forming the *fcc* lattice (R) is $r^{iO}/R = 0.414$ for iO position and $r^{iT}/R = 0.225$ for iT position. It means that octahedral site is much larger than the tetrahedral one in *fcc* metals.

Further, the tilt Σ5(210) grain boundary in both a clean state and with impurities and vacancies (Figure 1b) was analyzed. This grain boundary was created by means of the rotation of two standard *fcc* cells with 4 atoms around the [001] axis by 53°. The method used for the construction of this supercell is called the Coincident Site Lattice principle [20]. In case of the supercells with GB, the supercell with 60 atoms (consisting of 3 CSL cells) has the lattice parameters $\sqrt{5}a$, $3\sqrt{5}a$, a, and the supercell with 120 atoms (consisting of 6 CSL cells) has the lattice parameters $\sqrt{5}a$, $3\sqrt{5}a$, $2a$ (Figure 1b). Because of the periodicity reasons, our GB supercells contain two grain boundaries oriented in the opposite direction, which is obvious from Figure 1b.

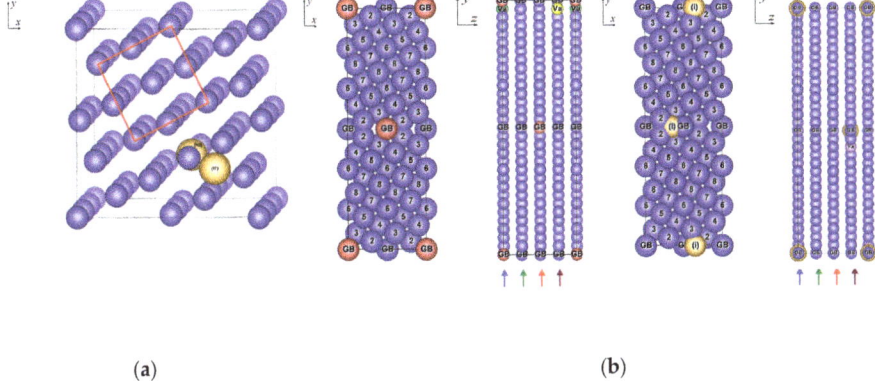

Figure 1. Supercells employed in the present calculations: (**a**) The fcc Ni$_{60}$ supercell (Ni$_{60}$–3 CSL cells behind each other) in a typical (3 CSL cells) arrangement having the lattice parameters $\sqrt{5}a$, $\sqrt{5}a$, $3a$. The orange spheres denoted as iT (iO) occupy the tetrahedral (octahedral) positions of interstitial impurity atoms. The red square shows the position of the conventional fcc cell with 4 atoms. (**b**) The Σ5(210) grain boundary in fcc Ni supercell (GB Ni$_{120}$, consisting of 6 CSL cells, 1×3×2 next to each other) having the lattice parameters $\sqrt{5}a$, $3\sqrt{5}a$, $2a$. The blue spheres correspond to Ni atoms, the spheres labelled by "GB" denote the GB plane, the red spheres occupy the position of substitutional impurity atoms, the orange spheres denoted as (i) mark the positions of interstitial impurity atoms. The green, yellow and violet spheres with label Va show the positions of vacancies in the structure with the substitutional Al$^{(s)}$, substitutional Si$^{(s)}$ and interstitial Si$^{(i)}$ impurity, respectively. The numbers on atoms mark the layers counted from the GB plane. The planes with $z = 0$ and $z = 0.5$ ($z = 0.25$ and $z = 0.75$) are denoted by blue and red (green and violet) arrows.

The above-mentioned structure is called B'.B' [21] and is one of the known Σ5(210)[001] GB configurations. The other one (differing in the shape of interstitial sites) is the B.B structure. For this study, the B'.B' configuration was chosen because it is supposed to have a lower GB energy, as deduced from atomistic studies [22,23].

All our calculations were performed at 0 K temperature within the Density Functional Theory (DFT) using the Vienna Ab initio Simulation Package (VASP) code [24–26] with Projector-Augmented-Wave – Perdew–Burke–Ernzerhof (PAW–PBE) potentials [27–29], i.e., in the Generalized Gradient Approximation (GGA). The optimum setting of computational parameters found by test calculations was as follows: the ENCUT parameter (the cut off energy defining the number of plane waves in the basis set) was 500 eV and the KSPACING parameter (which defines the number of k-points in the irreducible part of the Brillouin zone) was 0.1 Å^{-1}. If not mentioned otherwise, the structure relaxations were performed as follows. The total energy was minimized with respect to the lattice parameters and atomic positions. At first, the conjugate–gradient method (using the Hellmann–Feynman forces acting on the atoms) and then the quasi–Newton method was employed to reach the required force limit. The Brillouin zone was sampled by the Monkhorst–Pack scheme and its integration was performed by Methfessel–Paxton or by the tetrahedron method, depending on whether relaxation or static calculation was carried out.

3. Results

3.1. Bulk Material

3.1.1. Elemental Bulk Material

As the ferromagnetic fcc structure of elemental Ni is taken as the reference state in this study, its equilibrium properties (lattice parameters, atomic positions and total energy) were determined at first with the help of a sixty-atom supercell (Figure 1a). Similar analyses were performed for nonmagnetic fcc Al (4 atoms per cell) and nonmagnetic Si in diamond structure (8 atoms per cell) for the same reason. These three structures are taken here as the Standard Element Reference state (SER). The results obtained are summarized in Table 1.

Table 1. Equilibrium properties of the bulk material of all three elements studied and fcc Ni containing impurities. Here, a, b and c stand for lattice parameters, E^f is the energy of formation related to the standard element reference states: ferromagnetic fcc Ni, nonmagnetic fcc Al and nonmagnetic Si in the diamond structure (Equation (1)). μ_{Ni} stands for the magnetic moment per atom obtained by averaging over all Ni atoms in the structure and $\mu_{Ni,NN}$ denotes the magnetic moment of Ni atom that is the nearest neighbor of the impurity. Superscripts (s) and (iT) or (iO) mark the substitutional or interstitial impurity position. The fcc supercell lattice parameters obtained from literature were derived from the lattice parameters of fcc Ni unit cell a = 3.498 Å [30] and a = 3.5138 Å [31].

Configuration	a (Å)	b (Å)	c (Å)	V_{at} (Å³)	E^f (eV.atom⁻¹)	μ_{Ni} (μ_B)	$\mu_{Ni,NN}$ (μ_B)	Reference
Si	5.469	5.469	5.469	20.443	0	-	-	This work
Al	4.040	4.040	4.040	16.481	0	-	-	This work
Ni$_{60}$	7.866	7.866	10.552	10.882	0	0.632	-	This work
	7.824	7.824	10.497	10.710	-	0.628	-	[32]
	7.821	7.821	10.494	10.669	-	0.607	-	[30] [a]
	7.857	7.857	10.541	10.846	-	-	-	[31] [a]
Ni$_{59}$Al$^{(s)}$	7.872	7.872	10.562	10.909	−0.026 **	0.608	0.525	This work
	-	-	-	10.735	-	-	0.528	[32]
	-	-	-	-	−0.029	-	-	[33,34] [b]
Ni$_{59}$Si$^{(s)}$	7.866	7.866	10.550	10.878	−0.027 **	0.624	0.499	This work
	-	-	-	10.699	-	-	0.484	[32]
	-	-	-	-	−0.031	-	-	[33,34] [b]
Ni$_{60}$Si$^{(iT)}$ *	7.945	7.929	10.654	11.003	0.024	0.579	0.278	This work
Ni$_{120}$Si$^{(iT)}$	7.899	7.899	21.245	10.956	0.020	0.600	0.251	This work
Ni$_{60}$Si$^{(iO)}$ *	7.938	7.938	10.670	11.021	0.026	0.598	0.302	This work
Ni$_{120}$Si$^{(iO)}$	7.903	7.903	21.218	10.951	0.014 **	0.623	0.302	This work

* The interstitial impurity migrated away from the ideal tetrahedral (octahedral) position. ** From these values, the energies of entering of one impurity into the pure fcc nickel (E^f (X)) used in Equation (20) were evaluated by multiplying these values by the total number of atoms in the structure. These values are $E^f(Al^{(s)})$ = −1.574 eV.X⁻¹, $E^f(Si^{(s)})$ = −1.608 eV.X⁻¹ and $E^f(Si^{(i)})$ = −1.669 eV.X⁻¹. [a] The data were calculated for Ni$_4$ unit cell. [b] The formation energy was obtained by the extrapolation of the formation energies for Ni, Ni$_3$Al, Ni$_3$Si by Open Quantum Mechanics Database tools.

Comparing the lattice parameters of the Ni$_{60}$ structure with experimental data published in References [30,31], only a very small deviation within 0.6% and 0.1%, respectively, was found. The magnetic moment differs by 5.6%.

3.1.2. Bulk Material with Impurities

The supercells including both substitutional and interstitial impurities were investigated using the following configurations: **Ni$_{59}$Al$^{(s)}$** (a structure with 59 atoms of Ni and one atom of Al in a substitutional position), **Ni$_{59}$Si$^{(s)}$** (a structure with 59 atoms of Ni and one atom of Si in a substitutional position) and **Ni$_{60}$Si$^{(i)}$** (a structure with 60 atoms of Ni and one atom of Si in an interstitial tetrahedral and octahedral position). The study of the configuration with the interstitial Al atom (Ni$_{60}$Al$^{(i)}$) was omitted as this structure is known as very unstable [35]. For comparison, we also studied the structure of **Ni$_{120}$Si$^{(i)}$** (a structure with 120 atoms of Ni and one atom of Si in a tetrahedral and octahedral interstitial position). The results of calculations of four fcc Ni supercells with interstitial Si (Ni$_{60}$Si$^{(iT)}$, Ni$_{120}$Si$^{(iT)}$, Ni$_{60}$Si$^{(iO)}$, Ni$_{120}$Si$^{(iO)}$) provide almost the same results: lattice parameters a, b and c differ by 0.57% at most and the volume per atom by 0.64%. The parameter c is two times larger in $Ni_{120}Si^{(i)}$ than in $Ni_{60}Si^{(i)}$, which is given by the increase of the number of atoms in the supercell.

The equilibrium lattice parameters and the total energies of all configurations were determined using the full relaxation of structure parameters. At first, the conjugate-gradient method and then the quasi-Newton method was used. The results obtained are summarized in Table 1.

The equilibrium lattice parameters and volume per atom (Table 1) show that the interstitial Si atom causes the largest volume increase in comparison with Ni_{60}. Concerning the structures with substitutional Al and Si impurities, $Ni_{59}Al^{(s)}$ and $Ni_{59}Si^{(s)}$, the increase in volume per atom is larger for $Ni_{59}Al^{(s)}$. This can be explained by the fact that the atomic radius of the Al impurity is larger than that of the Si impurity.

Subsequently, the energies of formation of particular structures were calculated according to Equation (1):

$$E^f = \left(E_{A_xB_y} - \left(x\frac{E_A}{N_A} + y\frac{E_B}{N_B}\right)\right)/N_{A_xB_y}, \qquad (1)$$

where $E_{A_xB_y}$ corresponds to the total energy of studied configuration, x denotes the number of atoms A and y denotes the number of atoms B. E_A/N_A is the ground state energy per one atom A and E_B/N_B is the ground state energy per one atom B. $N_{A_xB_y}$ is the total number of atoms in the compound A_xB_y.

For the studied structures in equilibrium arrangement, the following decreasing trend in the energy of formation related to fcc Ni, fcc Al and Si in diamond structure was found: $Ni_{60}Si^{(iO)} > Ni_{60}Si^{(iT)} > Ni_{120}Si^{(iT)} > Ni_{120}Si^{(iO)} > Ni_{60} > Ni_{59}Al^{(s)} > Ni_{59}Si^{(s)}$, where the formation energy per atom of $Ni_{59}Al^{(s)}$ and $Ni_{59}Si^{(s)}$ differs only by 2% and is negative while the energies of formation of structures with interstitial Si are positive. This trend in energies of formation was confirmed by the calculations of the total energy of smaller cells, where the formation energies decrease in the sequence $Ni_4Si^{(iT)} > Ni_3Al^{(s)} > Ni_4 > Ni_3Si^{(s)}$. However, these structures are not used for the study of impurities in this work as the concentration of impurities is too high in comparison with real materials. Based on the above-mentioned results, one can expect that both Si and Al would prefer the substitutional positions in bulk fcc Ni, which is in agreement with the results of a previous study [35].

Significant structural changes occurred during the relaxation of structures $Ni_{60}Si^{(iT)}$ and $Ni_{60}Si^{(iO)}$. The change in a position of atoms is the largest for structure $Ni_{60}Si^{(iT)}$. Here, the interstitial Si atom moves from the position between two planes towards one plane and changes the position of surrounding Ni atoms. After the relaxation, the new position of impurity cannot be considered as tetrahedral. The fractional coordinates of impurity in the 61 atomic unit cell (Figure 1a) has changed from $x = 0.25$, $y = 0.25$, $z = 0.08$ determining the tetrahedral position to nonspecific $x = 0.23$, $y = 0.13$, $z = 0.17$. The same trend can be observed for the structure $Ni_{60}Si^{(iO)}$. Here, the silicon impurity atom moves from the plane of atoms to the interplanar space and affects the surrounding Ni atoms and the relaxed position of the impurity atom is not octahedral. The fractional coordinates of this impurity in the 61 atomic unit cell (Figure 1a) has changed from $x = 0.30$, $y = 0.40$, $z = 0.33$ determining the octahedral position to nonspecific $x = 0.40$, $y = 0.29$, $z = 0.38$. The position of the Si impurity atom in structures $Ni_{120}Si^{(iT)}$ and $Ni_{120}Si^{(iO)}$ remains unchanged even after the relaxation. These structures can be considered as the structures with interstitial impurity atom occupying the tetrahedral or octahedral site. Comparing the larger structures with the smaller ones, it was concluded that the larger structures require smaller energies of formation (Table 1) which results in their larger stability. Comparing the structures $Ni_{120}Si^{(iT)}$ and $Ni_{120}Si^{(iO)}$, we can confirm the fact that the octahedral site is larger than the tetrahedral one, which results in greater stability of structure with the interstitial impurity in octahedral site where the smaller deformation of structure (displacement of adjacent Ni atoms) is needed to accommodate the impurity atom. The energy difference between $Ni_{120}Si^{(iT)}$ and $Ni_{120}Si^{(iO)}$ is 0.5886 kJ.mol^{-1}, which predicts $Ni_{120}Si^{(iO)}$ as more stable.

Our results on energy of formation of $Ni_{59}Al^{(s)}$ and $Ni_{59}Si^{(s)}$ can be compared with the values obtained from References [33,34]. For this comparison, the values calculated as the linear combination of the literature data on the energies of formation of the cells with compositions Ni_3Al, Ni_3Si and pure Ni were employed. Hence, these results are not directly comparable with the data calculated for $Ni_{59}Al^{(s)}$ and $Ni_{59}Si^{(s)}$ structures. The energy of formation for $Ni_{59}Al^{(s)}$ calculated in this work and that obtained from the literature differ by 11.5%. In the case of $Ni_{59}Si^{(s)}$ structure, the difference between the calculated and reported energy of formation is 17.0%. In comparison with literature, our data show lower stability of studied structures.

It turns out that impurities significantly influence the magnetic moments of their neighboring Ni atoms (see Table 1) and they themselves obtain a small magnetic moment. The induced magnetic moment μ on the substitutional Al atom is slightly negative ($\mu = -0.018\ \mu_B$) with respect to Ni atoms and causes the decrease of magnetic moments of neighboring Ni atoms by 0.13 μ_B. Substitutional Si ($\mu = -0.018\ \mu_B$) causes the decrease of magnetic moment of neighboring Ni atoms by 0.16 μ_B and interstitial Si in $Ni_{60}Si^{(i)}$ ($\mu = -0.021\ \mu_B$) induces the decrease of magnetic moment of neighboring Ni atoms by 0.35 μ_B. In case of the $Ni_{120}Si^{(i)}$ ($\mu = -0.022\ \mu_B$), the decrease of magnetic moment of neighboring Ni atoms reaches 0.39 μ_B. This significant impact of interstitial Si can be explained by the fact that in this case the neighboring atoms are much closer to the impurity than in the case of structures with substitutional impurity. Here, the largest decrease is caused by the interstitial atom of Si and the smallest decrease is due to the substitutional Al atom. Going further from the impurity, the magnetic moments of Ni atoms converge to the bulk value.

A similar study was also presented in References [2,32], where the change of magnetic moment of the 1st nearest neighbor (NN) nickel of $Al^{(s)}$ ($\mu = -0.02\ \mu_B$) and $Si^{(s)}$ ($\mu = -0.01\ \mu_B$) impurity are $-0.10\ \mu_B$ and $-0.14\ \mu_B$, respectively. These findings correlate very well with our values. Further, the lattice parameters a, c and the volume per atom found in the present and above-mentioned studies differ only by -0.45%, -0.64% and -1.54% (expressed in % of data found in this work) for $Ni_{59}Al^{(s)}$, respectively, and by -0.48%, -0.64% and -1.60% for $Ni_{59}Si^{(s)}$.

3.1.3. Bulk Material with Impurity and Vacancy

To study the effect of vacancies on the material properties, the basic quantities such as the vacancy-formation energy and the tendency to vacancy-impurity binding have to be evaluated. Hence, the relaxed configurations Ni_{60}, $Ni_{59}Al^{(s)}$, $Ni_{59}Si^{(s)}$, $Ni_{60}Si^{(iT/iO)}$, and $Ni_{120}Si^{(iT/iO)}$ were relaxed again with one additional vacancy in their structure. The relaxation procedure was the same as for the systems without vacancy. The equilibrium lattice parameters, the volume per atom V_{at}, the energies of formation of the structure E^f and of the vacancy $E^f(Va)$ as well as the binding energies between impurity X and vacancy $E^b(X;Va)$ are summarized in Table 2. Subsequently, the effect of distance between the vacancy and impurity D on structural parameters, energetics and magnetic moments was analyzed.

Discussing the structure of configurations with and without vacancies, it is necessary to point out that all configurations of $Ni_{59}Si^{(iT/iO)}$+Va and $Ni_{119}Si^{(iT/iO)}$+Va structure (except for the fourth one of $Ni_{119}Si^{(iT)}$+Va and the second ones of $Ni_{119}Si^{(iO)}$+Va and $Ni_{59}Si^{(iO)}$+Va, all of them denoted by bold text in Table 2) are not stable and they transform to $Ni_{59}Si^{(s)}$ or $Ni_{119}Si^{(s)}$ structure. In such case, the vacancy is occupied by interstitial Si atom or Ni atom and disappears. In this way, the interstitial Si atom becomes substitutional and the structure achieves the properties comparable with the $Ni_{59}Si^{(s)}$ structure. In case of the third configuration of both structures, the values of D parameters differ. This comes from the fact that one of them is the starting value (denoted by *) and the other is the equilibrium one.

The lattice parameters almost do not depend on the position of the vacancy (not discussing the structures $Ni_{119}Si^{(iT)}$+Va where the equilibrium lattice angles are not 90°). They vary within one hundredth, exceptionally one tenth, of percent. The lattice parameters of the configurations, where the "recombination" has occurred, exhibit the differences amounting up to 1% from values calculated for structures without vacancy.

In structures with a stable vacancy, the vacancy causes a decrease of the cell volume but an increase of the volume per atom, which is obvious because the structure relaxation cannot cause the decrease of volume in such extent, which would correspond to the volume per atom of missing particles.

Table 2. Equilibrium properties of bulk material with vacancies (both elemental fcc Ni and fcc Ni containing impurities). Va means one vacancy in the structure, D denotes the equilibrium distance of Va from the impurity, a, b and c stand for lattice parameters, V_{at} denotes the volume per atom, E^f is the energy of formation related to the ground-state of elemental ferromagnetic fcc Ni, fcc Al and diamond Si. $E^f(Va)$ is the energy of vacancy formation and $E^b(X;Va)$ is the binding energy between impurity X and vacancy. The values written in bold correspond to the most stable configuration.

Config.	D (Å)	a (Å)	b (Å)	c (Å)	V_{at} (Å³)	E^f (eV.atom⁻¹)	$E^f(Va)$ (eV.Va⁻¹)	$E^b(X;Va)$ (eV.X⁻¹Va⁻¹)
Ni₅₉+Va	-	7.8496	7.8497	10.5332	11.0003	0.0239	1.4101	-
Ni₅₈Al⁽ˢ⁾+Va	2.4594	7.8541	7.8652	10.5256	11.0206	−0.0035	1.3664	−0.0451
	3.5249	7.8566	7.8561	10.5375	11.0236	−0.0022	1.4431	0.0316
	4.2946	**7.8595**	**7.8607**	**10.5340**	**11.0307**	**−0.0023**	**1.4372**	**0.0257**
	4.9448	7.8562	7.8615	10.5326	11.0255	−0.0029	1.4018	−0.0097
Ni₅₈Si⁽ˢ⁾+Va	2.4263	7.8489	7.8409	10.5318	10.9856	−0.0052	1.3005	−0.1110
	3.5024	7.8490	7.8493	10.5229	10.9883	−0.0036	1.3949	−0.0166
	4.2894	**7.8485**	**7.8495**	**10.5259**	**10.9910**	**−0.0030**	**1.4303**	**0.0188**
	4.9468	7.8462	7.8470	10.5296	10.9881	−0.0034	1.4067	−0.0048
Ni₅₉Si⁽ⁱᵀ⁾+Va (Ni₅₉Si⁽ˢ⁾*)	1.5092 *	7.8628	7.8630	10.5499	10.8050	−0.0269	−3.0488 *	—*
	2.9128 *	7.8630	7.8628	10.5500	10.8050	−0.0269	−3.0488 *	—*
	3.8100 *	7.8630	7.8630	10.5498	10.8050	−0.0269	−3.0492 *	—*
	4.5578 *	7.8643	7.8645	10.5406	10.8654	−0.0269	−3.0485 *	—*
	5.2036 *	7.8630	7.8628	10.5499	10.8709	−0.0269	−3.0488 *	—*
Ni₁₁₉Si⁽ⁱᵀ⁾+Va (Ni₁₁₉Si⁽ˢ⁾*)	1.5233 *	7.8639	7.8644	21.0950	10.8718	−0.0136	−4.0395 *	—*
	2.9169 *,ᵃ	8.2455	10.8490	14.9017	11.1086	−0.0141	−4.1043 *	—*
	4.5700 *,ᵃ	9.5257	9.2313	14.8911	10.9022	0.0098	−1.2291 *	—*
	5.1815 ᵇ	**7.8916**	**7.8899**	**21.2043**	**11.0021**	**0.0244**	**0.5153**	**−0.1578**
Ni₅₉Si⁽ⁱᴼ⁾+Va (Ni₅₉Si⁽ˢ⁾*)	1.7603 *	7.8632	7.8633	10.5498	10.8717	−0.0271	−3.2177 *	—*
	3.4306 ᶜ	**7.9078**	**7.9159**	**10.6642**	**11.1259**	**0.0461**	**1.1772**	**−0.2343**
	3.9362 *	7.8629	7.8629	10.5494	10.8703	−0.0271	−3.2181 *	—*
	5.2802 *	7.8625	7.8630	10.5499	10.8704	−0.0271	−3.2192 *	—*
Ni₁₁₉Si⁽ⁱᴼ⁾+Va (Ni₁₁₉Si⁽ˢ⁾*)	1.7436 *	7.8631	7.8631	21.1068	10.8749	−0.0138	−3.3306 *	—*
	3.1287	**7.9000**	**7.8963**	**21.2041**	**11.0227**	**0.0257**	**1.4128**	**0.0012** **
	3.9332 *	7.8623	7.8629	21.1038	10.8720	−0.0138	−3.3306 *	—*
	5.2770 *	7.8669	7.8669	21.1127	10.8886	−0.0268	−4.8852 *	—*

* The structure in parentheses denotes the final equilibrium arrangement after the relaxation and the D values correspond to the distances before relaxation. E^f corresponds to the energy of formation of Ni₅₉Si⁽ˢ⁾ or Ni₁₁₉Si⁽ˢ⁾. In this case, $E^f(Va)$ is the energy difference between structures with interstitial and substitutional Si and $E^b(X;Va)$ cannot be evaluated as Va disappeared during the structure relaxation. ** In case of the structure Ni₁₁₉Si⁽ⁱᴼ⁾+Va, the value of $E^b(X;Va)$ was calculated with respect to structures Ni₅₉+Va and Ni₁₂₀Si⁽ⁱᴼ⁾. Hence, the value −0.2343 eV.X⁻¹Va⁻¹ should be considered more reliable as it was calculated using the energies of structures with comparable size. ᵃ The shape of the cell has changed and the equilibrium angles are not $\alpha = \beta = \gamma = 90.00°$. ᵇ The value of D before relaxation was 3.8336 Å. ᶜ The value of D before relaxation was 3.0489 Å.

In case of Ni₅₈Al⁽ˢ⁾+Va and Ni₅₈Si⁽ˢ⁾+Va structure, the minimum-energy configurations are those with the vacancy and impurity being the nearest neighbors. In case of the structure Ni₅₉Si⁽ⁱᵀ⁾+Va, the relaxation of all starting configurations results in the same final arrangement, namely Ni₅₉Si⁽ˢ⁾. In Ni₁₁₉Si⁽ⁱᵀ⁾+Va, the only stable configuration with Va is that one with the distance between the vacancy and impurity being 5.2 Å, i.e. Si and Va are in the 4th nearest neighbor position. In Ni₅₉Si⁽ⁱᴼ⁾+Va, the stable configuration is that one with the distance between the vacancy and impurity being 3.5 Å and in Ni₁₁₉Si⁽ⁱᴼ⁾+Va being 3.1 Å, i.e. Si and Va are in the 2nd nearest-neighbor's position. Analyzing the stability of vacancy in the vicinity of the impurity, it must be concluded that substitutional impurities prefer to be in the vicinity of the vacancy. On the other hand, the Si interstitial must be more distant from the vacancy to prevent the recombination of defects resulting in disappearance of vacancy and change of the impurity site from interstitial to substitutional (only three configurations were able to retain Va in their structure).

According to the values of the energies of formation provided in Table 2, it is straightforward that the interstitial Si atoms with a vacancy in their neighborhood prefer the recombination of defects resulting in a structure without a vacancy and with substitutional Si. If the vacancy is too distant from the interstitial Si atoms, the interstitial position of Si can be retained which results in the positive value of the energy of formation which are comparable for both types of the interstitial position in large cells.

The energies of vacancy formation of structures $Ni_{59}+Va$, $Ni_{59}Al^{(s)}+Va$, $Ni_{59}Si^{(s)}+Va$, $Ni_{59}Si^{(iT/iO)}+Va$ and $Ni_{119}Si^{(iT/iO)}+Va$ were computed according to Equation (2):

$$E^f(Va) = (E_{bulk+Va} + E_{1fccNi}) - E_{bulk} \qquad (2)$$

and are listed in Table 2. It is obvious that the vacancy formation causes the structure destabilization characterized by positive values of $E^f(Va)$. For the structures $Ni_{59}+Va$, $Ni_{58}Al^{(s)}+Va$, and $Ni_{58}Si^{(s)}+Va$, the increase in the total energy due to the $E^f(Va)$ contributions amounts to about 1.4 eV.Va^{-1}. In case of the structures with interstitial Si, the situation is more complicated. For the $Ni_{59}Si^{(iT)}+Va$ and the first two configurations of $Ni_{119}Si^{(iT)}+Va$, a decrease in the formation energy of -3.05 eV.Va^{-1} and of about -4 eV.Va^{-1} was found. The third configuration ($D = 4.5$ Å) of $Ni_{119}Si^{(iT)}$ exhibits even the formation energy of -1.2 eV.Va^{-1}, which can be caused by the change of structure during the relaxation where the equilibrium angles are $\alpha = 90.00°$, $\beta = 89.99°$ and $\gamma = 92.43°$. The formation energies of the structures with octahedral impurity ($Ni_{59}Si^{(iO)}+Va$ and $Ni_{119}Si^{(iO)}+Va$), except the second configuration, are in the range from -3.2 to -4.8 eV.Va^{-1}. However, the stabilization in all above-mentioned configurations with interstitial impurities is caused by the formation of the $Ni_{59}Si^{(s)}$ or $Ni_{119}Si^{(s)}$ structure rather than by the Va formation itself. In case of the fourth configuration of $Ni_{119}Si^{(iT)}+Va$ and the second of $Ni_{59}Si^{(iO)}+Va$ and $Ni_{119}Si^{(iO)}+Va$, the energy of vacancy formation reaches low positive values of 0.5, 1.1, and 1.4 eV.Va^{-1}. This is natural because here the created empty (Va) space compensates the destabilization caused by the tension originating from the interstitial impurity. Simultaneously, the most stable configurations of all structures reveal the lowest values of $E^f(Va)$.

The influence of the position of vacancy in a particular configuration on its energy of formation E^f and on the vacancy-formation energy $E^f(Va)$ is very small except for the cases with interstitial Si. For the structures represented by the supercells $Ni_{58}Al^{(s)}+Va$ and $Ni_{58}Si^{(s)}+Va$, the scatter in the energy of formation of vacancy is within several hundredths of eV.Va^{-1}. This means that there is no special interaction between the substitutional impurity and vacancy in comparison with bulk Ni (this is confirmed by similar values of $E^b(X;Va)$ as discussed below). For all configurations of structures with interstitial impurity (except for the fourth one of $Ni_{119}Si^{(iT)}+Va$ and second ones in $Ni_{59}Si^{(iO)}+Va$ and $Ni_{119}Si^{(iO)}+Va$), the formation energy of Va cannot be evaluated (it corresponds to the energy of structure transformation).

Regarding the literature data, our results are comparable with experimental vacancy-formation energies of 1.7 eV [36], 1.78 eV and 1.79 eV and with other calculated data (see Reference [37] and references therein). Another experimental value of formation energy of vacancy in bulk Ni is 1.73 eV [38]. Theoretical values found in literature amount to 1.41 eV (VASP, GGA, Perdew–Wang) [39], 1.379 eV [40], 1.48 eV (QE GGA-PBE, mag.) [41], and 1.48 eV (GGA-PBE) and 1.63 eV (Local Density Approximation (LDA)) (both Reference [42]) followed by similar data of 1.57 eV [43], 1.58 eV [44], 1.31 eV [45], 1.6 eV [46], 1.279 eV [47], 1.77 eV [37] and 1.39 eV [35].

From the energetical point of view, the further quantity – the binding energy of two defects (impurity and vacancy) – can be evaluated by Equation (3):

$$E^b(X;Va) = (E_{bulk+X+Va} + nE_{1fccNi}) - (E_{bulk+X} + E_{bulk+Va}), \qquad (3)$$

where the first term describes the energy of material with interacting defects, the second term is the correction by means of the energy of bulk Ni and the third and fourth term correspond to the energy of noninteracting defects placed in different computational cells. The calculated values are given in Table 2.

In case of the most stable structures with substitutional impurities, the binding energy $E^b(X;Va)$ reaches the values of -0.0451 and -0.1110 eV.X^{-1}Va^{-1}. In case of interstitial Si, these values are much lower: −0.1578 eV.X⁻¹Va⁻¹ for Si in iT position and −0.2343 and 0.0012 eV.X⁻¹Va⁻¹ for Si in iO position. Nevertheless, the most negative values are obtained for the most stable configurations (1st NN for substitutional impurity, 3rd NN for iT impurity and 2nd for iO impurity). The negative values can be interpreted as the tendency to the binding between vacancy and impurity.

It was found that the presence of a vacancy causes the increase of magnetic moments of neighboring Ni atoms by 0.01–0.03 μ_B. The size of this effect depends on the mutual position and interplay of all three "particles": Ni atom, impurity atom and vacancy. Usually the "magnetic-moment-decreasing" effect of antiferromagnetic impurity (which is induced by surrounding Ni atoms because originally the Si and Al impurities are nonmagnetic) is bigger than the "magnetic-moment-increasing" influence of vacancy (Figure 2). It was proved that the influence of impurities and vacancies on Ni magnetism gradually decreases with distance and the magnetic moment converges to bulk-material value. The bulk values were reached when the distance of a Ni atom from the impurity was equal to 3.5–5.0 Å.

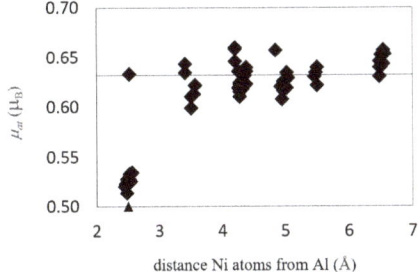

Figure 2. The dependence of the magnetic moment of Ni atom on its distance from Al$^{(s)}$ in the structure Ni$_{58}$Al$^{(s)}$+Va (the vacancy is situated in the first nearest neighbor position of the Al atom (D = 2.4594 Å)). The line corresponds to the calculated value of magnetic moment of bulk fcc Ni 0.632 μ_B.

There can be a difference in the magnetic moments of atoms with the same distance from the impurity atom, which is caused by the presence of vacancy in the structure (Figure 2). Here, the Ni atoms lying closer to the vacancy possess higher magnetic moments than those lying further from it but having the same distance from Al impurity.

3.2. Clean Grain Boundary

3.2.1. Structure Arrangement

After the studies of bulk material, a supercell with Σ5(210) grain boundary including sixty atoms of nickel (GB-Ni$_{60}$) (Figure 1b) and also the derivatives of the bigger cell (GB-Ni$_{120}$) were studied. Because of actual scientific discussions, how to properly optimize structural parameters in materials with grain boundaries [48] detailed tests of two relaxation methods were performed to confirm the reliability of our further approach. These tests are significant as their conclusions are transferable to other systems. The first method (Method 1) determines the equilibrium lattice parameters and total energy from the automatic full relaxation of structure parameters based on the minimization of forces acting on particular atoms. The other one (Method 2) consists in the evaluation of a set of calculations where the two lattice parameters (in the plane of the GB a and c) are kept fixed, one parameter (dimension perpendicular to the boundary, b) is changed to various values and the simultaneous relaxation of atomic positions is performed for each setting of lattice parameters. In this case, the equilibrium structure and energy were determined with respect to a) minimum energy (Method 2a) or b) the zero stress along the dimension perpendicular to the boundary (Method 2b).

The full relaxation corresponds to a situation in small grains, where also the bulk region is affected by the grain boundary, which plays a major role. The situation in larger grains is different. Here, the bulk interiors keep their own bulk values (lattice constant) so that a restricted relaxation (when the dimension perpendicular to the GB is relaxed and the interface dimensions in the GB plane are kept equal to the bulk values) should provide a better description.

Both above mentioned relaxation approaches (Method 1, Method 2) were used for the GB-Ni$_{60}$, GB-Ni$_{120}$, GB-Ni$_{118}$Al$_2^{(s)}$, GB-Ni$_{118}$Si$_2^{(s)}$ and GB-Ni$_{120}$Si$_2^{(i)}$ supercells. In case of the structures with an impurity, the large cells were used only to ensure the conservation of bulk-material properties between the grain boundaries. It was confirmed that the second method provides almost the same lattice parameters as the first one regardless of the method of determination of equilibrium structure. There is a difference of 0.7% between the equilibrium cell volumes obtained using the full relaxation and using the second approach for GB-Ni$_{60}$. The b/a ratio and the free volume also do not change very much (see Table 3 in next chapter). Because of these insignificant differences, the both relaxation methods can be considered as equivalent and the Method 1 was used in the subsequent calculations.

For better orientation in the following text, we will use the word "layer" for layers of the (210) type (those parallel with the GBs) and the word "plane" for planes of the (001) type (those perpendicular to GBs and lying in the plane of the paper in Figure 3, i.e. those with GB image).

The dependence of interlayer distances in a supercell containing a clean GB with respect to the type of atomic layers is displayed in Figure 4a (next section) and numerical data are provided in Table S1 in the Supplemental Material. The first interlayer distance GB/2 (the distance between the grain boundary and the second atomic layer) expanded to 1.10 Å and is by about 40% larger than the calculated bulk value of 0.79 Å. On the other hand, the second interlayer distance 2/3 (the distance between the second and the third atomic layer) contracted to 0.57 Å and is by about 27% smaller than the value for the bulk *fcc* nickel. The interlayer distance 3/4 is expanded again, namely to 0.86 Å, i.e. it is about 9% larger than the bulk value, and so on. From the sixth layer on, the interlayer distances closely oscillate around the calculated bulk value. This proves that the supercell used in the calculations is large enough to avoid the interactions between the present two grain boundaries. Our results are also in agreement with the data calculated by Všianská and Šob [2].

3.2.2. Energetics

The GB energy (energy of formation of a GB) γ_{GB} is defined as the energy needed to create 1 m^2 of GB in the bulk material. Within the ab initio approach, it can be calculated as the difference of the total energies of two supercells: one with the grain boundary (E_{GB}) and the other one without it (E_{bulk}). This difference is then divided by the doubled area of the GB in the cell S (because there are two boundaries per cell in the periodic system) [49]. This results in the following Equation (4):

$$\gamma_{GB} = \frac{E_{GB} - E_{bulk}}{2S}. \qquad (4)$$

The GB energy γ_{GB} (see Table 3 in the next chapter) calculated with the help of clean GB-Ni$_{60}$ and GB-Ni$_{120}$ supercell is in a very good agreement with data obtained from literature [2,50]. The energy per atom of Ni in the supercell with grain boundary is higher by 1.34% than the energy per atom in bulk Ni for both GB-Ni$_{60}$ and GB-Ni$_{120}$, which means that the grain boundary is unstable with respect to the bulk material.

3.2.3. Magnetism

Concerning the magnetism, it was found that the magnetic moment of Ni is strongly influenced by the presence of the grain boundary as it increases up to its maximum 0.677 μ_B in the layer next to the GB in case of GB-Ni$_{60}$ supercell. This value is by 7% higher than the bulk calculated value of 0.632 μ_B, which is in agreement with the enhancement found in [2]. Then, the magnetic moments of Ni atoms located farther from the grain boundary gradually decrease to the bulk value (Figure 5a in the next chapter).

In case of GB-Ni$_{120}$ supercell (Figure 5a), the magnetic moment of Ni atoms situated at the grain boundary (0.656 μ_B) is by 3.7% higher than the calculated magnetic moment of Ni atom in the bulk Ni

(0.632 μ_B). And again, the atoms in the layer 2 exhibit a larger enhancement of magnetic moment that is higher by 7.1% in comparison with the bulk value. Atoms in the 3rd layer have magnetic moment higher by 2.8%. However, the Ni atoms in the 4th layer exhibit a value by 2.0% lower than the magnetic moment of Ni atoms in the bulk. Then, the values of the magnetic moments of atoms located in further layers decrease and oscillate around the bulk value of 0.632 μ_B. These values are also in a very good agreement with those published in Reference [2].

3.3. Grain Boundary with Impurities

3.3.1. Structure Arrangement

To find the most preferable positions of impurities in Ni material, the properties of the supercells with $\Sigma 5(210)$ grain boundary with two impurity atoms were investigated. For this purpose, three different configurations of grain boundary were proposed: GB-Ni$_{118}$Al$_2$(s) (with two substitutional Al atoms), GB-Ni$_{118}$Si$_2$(s) (with two substitutional Si atoms) and GB-Ni$_{120}$Si$_2$(i) (with two interstitial Si atoms in position at GB layer visualized in Figure 1b). The equilibrium structure data are provided in Table 3 and the corresponding structures are displayed in Figure 3.

Table 3. Properties of structures with clean and impurity-segregated GBs (equilibrium b/a parameter, volume per atom V_{at}, excess free volume V^f and GB energy γ_{GB}) obtained by different methods of relaxation. Method 1 is the automatic full relaxation. Method 2 keeps the two lattice parameters (in the plane of the GB) fixed and the lattice parameter perpendicular to the GB is changed by the simultaneous relaxation of atomic positions. The equilibrium structures were determined with respect to (a) minimum energy (Method 2a) or (b) the zero stress along the dimension perpendicular to the GB (Method 2b). Here, b is the supercell lattice constant perpendicular to the GB plane and a is the lattice constant in the plane of the GB (see Figure 1b).

Method of relax.	GB-Ni$_{60}$				GB-Ni$_{120}$			
	b/a	V_{at} (Å^3)	V^f (Å^3·Å^{-2})	γ_{GB} (J·m^{-2})	b/a	V_{at} (Å^3)	V^f (Å^3·Å^{-2})	γ_{GB} (J·m^{-2})
1	3.13	11.12	0.26	1.29 1.23 [2]	3.13	11.12	0.26	1.29
2a	3.09	11.20	0.34	1.31 1.43 * [50]	3.09	11.20	0.34	1.30
2b	3.09	11.20	0.34	1.31	3.09	11.21	0.35	1.30

Method of relax.	GB-Ni$_{118}$Al$_2$(s)				GB-Ni$_{118}$Si$_2$(s)				GB-Ni$_{120}$Si$_2$(i)			
	b/a	V_{at} (Å^3)	V^f (Å^3·Å^{-2})	γ_{GB} (J·m^{-2})	b/a	V_{at} (Å^3)	V^f (Å^3·Å^{-2})	γ_{GB} (J·m^{-2})	b/a	V_{at} (Å^3)	V^f (Å^3·Å^{-2})	γ_{GB} (J·m^{-2})
1	3.11	11.14	0.26	2.50	3.12	11.11	0.26	2.71	3.15	11.10	0.11	0.64
2a	3.08	11.22	0.29	2.53	3.08	11.19	0.28	2.75	3.13	11.13	0.13	0.65
2b	3.09	11.23	0.29	2.53	3.08	11.19	0.28	2.75	3.13	11.13	0.14	0.65

* The computational cell was GB-Ni$_{40}$.

Comparing the values of V_{at} for various GB configurations, it is obvious that they are comparable, which means that the introduction of impurity does not influence significantly the size of the structure. The lowest value was obtained for the structure with interstitial impurities. It may be seen that there is an increase of volume related to the GB introduction. This effect can be quantified by means of the excess free volume:

$$V^f = (V_{GB} - V_{bulk})/2S, \qquad (5)$$

which is a GB specific property [51,52] like the GB energy discussed below. The value of V^f for clean GB obtained from automatic relaxation is 0.26 Å^3·Å^{-2} (see Table 3). This value is identical with the values obtained by the same method for structures with substitutional impurities. This means that the structures need some additional volume to form the GB. The value of V^f for structure with interstitial impurity is 0.11 Å³·Å⁻² and it is much lower than the values corresponding to other structures. This means that the big change in volume of bulk structure caused by introduction of the interstitial impurity (see Table 1) is consequently also used for the formation of GB. In case of the GB-Ni$_{60}$ supercell, the values of V^f (Table 3) are fully comparable with the values of other structures. This is caused by the fact that both the V_{GB}-V_{bulk} and 2S for GB-Ni$_{120}$ modifications are two times larger than for GB-Ni$_{60}$. The comparability between the results of Method 1 and 2 is not ideal, as the two lattice parameters (in the plane of the GB) and hence also the value of 2S are kept fixed in Method 2.

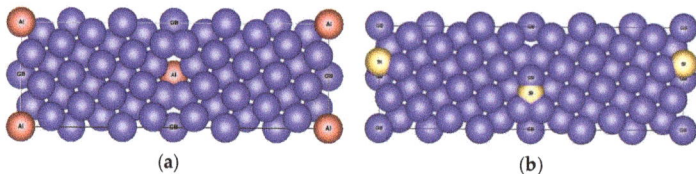

Figure 3. Equilibrium GB supercells containing impurities: (**a**) GB-Ni$_{118}$Al$_2$$^{(s)}$, (**b**) GB-Ni$_{120}Si_2$$^{(i)}$. The blue spheres correspond to Ni atoms, the spheres labelled by "GB" denote the GB plane, the red spheres occupy the position of substitutional impurity atoms, the orange spheres denoted as Si mark the positions of interstitial impurity atoms.

The first coordination sphere of the substitutional impurity is very similar to that of atoms in elemental bulk *fcc* Ni. Nevertheless, the interatomic distances are scattered a little bit and increased with respect to ideal value (2.475 Å). In case of Al$^{(s)}$, they reach values in the range from 2.585 to 2.640 Å. The first coordination sphere of substitutional Si$^{(s)}$ is even more scattered and slightly enlarged (2.504–2.645 Å). On the other hand, in the structure with interstitial Si, several Ni atoms in the neighborhood of Si atom changed their positions to make more space for the impurity. The coordination sphere is slightly compressed, and the scatter exhibits the largest range here (2.177-2.444 Å) here. The Ni atoms in the third layer of GB with the same z coordinate as Si atom are most affected. The equilibrium position of these Ni atoms differs from the initial position by 0.012% in y direction. This movement causes the increase of the distance between Ni and Si atom.

Analyzing the interlayer distances (see Figure 4 and Table S1 in Supplemental Material), it is obvious that they are comparable with the results presented in the papers [2,49], i.e. there is a big increase in the distance between the GB and layer 2 (GB(X,Ni)/2, GB(Ni,Ni)/2) (with a maximum for the GB-Ni$_{120}$Si$_2$$^{(i)}$ structure) and a significant decrease in the distance between the layers 2 and 3 (2/3) in all studied configurations. The distances between next layers are not so much affected by the presence of the impurity and they converge with small oscillations to the bulk value. As there are two types of planes (without and with impurity) in the supercell, there are also two types of lines (blue with squares and red with circles) depicted in Figure 4. In case of the substitutional impurities (Figure 4b), these curves mostly overlap. The red one (plane without impurities) is symmetric with respect to the center of the picture. The blue curve (plane with impurities) reveals some deviations from this symmetry, which reflects the fact that the GBs are not fully equivalent. However, as the deviations are very small, it can be concluded that the type of GB (clean or with impurity) has a negligible impact on the interlayer distances. These conclusions are valid also for the system GB-Ni$_{118}$Si$_2$$^{(s)}$ which is not displayed in Figure 4. In case of the interstitial Si (Figure 4c), the figure looks different. This is caused by the fact that the two impurity atoms are positioned in the supercell asymmetrically (see Figure 1b). The most obvious difference may be seen in the case of the blue line in the left part of the figure, which reveals a strong asymmetry. This reflects the fact that the interstitial impurities influence the structure much stronger than the substitutional ones.

Again, our findings are in a good agreement with the data calculated by Všianská and Šob [2] and we recalculated them to have a benchmark. Nevertheless, the present results are not fully comparable with those in [2] as the concentrations of the impurities in the layers are not the same. The ratio of the number of atoms Ni:X$^{(s)}$ (Ni:X$^{(i)}$) in one layer is 3:1 (4:1) in our work and 0:2 (2:2) in Reference [2], which means that our supercells exhibit a considerably lower concentrations of impurities which is closer to real systems. In the case of structure with a substitutional impurity in Reference [2], there are 56 atoms of Ni and 4 atoms of impurity. The impurity atoms are forming the monolayer at the GB layer including 2 impurity atoms at each GB. For the structure with an interstitial impurity in Reference [2], there are 60 atoms of Ni and 4 atoms of impurity. Each GB is occupied by 2 atoms of Ni and 2 atoms of impurity. We also have to take into account that the GB cells used by Všianská and Šob [2] are usually smaller than the cells presented in this work.

The differences may be characterized as follows. In comparison of the interlayer distances of structures with GB and impurity calculated by Všianská and Šob [2] and presented in this work, it can be said that the interlayer distances calculated for structures GB-Ni$_{120}$ and GB-Ni$_{118}$Al$_2^{(s)}$ differ in maximum by 2%. For the structure GB-Ni$_{118}$Si$_2^{(s)}$ the 2/3 interlayer distance is by 26% larger and 3/4 interlayer distance is by 12% smaller than the same interlayer distances in the paper [2]. Additionally, for the structure GB-Ni$_{120}$Si$_2^{(i)}$, there are significant differences in GB/2 interlayer distance and 2/3 interlayer distance. Both values are smaller than data published in [2]; the difference is 10% and 15%, respectively.

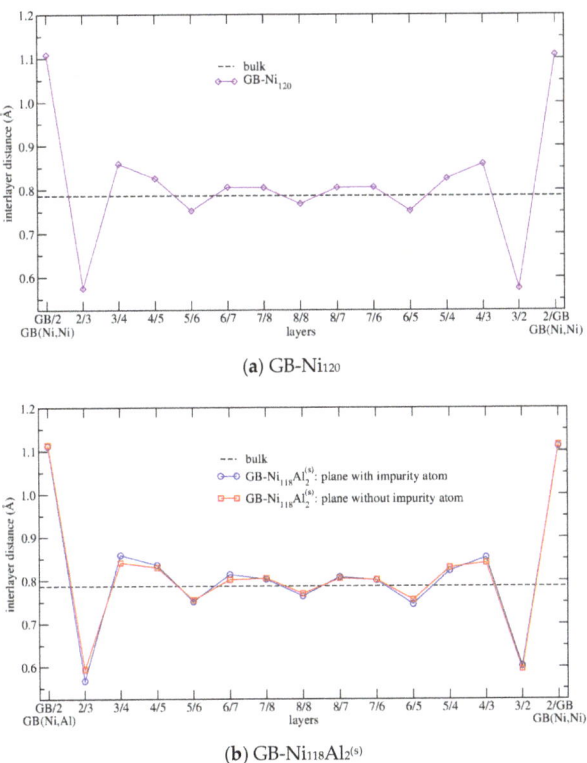

(a) GB-Ni$_{120}$

(b) GB-Ni$_{118}$Al$_2^{(s)}$

Figure 4. Cont.

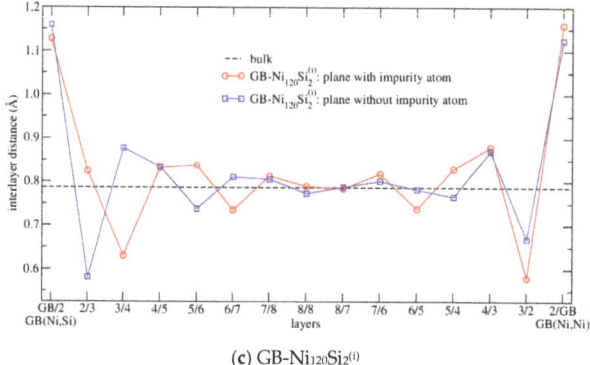

(c) GB-Ni$_{120}$Si$_2^{(i)}$

Figure 4. The interlayer distances in supercells with a GB: (a) GB-Ni$_{120}$, (b) GB-Ni$_{118}$Al$_2^{(s)}$, (c) GB-Ni$_{120}$Si$_2^{(i)}$. The violet line with diamonds corresponds to a clean GB in elemental *fcc* Ni. The blue (red) line with squares (circles) characterizes the planes without (with Al or Si) impurity, which are parallel to the plane of paper in structures shown in Figure 3. The exact meaning of expression plane is defined in the text. GB(X,Ni) denotes the GB with impurity X (X = Al$^{(s)}$, Si$^{(i)}$) and GB(Ni,Ni) the clean GB. The lines do not reflect any physical meaning; they are used to guide the eye only. The dashed line corresponds to the interlayer distance in ferromagnetic *fcc* Ni (this work). For a better demonstration of results in a case of (b) and (c), the plane with impurity atom is always presented so that there is an impurity on the left part of the figures. This was achieved by the shift $\frac{1}{2}$ of the lattice parameter *b* of lines representing the results of some planes with impurities (the middle plane of GB-Ni$_{118}$Al$_2^{(s)}$ or the second plane of GB-Ni$_{120}$Si$_2^{(i)}$) and one without impurity (the middle plane of GB-Ni$_{120}$Si$_2^{(i)}$) see Figure 1. This figure does not contain the system GB-Ni$_{118}$Si$_2^{(s)}$ as it provides results similar to data presented for system GB-Ni$_{118}$Al$_2^{(s)}$ in part (b).

3.3.2. Energetics

Based on the total energies, we can estimate the preference of Si to occupy the substitutional or interstitial positions at the studied GB using the following equation:

$$\Delta E = \frac{\left(E_{GB-Ni_{118}Si_2^{(s)}} + 2E_{1fccNi}\right) - E_{GB-Ni_{120}Si_2^{(i)}}}{2}, \quad (6)$$

where the negative value of the ΔE corresponds to the preference of the substitutional position. This is not the case of Si impurity at the GB, where $\Delta E = 0.4798$ eV·atom^{-1} of Si. Hence, in case of the Si the occupational preference is changed to interstitial positions. The fraction of Si atoms located in substitutional positions at the Σ5(210) GB will be therefore negligible.

The energy per one atom in the supercell with a grain boundary including the impurity atoms is higher by 0.3–1.4% than the energy of bulk structure with impurities. The energy difference (E_{GB}-E_{bulk}) is lowest in case of the GB-Ni$_{120}$Si$_2^{(i)}$ supercell and the largest in GB-Ni$_{118}$Si$_2^{(s)}$.

Furthermore, we can also analyze the GB energies γ_{GB} (Equation (4), Table 3). Here, the relaxation methods 1, 2a and 2b give almost the same results for structures with impurities, which differ within 2.2% for Ni$_{120}$Si$_2^{(i)}$. In a case of structures without impurity the GB energies given by the three types of relaxation differs less. It is 1.2% for GB-Ni$_{60}$ and even 1.0% for GB-Ni$_{120}$. Because the differences are not significant, this is a further argument supporting the statement that all methods of relaxation used here are equivalent. In case of the GB-Ni$_{60}$ and GB-Ni$_{120}$ supercell, the values of γ_{GB} (Table 3) are fully comparable because both the V_{GB}-V_{bulk} and $2S$ for GB-Ni$_{120}$ are two times larger than for GB-Ni$_{60}$. According to the presented values of GB energy, it can be said that the energy related to the formation of GB is increasing in this sequence: GB-Ni$_{120}$Si$_2^{(i)}$ < GB-Ni$_{60}$ = GB-Ni$_{120}$ < GB-Ni$_{118}$Al$_2^{(s)}$ <

GB-$Ni_{118}Si_2^{(s)}$, which means that the formation of the GB is easiest in the structure with interstitial Si, even easier than in elemental *fcc* nickel, and that substitutional impurities make the GB formation most difficult. The γ_{GB} for GB-$Ni_{120}Si_2^{(i)}$ configuration is very small, which supports the stability of the structure with interstitial Si with respect to the structure containing substitutional Si. Our calculated GB energy γ_{GB} in elemental Ni (1.29 J.m^{-2}, Table 3) agrees very well with the value of 1.23 J.m^{-2} published in Reference [2], which was obtained by the VASP code within the same GGA approach. We also obtained a reasonable agreement with a value of 1.43 J.m^{-2} reported in Reference [50], where the WIEN2k code with GGA approach in the PBE96 parametrization was employed, the $\sqrt{5}a$, $2\sqrt{5}a$, a cell was used but the spin polarization was omitted. The GB energy was also studied by atomistic simulations [53] providing the values of 1.23 J.m^{-2} and 1.28 J.m^{-2}. These values also compare well with our data. To the best of our knowledge, no experimental value for GB energy of the Σ5(210) GB in Ni is available, and hence a comparison of theoretical calculations with experiment is not possible yet.

The other energetic quantity, which can characterize the properties of the GB and/or impurities is the segregation energy (the binding energy between impurity and GB). The segregation energy of an impurity at a GB can be measured [3] and is defined as the lowering of the energy of the system when solute atoms go from the bulk material to the GB. It can be expressed by the Equation (7):

$$E^{seg}(X) = \frac{(E_{GB+X} - E_{GB})}{2} - (E_{bulk+X} - E_{bulk}) = E_{XatGB} - E_{Xinbulk}, \qquad (7)$$

where E_{GB+X} and E_{GB} are the energies of the GB structure with and without impurity atoms (both containing the same number of Ni atoms) and E_{bulk+X} and E_{bulk} are the total energies of a bulk material with and without impurity (both containing the same number of Ni atoms). The energy difference for bulk material has to be subtracted here to eliminate the energetics of adding impurity to the bulk material [49,54]. E_{XatGB} and $E_{Xinbulk}$ are the total energies of one atom of impurity at GB and in a bulk material. The results are summarized in Table 4.

Table 4. Segregation energies obtained by different methods of relaxation. Here eV.X^{-1} stands for eV per atom of impurity.

Structure	$E^{seg}(X)$ (eV.X^{-1})		
	GB-$Ni_{118}Al_2^{(s)}$	GB-$Ni_{118}Si_2^{(s)}$	GB-$Ni_{120}Si_2^{(i)}$
Method of relaxation			
1	−0.144 −0.22 * [2] 0.05 * [8]	0.222	−0.258 −0.76 * [2] −0.71 * [8]
2a	−0.132 −0.19 * [50]	0.252	−0.281 −0.83 * [50]
2b	−0.132	0.252	−0.275

* The values of segregation energy are not directly comparable with our results because of various sizes of unit cells and number of impurities. In Reference [2], the unit cells are GB-$Ni_{56}Al_4^{(s)}$ and GB-$Ni_{60}Si_4^{(i)}$; Reference [8] deals with the unit cells GB-$Ni_{79}Al_1^{(s)}$ and GB-$Ni_{80}Si_1^{(i)}$ and in Reference [50], the unit cell corresponds to GB-$Ni_{36}Al_4^{(s)}$ and GB-$Ni_{40}Si_4^{(i)}$.

The segregation energies of substitutional Al$^{(s)}$ atom and interstitial Si$^{(i)}$ atom are negative for all methods of relaxation, which means that both impurities prefer a location in the neighborhood of GB to that in the bulk material. These values also support the stability of these configurations. The segregation energy of substitutional Si$^{(s)}$ atom is positive, so that substitutional silicon prefers to stay in the bulk. However, this arrangement is less stable than the structure with interstitial Si$^{(i)}$ as stated above. Equation similar to Equation (7) was also used by Yamaguchi et al. in Reference [55].

Unfortunately, there are no experimental data available in literature regarding the segregation enthalpies of the Σ5(210) GB for Si and Al. The only data found are those for Al at specific GB types where the values −2, −4 and −5 kJ.mol^{-1} were obtained for the surface (001), (123) and (015), respectively (see Reference [3] and references therein). Thus, our calculated segregation energies are

6.9, 3.5 and 2.8 times higher than those obtained from experiments. It should be mentioned that these values correspond to different types of GBs and therefore our comparison of calculations and experiment is somewhat limited. It should also be considered that other configurations of the Σ5(210) GB with Si and Al atoms may occur in reality and in the present work we examined only one of them.

The obtained results for the segregation of Al$^{(s)}$ and Si$^{(i)}$ at the Σ5(210) GB can be also compared with other computational studies [2,50]. The calculated values from Reference [2] and [50] correspond to ours very well as they differ only by hundredths and tenths of eV in case of Al$^{(s)}$ and Si$^{(i)}$, respectively.

The calculations in Reference [2] were performed using the same exchange-correlation energy but a smaller cell with a higher concentration of impurities at the GB (impurity monolayer). In reference [8] norm-conserving pseudopotentials and local density approximation for the exchange-correlation potentials were employed. The method used in Reference [50] is based on the full-potential linearized augmented plane wave method with the generalized gradient approximation of PBE96 parametrization [29] and, again, a smaller cell with even higher concentration of impurities than in [2] (impurity monolayer at GB) is applied. Moreover, in [50], the calculations were performed as nonmagnetic. Considering the above-mentioned differences in the approach in Reference [50] and configurations [2], the agreement is reasonable and confirms the behavior of segregating elements at the GB, i.e. Al as substitutional and Si as interstitial.

3.3.3. Magnetism

As it was shown in previous chapters, the presence of impurities and grain boundaries has a very strong impact on the value of magnetic moment of neighboring Ni atoms. The calculated magnetic moments of Ni atoms are presented in Figures 5 and 6 as well as in Table 5 (in all cases the configurations with the impurity localized at GB are considered).

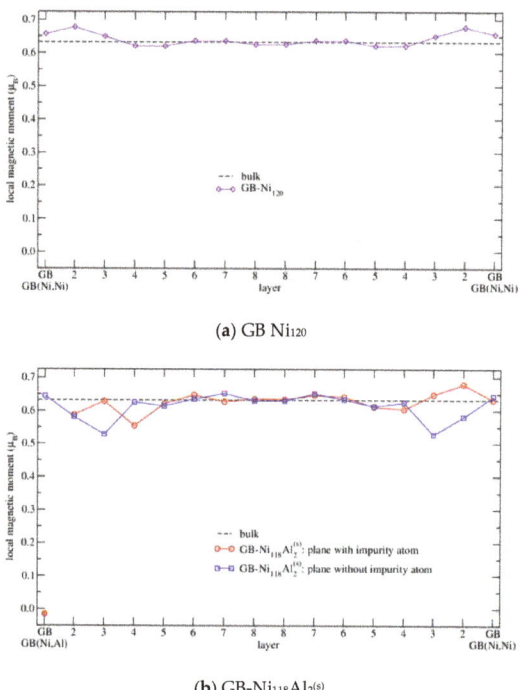

(a) GB Ni$_{120}$

(b) GB-Ni$_{118}$Al$_2^{(s)}$

Figure 5. Cont.

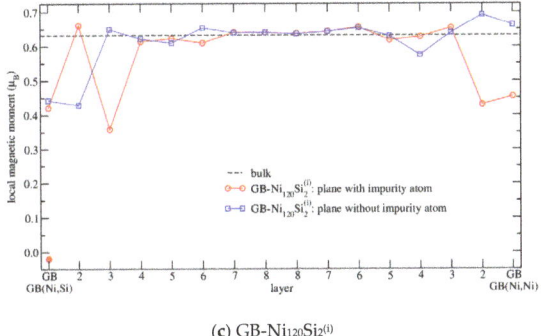

(c) GB-Ni$_{120}$Si$_2^{(i)}$

Figure 5. The dependence of magnetic moment of Ni atom on the number of the atomic layer: (**a**) GB-Ni$_{120}$, (**b**) GB-Ni$_{118}$Al$_2$ $^{(s)}$, (**c**) GB-Ni$_{120}$Si$_2$ $^{(i)}$. The violet line with diamonds corresponds to the Σ5(210) GB in elemental fcc Ni. The blue (red) line with squares (circles) corresponds to the planes without (with Al or Si) impurity, which are parallel to the plane of paper in structures shown in Figure 3. The exact meaning of expression plane is defined in the text above. GB(X,Ni) denotes the GB with impurity X (X = Al$^{(s)}$, Si$^{(i)}$) and GB(Ni,Ni) the clean GB. The lines do not reflect any physical meaning; they are used to guide the eye only. The dashed line shows the magnetic moment of fcc bulk Ni of 0.632 μ_B calculated in this work. The full red circle corresponds to the magnetic moment of Al (Si) impurity. This figure does not contain the system GB-Ni$_{118}$Si$_2$ $^{(s)}$ as it provides results similar to data presented for system GB-Ni$_{118}$Al$_2$ $^{(s)}$ in part (**b**). For a better demonstration of results in case of (**b**) and (**c**), the plane with impurity atom is always presented so that there is an impurity on the left part of the figures. This was achieved by the shift of lines representing the results of some planes with impurities (the middle plane of GB-Ni$_{118}$Al$_2$ $^{(s)}$ or the second plane of GB-Ni$_{120}$Si$_2$ $^{(i)}$) and one without impurity (the middle plane of GB-Ni$_{120}$Si$_2$ $^{(i)}$) by $\frac{1}{2}$ of the lattice parameter b.

In case of a clean grain boundary (Figure 5a), we mentioned the oscillations in magnetic moments of Ni atoms in the previous chapter; they are enhanced in the GB region. This enhancement is in agreement with literature [2,33,34,49,50,54], the maximum magnetic moment was found in layer 2.

Further, let us discuss the results concerning the magnetism of GBs with impurities. Here, the "magnetic-moment-decreasing" effect of impurity is suppressing the "magnetic-moment-increasing" effect of grain boundary. In case of GB-Ni$_{118}$Si$_2$ $^{(s)}$ and GB-Ni$_{118}$Al$_2$ $^{(s)}$ (Figure 5b) supercells, there is a very similar trend in the distribution of magnetic moments. However, the "magnetic-moment-decreasing" effect is a little bit stronger for substitutional silicon than for substitutional aluminum. This effect is most obvious for atoms in the 2nd (the strongest effect), 3rd and 4th layer. The magnetic moments in the 5th layer are still a little bit below the bulk value. Then, the moments of atoms from further (6th–8th) layers oscillate around the bulk value. In case of the structure GB-Ni$_{120}$Si$_2$ $^{(i)}$ (Figure 5c), the largest decrease in magnetic moment can be observed near to the impurity. The reason for this is the small free volume caused by interstitial impurity. Here, the magnetic moment distribution is different from the case of structures discussed previously. In particular, the atoms at the boundary are strongly influenced by the interstitial impurity.

The magnetic moments in structures containing GB and impurity exhibit a scattered character in the same layer. A more detailed study reveals that they increase rather monotonically with the distance from the impurity atom at the GB (Figure 6), which gives rise to scattering as Ni atoms in the same layer have different distances from the impurity. For more distant planes, the magnetic moment of Ni atoms converges to the bulk value. The only case where the magnetic moments are increased is that of the planes without impurity atoms in the part of the GB(Ni,Ni). In this case, the Ni atoms do not have any impurity in their neighborhood and therefore the "magnetic-moment-increasing" effect of grain boundary prevails. In Figure 6, the scatter of values corresponding to atoms with the same distance from X is caused by their different distance from GB.

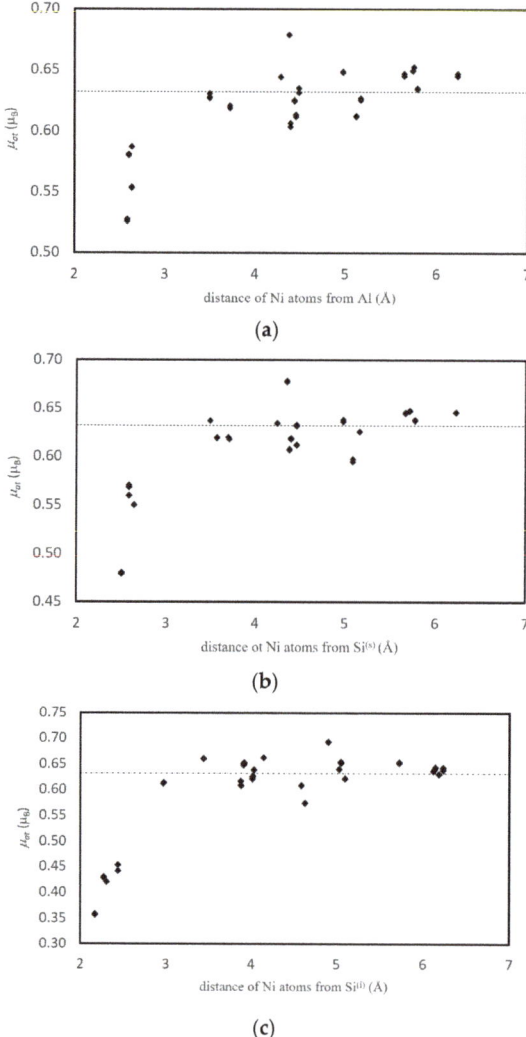

Figure 6. The dependence of magnetic moment of Ni atom on the distance from the impurity: (a) GB-Ni$_{118}$Al$_2^{(s)}$, (b) GB-Ni$_{118}$Si$_2^{(s)}$, (c) GB-Ni$_{120}$Si$_2^{(i)}$. The dashed line corresponds to the magnetic moment of elemental *fcc* bulk Ni of 0.632 μ_B calculated in this work.

Table 5 presents more details on the nearest Ni atoms of studied impurities. The magnetic moments of Ni atoms lying in the nearest neighborhood of substitutional Al atom (about 0.528–0.588 μ_B) are lower in comparison to the μ of atoms lying in the same layers but further from impurity. The values for Ni atoms more distant from Al are shown in Figure 6 and their magnetic moments are higher than 0.6 μ_B.

Table 5. Magnetic moment of Ni atoms μ_{at} in the neighborhood of impurity X. R stands for the distance from impurity.

Structure	Plane	R (Å)	μ_{at} (μ_B)	Layer
GB-Ni$_{118}$Al$_2$ $^{(s)}$ GB (Ni,X)	clean	2.586	0.528	3
	clean	2.599	0.581	2
	with X	2.634	0.588	2
	with X	2.639	0.554	4
GB-Ni$_{118}$Si$_2$ $^{(s)}$ GB (Ni,X)	clean	2.504	0.483	3
	clean	2.588	0.572	2
	with X	2.592	0.563	2
	with X	2.645	0.552	4
GB-Ni$_{120}$Si$_2$ $^{(i)}$ GB (Ni,X)	with X	2.177	0.357	3
	clean	2.278	0.429	2
	with X	2.312	0.420	GB (Ni,X)
	clean	2.444	0.453	GB (Ni,X)

The trends of magnetic moments of Ni atoms in both structures with substitutional impurities are similar. The lowest magnetic moment is found at the nearest Ni atoms. The further two Ni atoms reveal the increase of magnetic moment as they lay in the 2nd layer and are in the vicinity of GB. The last (4th) Ni atoms possess the lowered magnetic moments again as they are accommodated in the 4th layer.

In case of the structure with interstitial Si, the situation is different. The magnetic moment of Ni increases (with small deviation for the third Ni atom) with increasing distance from Si atom (column 3 of Table 5) and it is simultaneously approaching to the GB (column 5 of Table 5).

From the previous studies, it is apparent that the segregation of nonmagnetic impurities has a detrimental effect on GB magnetism [2,56]. Another example of a reduction of magnetic moments due to impurities and other structural imperfections may be found in the paper [57].

Therefore, it may be supposed that the effect of impurities on magnetic properties will be more significant in materials with high GB concentrations, e.g. in nanocrystalline materials. It was demonstrated that nanocrystalline Ni does not exhibit any substantial change in the saturation magnetization compared to a polycrystalline sample [58] but, probably, because of exposition to air, the saturation magnetization is lowered due to a GB contamination by oxygen. The effect of segregation of nonmagnetic elements on the magnetic properties of GBs was found experimentally for tungsten segregating at GBs of nanocrystalline Ni in Reference [59]. Some other works dealing with magnetism on GBs in nanomaterials were also published, e.g. in Reference [60] dealing with Fe. To the best of our knowledge, no such study was made for nanocrystalline Ni with Al or Si.

Comparing the magnetic moment of bulk Ni atom calculated by Všianská and Šob [2] and published in this work, the magnetic moment of Ni atom presented in this work is by 2% larger. For the structure GB-Ni$_{120}$, the differences in magnetic moments are not larger than 2%. There are larger differences in magnetic moments of atoms in structures GB-Ni$_{118}$Al$_2$$^{(s)}$, GB-Ni$_{118}Si_2$$^{(s)}$ and GB-Ni$_{120}$Si$_2$$^{(i)}$ calculated in [2] and presented in this work.

The largest differences can be observed for magnetic moments of Ni atoms in the layers GB, 2, 3, 4 and for magnetic moments of impurities. In these layers, the magnetic moments of Ni atoms calculated in this work are always larger than the magnetic moments calculated in Reference [2]. For structures containing substitutional Al and Si the differences are 34% and 54% for the 3rd layer. For the structure GB-Ni$_{120}$Si$_2$$^{(i)}$, the difference in magnetic moment of Ni atoms attains its maximum for Ni atoms at the GB layer, the value is almost 93%.

The above-mentioned differences are caused by the differences in calculations discussed in the section devoted to the interlayer distances and, of course, by a different concentration of impurities at GBs.

According to [61], the solubility of Al in bulk Ni is about 7 at.% at 400 °C, our concentration in bulk and in structure with GB is equal to 1.67 at.%. In [62], the solubility of Si in bulk Ni is found to be

10 at.% at 700 °C. Our concentration is 1.64 at.% in GB-Ni$_{120}$Si$_2$$^{(i)}$, 0.83 at.% in Ni$_{120}Si^{(i)}$ and 1.64 at.% in Ni$_{60}$Si$^{(i)}$. Despite the temperature effect, our concentrations are lower than the experimental ones and hence, according to Lejček et al. [63], we can consider our values of segregation energies reliable.

3.4. Grain Boundary with Impurities and Vacancies

In this chapter, the effect of vacancy in the structures GB-Ni$_{120}$, GB-Ni$_{118}$Al$_2$$^{(s)}$, GB-Ni$_{118}Si_2$$^{(s)}$ and GB-Ni$_{120}$Si$_2$$^{(i)}$ is analyzed. The interactions between the grain boundary, the impurity and the vacancy are studied as well as the effect of the distance between vacancy and impurity atom at the GB. These phenomena are investigated from both the energetic and structural point of view.

3.4.1. Structure Arrangement

In case of impurities, the presence of vacancies is crucial for their diffusion as they enable an easier exchange of atomic positions. In Table 2, the preference of Al and Si to bind to vacancies in bulk Ni material was presented, which further emphasizes the importance of vacancy diffusion mechanism of these two impurities in Ni. In principle, there is a large variety of possible impurity-vacancy configurations at the studied GB (or in its nearest vicinity) which can be investigated.

Here, the vacancies on GBs without impurities are studied at first. In the bulk material, the presence of a vacancy can be characterized by a free volume. This volume can disappear only when the vacancy meets the interstitial atom or another lattice defect, e.g. GB. In this case the vacancies behave unusually [64–66], i.e. they can be delocalized or become instable [65]. The volume of delocalized vacancy disappears by rearrangement of neighboring atoms. On the other hand, the instability is related to the movement of the vacancy to another site where the delocalization can occur again.

In this part of study, we have investigated all four above mentioned configurations with vacancies accommodated in various layers. The results concerning the vacancy formation are summarized in Table 6.

Table 6. Energy and volume effects associated with vacancy formation in various layers of GB-Ni$_{119}$+Va, GB-Ni$_{117}$Al$_2$$^{(s)}$+Va, GB-Ni$_{117}Si_2$$^{(s)}$+Va and GB-Ni$_{119}Si_2$$^{(i)}$+Va structure. The quantity $E^f(Va)$ is defined as the formation energy of the vacancy. V^f is the change of volume of cell associated with the formation of vacancy and it is defined as the difference of the volume of the structure with and without a vacancy. The values written in bold correspond to the most stable configurations.

Structure	GB-Ni$_{119}$+Va		GB-Ni$_{117}$Al$_2$$^{(s)}$+Va		GB-Ni$_{117}$Si$_2$$^{(s)}$+Va		GB-Ni$_{119}$Si$_2$$^{(i)}$+Va	
Layer with Va	$E^f(Va)$ (eV.Va^{-1})	V^f (Å^3.Va^{-1})	$E^f(Va)$ (eV.Va^{-1})	V^f (Å^3.Va^{-1})	$E^f(Va)$ (eV.Va^{-1})	V^f (Å^3.Va^{-1})	$E^f(Va)$ (eV.Va^{-1})	V^f (Å^3.Va^{-1})
GB	1.5727	−1.2460	1.5808	6.3510	1.4961	2.6906	1.4304	−5.6628
2	**0.5791**	**−9.7127**	**0.6169**	**−3.3396**	**0.2466**	**−5.7481**	0.9106	−10.8557
3	1.2866	−4.0505	1.2833	3.2737	1.2105	−1.3837	**0.5746**	**−14.3407**
4	1.3088	−4.9483	1.3428	1.5724	1.1510	−5.5262	1.2894	−8.8176
5	1.2874	−4.7500	1.3190	1.8658	1.2819	−0.4996	1.3253	−9.1865

For GB-Ni$_{119}$+Va, GB-Ni$_{117}$Al$_2$$^{(s)}$+Va and GB-Ni$_{117}Si_2$$^{(s)}$+Va (GB-Ni$_{119}Si_2$$^{(i)}$+Va) structures, the configuration with vacancy in layer 2 (layer 3) reveals the highest decrease of the volume per unit cell and the lowest but still positive energy of vacancy formation. It can be said that the largest decrease in volume per unit cell always corresponds to the most easily but still unfavorably formed vacancy and to the most stable arrangement with the lowest energy of formation with respect to the standard element reference states (Table 7). However, as all structures possess the positive values of the energy of formation they must be considered as unstable at 0 K. The structures with the lowest energies are shown in Figure 7.

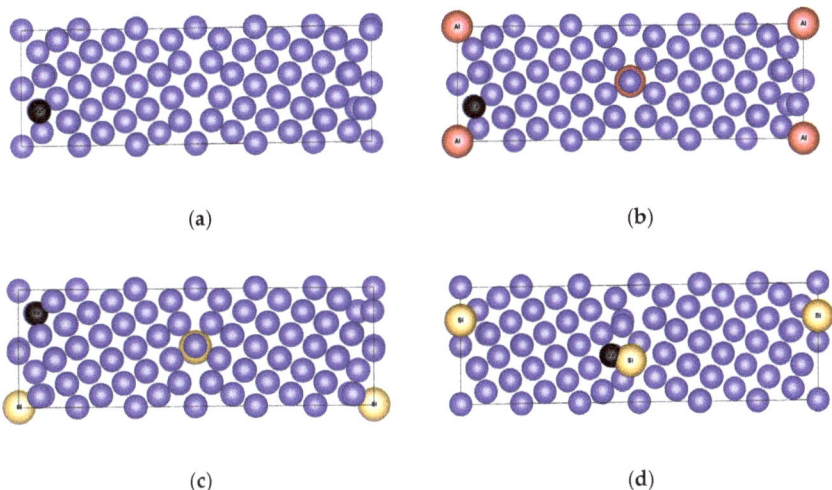

Figure 7. The structures with the lowest energies: (a) GB-Ni$_{119}$+Va^{L2}, (b) GB-Ni$_{117}$Al$_2$$^{(s)}$+Va^{L2}, (c) GB-Ni$_{117}Si_2$$^{(s)}$+Va^{L2}, (d) GB-Ni$_{119}Si_2$$^{(i)}$+Va^{L3}. Here, the denoted positions of impurities correspond to the general positions shown in Figure 1b. The dark particles denote the positions of vacancies.

In case of the structure GB-Ni$_{119}$+Va^{L2}, there occur changes in positions of atoms in the closest neighborhood of the vacancy, as they tend to occupy the free space provided by this lattice defect. The vacancy even attracts the Ni atom on the opposite side of the grain boundary, which corresponds to the vacancy mirror image.

The vacancy in the GB-Ni$_{117}$Al$_2$$^{(s)}$+Va^{L2} and GB-Ni$_{117}Si_2$$^{(s)}$+Va^{L2} structure also causes the attraction of Ni atoms lying in the nearest GB layer towards the vacancy site. The effect is largest for the Ni atoms with the same z coordinate as the vacancy.

The vacancy in the structure GB-Ni$_{117}$Al$_2$$^{(s)}$+Va^{L2} attracts the Ni atom situated on the opposite side of the GB and in the same plane as vacancy (z = 0.0). In the lowest-energy structure, this atom is situated almost at the GB. This effect is not as strong as in the case of GB-Ni$_{117}$Si$_2$$^{(s)}$+Va^{L2} structure but it is still very significant. As a consequence of this structural change, another two Ni atoms are repulsed by the approaching Ni atom. Their z coordinates are changing from the value of 0.75 to 0.77 and from 0.25 to 0.23 to make the space for the newcomer. Here, the Al atom closer to the vacancy is also slightly moving towards the vacancy.

In the structure GB-Ni$_{117}$Si$_2$$^{(s)}$+Va^{L2} with the vacancy in the second layer, the largest structure change occurs in the plane with z = 0.75. This is the plane where the vacancy is placed, i.e., in the plane next to the plane with impurity. Here, the most affected Ni atom is situated in the second layer on the opposite side of the grain boundary with respect to the vacancy. This atom moves towards the grain boundary. In the lowest-energy structure, this atom is located almost at the grain boundary. This position rearrangement causes the repulsion of the Ni atom on the grain boundary (z = 0.75). The final position of coming Ni atom causes other changes in the structure arrangement of Ni atoms in the plane z = 0.5. The Si atom closer to the vacancy is also affected. This atom is changing its z coordinate from z = 0 to z = 0.19.

The vacancy in the structure GB-Ni$_{119}$Si$_2$$^{(i)}$+Va^{L3} causes the largest changes in the arrangement of atoms. The free volume of vacancy is partially occupied by the interstitial Si atom that is situated in the same plane as the vacancy (z = 0.75). This means that the interstitial Si in the vicinity of the vacancy changes its position and moves away from the grain boundary towards the bulk region. Its equilibrium position is between the GB and the 2nd layer. The second most affected atom is the Ni atom located in the second layer and in the same plane as vacancy and silicon impurity. This atom is also moving towards the position where the vacancy is.

Despite of the changes in the arrangement of the atoms in the lowest-energy structures, it can be concluded that the existence of the vacancy is preserved and no delocalization or instability has been observed.

The interlayer distances in all mentioned structures were also analyzed and the results are presented in Figure 8.

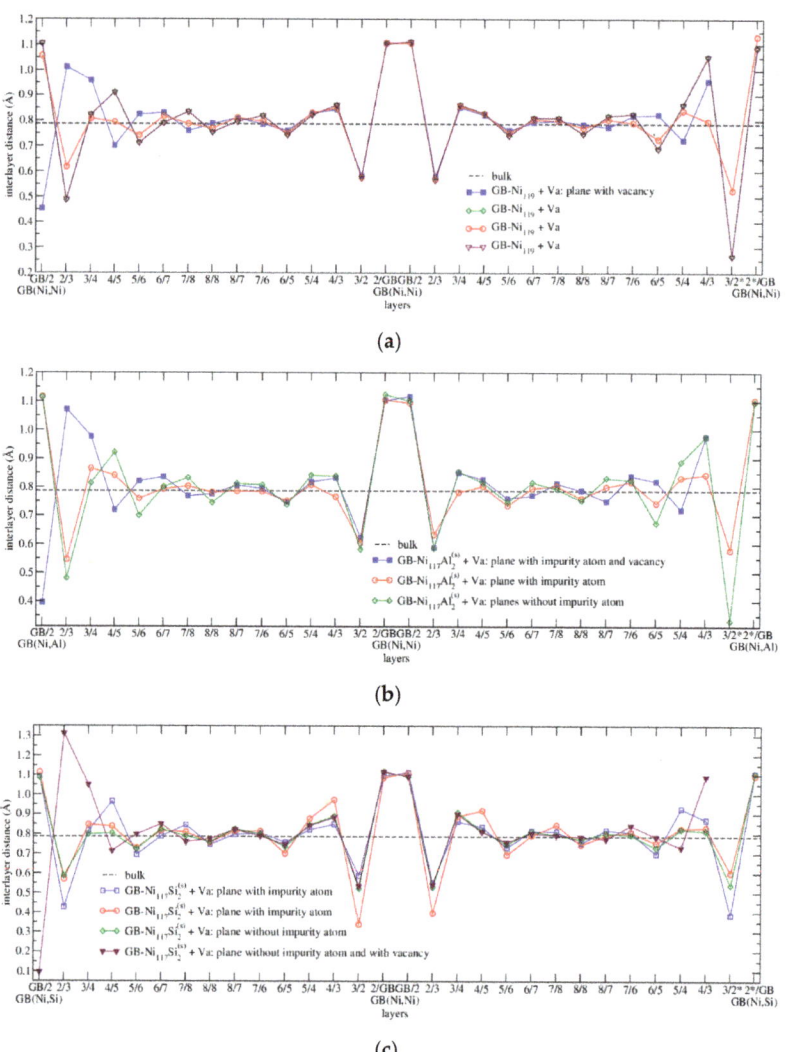

Figure 8. Cont.

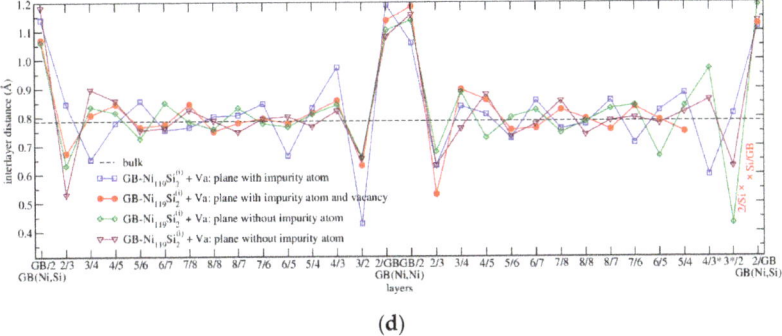

Figure 8. Interlayer distances in supercells with the GB, impurity and vacancy: (**a**) GB-Ni$_{119}$+Va^{L2}, (**b**) GB-Ni$_{117}$Al$_2$$^{(s)}$+Va^{L2}, (**c**) GB-Ni$_{117}Si_2$$^{(s)}$+Va^{L2}, (**d**) GB-Ni$_{119}Si_2$$^{(i)}$+Va^{L3}. The superscripts $L2$ and $L3$ denote the position of the vacancy in the second and third layer, respectively. × indicates the distances of interstitial Si in the layer with vacancy, which changed its position due to the structure relaxation. * denotes the layer with the vacancy. The dashed line corresponds to the interlayer distance in ferromagnetic fcc Ni (this work). The lines do not reflect any physical meaning; they are used to guide the eye only. For a better demonstration of results in the cases of Figures (b–d), the plane with impurity atom is always presented so that the impurity is on the edges of the figures. This was achieved by the shift of $\frac{1}{2}$ of the lattice parameter b of lines representing the results of some planes with impurities (the middle plane of GB-Ni$_{118}$Al$_2$$^{(s)}$ and GB-Ni$_{118}$Si$_2$$^{(s)}$ or the second plane of GB-Ni$_{120}$Si$_2$$^{(i)}$) and one without impurity (the middle plane of GB-Ni$_{120}$Si$_2$$^{(i)}$).

The situation in GB-Ni$_{119}$+Va^{L2} structure is the least complicated as this structure does not contain any impurity. The planes which do not contain the vacancy (green line with diamonds, red with circles and violet with triangles in Figure 8a, where the green and violet line are overlapping) reveal the typical shape of the curve describing the dependence of the interlayer distances on the type of layer. There is a large gap between the GB and second Ni layer (GB/2) followed by reduced spacing between layers 2 and 3 (2/3 and 3*/2). Some oscillations may be found in the interlayer distances between further layers, which converge to the interlayer distance typical for the bulk material. The reason why these curves are not ideally symmetric with respect to the center of the figure is that the interlayer distances are influenced by vacancy even if it is not placed directly in these planes. The plane in GB-Ni$_{119}$+Va^{L2} structure, which contains the vacancy (blue line with squares in Figure 8a), reveals untypical behavior. The Ni atom placed in the 2nd layer on the opposite side of the GB than vacancy (the mirror image of the vacancy) moves towards the GB in comparison with the plane without the vacancy and hence the value of interlayer distance GB/2 is very low. This shift is compensated by the increase of the interlayer distances 2/3 and 3/4 in the left part of Figure 8a. The missing points on the right-hand side of the curve correspond to the vacancy. The presence of the vacancy also causes the scatter of interlayer distances on the other curves at the edges of the figure and, as there are no impurities, the curves overlap in the middle.

The planes in GB-Ni$_{117}$Al$_2$$^{(s)}$+Va^{L2} structure which do not contain the vacancy (green with diamonds, describing behavior in two planes z = 0.25 and 0.75 and red line with circles) in Figure 8b reveal the same trends as described for the structure GB-Ni$_{119}$+Va^{L2}. The plane in the GB-Ni$_{117}$Al$_2$$^{(s)}$+Va^{L2} structure which contains the vacancy and impurity (blue line with squares in Figure 8b) reveals untypical behavior. The Ni atom placed in the 2nd layer on the opposite side of the GB than the vacancy (the mirror image of the vacancy) moves towards the GB by 0.72 Å in comparison with the plane without vacancy. This shift is compensated by the increase of the interlayer distances 2/3 and 3/4 in the most left part of the Figure 8b. The missing points on the right-hand side of the curve correspond to the vacancy.

The case of the GB-$Ni_{117}Si_2^{(s)}$+Va^{L2} structure is similar to GB-$Ni_{117}Al_2^{(s)}$+Va^{L2}, i.e. the planes without a vacancy (blue line with squares, green with diamonds and red with circles in Figure 8c) reveal the typical shape of the curves. The plane which contains the vacancy (violet line with triangles in Figure 8c) reveals the increase of the interlayer distances 2/3 and 3/4 in the most left part of the Figure 8c which is caused by the attraction of the Ni atom in the 2nd layer on the opposite side of GB (the mirror image of the vacancy) towards the vacancy. From this reason, the most left violet point possesses the lowest value of interlayer distance. The influence of the vacancy on the structure is much higher in this structure than in the structure containing aluminum which is demonstrated by a higher scatter of the points on the left-hand part of curves describing the planes with the vacancy.

In case of the GB-$Ni_{117}Si_2^{(i)}$+Va^{L3} structure, the situation is even more complicated, in particular in the case of the plane with an impurity and a vacancy (red line with circles in Figure 8d). On the right-hand side of this line, there are two points missing which correspond to the vacancy and there are two points × added which denote the distances of the Si atom from the GB and the 2nd layer. From their position, it is obvious that the Si atom moved away from the GB towards the 2nd layer. The green curve with diamonds and violet one with triangles in Figure 8d reveal typical trends with a maximum of the distance between the GB and layer 2 (GB/2) and a minimum of the distance between the layer 2 and 3 (2/3).

3.4.2. Energetics

The energy of formation of the structures with impurity, vacancy and GB with respect to the standard element reference states is evaluated using Equation (8):

$$E^f(GB - Ni_m Al_n Si_o + Va) = E_{GB-Ni_m Al_n Si_o+Va} - \left(mE_{1fccNi} + nE_{1fccAl} + oE_{1diamSi}\right). \quad (8)$$

Here, $E_{GB-Ni_m Al_n Si_o+Va}$ is the total energy of the structure with m Ni atoms, n Al atoms or o Si atoms, GB and one vacancy. The terms E_{1fccNi}, E_{1fccAl} and $E_{1diamSi}$ correspond to the total energy of one atom in its standard element reference state. The obtained results are summarized in Table 7.

Table 7. Energy of formation of structures GB-Ni_{119}+Va, GB-$Ni_{117}Al_2^{(s)}$+Va, GB-$Ni_{117}Si_2^{(s)}$+Va and GB-$Ni_{119}Si_2^{(i)}$+Va with three defects (GB, impurity, vacancy) for a vacancy in various layers. The quantity $E^f(GB - Ni_m Al_n Si_o + Va)$ is defined with respect to the standard element reference states. The values written in bold correspond to the lowest energy configuration.

Structure	GB-Ni_{119}+Va	GB-$Ni_{117}Al_2^{(s)}$+Va	GB-$Ni_{117}Si_2^{(s)}$+Va	GB-$Ni_{119}Si_2^{(i)}$+Va
Layer with Va	\multicolumn{4}{c}{$E^f(GB - Ni_m Al_n Si_o + Va)$ (eV.cell^{-1})}			
GB	10.33	6.94	7.52	6.50
2	**9.34**	**5.97**	**6.27**	5.98
3	10.05	6.64	7.24	**5.64**
4	10.07	6.70	7.18	6.36
5	10.05	6.68	7.31	6.39

According to the data in Table 7, we can see that the energy of formation of the structure with a vacancy and a GB with respect to the standard element reference states is the lowest for structures where the vacancy is situated in the second layer for GB-Ni_{119}+Va, GB-$Ni_{117}Al_2^{(s)}$+Va, and GB-$Ni_{117}Si_2^{(s)}$+Va and in the third layer for GB-$Ni_{119}Si_2^{(i)}$+Va. The position of the vacancy has the effect on the energy of formation up to 20%. Because of the positive values of $E^f(GB - Ni_m Al_n Si_o + Va)$, all structures must be considered as unstable at 0 K. Nevertheless, we can expect their existence at higher temperatures when also the simultaneous existence of more configurations of one structure having vacancies in various positions is probable as the energy differences between them are very small. In such a case, we expect the $Si^{(i)}$ to be more stable than $Si^{(s)}$. The lowest-energy configurations for structure GB-$Ni_{117}Al_2^{(s)}$+Va, GB-$Ni_{117}Si_2^{(s)}$+Va and GB-$Ni_{119}Si_2^{(i)}$+Va were used for the calculation of following quantities.

The key quantity for the vacancy formation energy is the difference of total energy for the defective (E_{GB+Va}) and perfect (E_{GB}) systems both in relaxed state [67]. Structural relaxation can reduce the vacancy formation energy by more than 30%. It is therefore important to include structural relaxation when calculating vacancy formation energies from first principles. Vacancies, created at the surface or at defects, exist at a certain concentration in all materials, depending exponentially on the formation energy of the vacancy and the temperature [68].

The stability of vacancy may be characterized by the vacancy formation energy which can be calculated according to the following Equation (9):

$$E^f(Va) = (E_{GB+X+Va} + E_{1fccNi}) - E_{GB+X}. \tag{9}$$

Here, $E_{GB+X+Va}$ and E_{GB+X} is the total energy of fcc Ni with GB, impurity X and Va and of fcc Ni with GB and impurity X. The value of $E^f(Va)$ provides the information whether there is a tendency towards the vacancy formation. The data obtained using Equation (9) are listed in Table 6.

In all configurations of studied structures, the energy of vacancy formation is positive, which means that the vacancies destabilize the structure. In a case of GB-Ni$_{119}$+Va structure, the lowest-energy configuration is that one with the vacancy situated in the 2nd layer (see Table 7). In this structure, the vacancy reveals the lowest positive energy of formation 0.5791 eV.Va^{-1}, i.e. is less unstable than in the other structures with the same composition. For this structure, we can also observe the largest decrease in the volume −9.7127 Å^3.Va^{-1}. The vacancy in the 2nd layer exhibits also the lowest energy for the GB-Ni$_{117}$Al$_2^{(s)}$+Va and GB-Ni$_{117}$Si$_2^{(s)}$+Va structures where the energy of vacancy formation is 0.6169 eV.Va^{-1} and 0.2466 eV.Va^{-1}, respectively. The formation volume is −3.3396 Å^3.Va^{-1} and −5.7481 Å^3.Va^{-1}. In the case of the last structure GB-Ni$_{119}$Si$_2^{(i)}$+Va, the lowest-energy configuration is that one with a vacancy situated in the 3rd layer. Here, the vacancy formation energy is 0.5746 eV.Va^{-1} and volume decrease is −14.3407 Å^3.Va^{-1}.

From the comparison of energies of vacancy formation in GB-Ni$_{119}$+Va, GB-Ni$_{117}$Al$_2^{(s)}$+Va, GB-Ni$_{117}$Si$_2^{(s)}$+Va and GB-Ni$_{119}$Si$_2^{(i)}$+Va configurations, it is obvious that the vacancy formation energy is the lowest for the structure GB-Ni$_{117}$Si$_2^{(s)}$+Va^{L2} (0.2466 eV.Va^{-1}). However, we have to take into account that the vacancy formation is energetically unfavorable process in all mentioned cases because the $E^f(Va)$ values are always positive. It can be said that the largest decrease in volume per atom always corresponds to the lowest-energy configuration. From all four configurations, the largest decrease was observed for the structure GB-Ni$_{119}$Si$_2^{(i)}$+Va^{L3} (14.3407 Å^3.Va^{-1}).

We have not found any literature data about the energetics of the Va formation at GBs in fcc Ni. The energies of vacancy formation at the Σ3(111) GB in vanadium were calculated in Reference [69]. Here, the formation energy of a vacancy at the GB is 2.51 eV, in the second layer 0.18 eV, in the third layer 1.65 eV and 2.54 eV in the fourth layer. It is in a good agreement with experimental data of 1.8–2.6 eV [70]. This trend in formation energy of vacancy with respect to its position is similar to the trend in vacancy formation energy for Σ5(210) GB in fcc Ni found in this work.

We can also evaluate further energy-related quantities, which are described in the following text and defined using Equations (10)–(15). The first such quantity is the binding energy of a vacancy to couple of lattice defects GB+X (the tendency of vacancy to move from the bulk to the GB with impurity) which is evaluated as:

$$E^b(Va; GB + X) = (E_{GB+X+Va} + E_{bulk}) - (E_{bulk+Va} + E_{GB+X}). \tag{10}$$

Here, $E_{bulk+Va}$ is the total energy of bulk fcc Ni with a vacancy. Based on the value of $E^b(Va; GB + X)$, we can predict whether the couple of defects GB+X facilitates the formation of vacancies in comparison with the bulk material. When we simplify the Equation (10) to the form:

$$E^b(Va; GB) = (E_{GB+Va} + E_{bulk}) - (E_{bulk+Va} + E_{GB}), \tag{11}$$

we get the binding energy of the vacancy to the clean GB, which is -0.831 eV for the most stable configuration GB-Ni$_{119}$+Va^{L2}.

We can also evaluate the binding energy of impurity to couple of lattice defects GB+Va (the tendency of impurity to move from the bulk to the GB with vacancy) as:

$$E^b(X;GB+Va) = (E_{GB+X+Va} + E_{bulk}) - \left(E_{bulk+X} + E_{GB+Va} + \frac{E_{GB+X}}{2} - \frac{E_{GB}}{2}\right). \quad (12)$$

Further, we can also study how two defects interact in the presence of the third one. In our research, we can find three such situations. The first one is the vacancy-impurity interaction at the Σ5(210) GB described with the help of the vacancy-impurity binding energy at GB. This quantity tells us how the GB simplifies or complicates the formation of X+Va couple from the separated X and Va both being at the GB. Here, the vacancies as well as impurities at the GB are used as a reference state. The vacancy-impurity binding energy is described for both the substitutional or interstitial impurity as follows:

$$E^b(GB+X;GB+Va) = (E_{GB+X+Va} + E_{GB}) - (E_{GB+X} + E_{GB+Va}). \quad (13)$$

Here, E_{GB+Va} is the total energy of bulk *fcc* Ni with GB and vacancy and the term E_{GB} is used to keep the balance in number of GBs and Ni atoms employed.

The vacancy-GB binding energy at the presence of impurity can be calculated using Equation (14):

$$E^b(X+Va;X+GB) = (E_{GB+X+Va} + E_{bulk+X}) - (E_{bulk+X+Va} + E_{GB+X}). \quad (14)$$

This energy provides the information how does the impurity simplify or complicate the formation of GB+Va couple from the separated GB and Va, in other words how simple the transfer of Va to GB in the presence of X in both places is.

The final (third) possibility of the above-mentioned interactions of two defects in the presence of the third one is characterised by the X-GB binding energy at presence of Va, which is defined by Equation (15):

$$E^b(Va+X;Va+GB) = (E_{GB+X+Va} + E_{bulk+Va}) - (E_{bulk+X+Va} + E_{GB+Va}). \quad (15)$$

This equation describes how does the Va simplify or complicate the formation of GB+X couple from the separated GB and X (i.e. the transfer of X to GB) in the presence of Va in both places. The data obtained by means of Equations (10) and (12)–(15) are summarized in Table 8.

Table 8. The energetic characteristics of interaction of three defects: the binding energy of a vacancy to the couple of GB+X ($E^b(Va;GB+X)$), the binding energy of an impurity to the couple of GB+Va ($E^b(X;GB+Va)$), the vacancy-impurity binding energy at the GB for substitutional or interstitial impurity ($E^b(GB+X^{(s/i)};GB+Va)$), the vacancy-GB binding energy in the presence of substitutional or interstitial impurity ($E^b(X^{(s/i)}+Va;X^{(s/i)}+GB)$), the X-GB binding energy in the presence of Va ($E^b(Va+X^{(s/i)};Va+GB)$) for GB-Ni$_{117}$Al$_2$$^{(s)}$+Va^{L2}, GB-Ni$_{117}Si_2$$^{(s)}$+Va^{L2} and GB-Ni$_{119}Si_2$$^{(i)}$+Va^{L3} structures.

Structure	$E^b(Va;GB+X)$ (eV)	$E^b(X;GB+Va)$ (eV)	$E^b(GB+X;GB+Va)$ (eV.X^{-1}Va^{-1})	$E^b(X+Va;X+GB)$ (eV.Va^{-1}GB^{-1})	$E^b(Va+X;Va+GB)$ (eV.X^{-1}GB^{-1})
GB-Ni$_{117}$Al$_2$$^{(s)}$+Va^{L2}	−0.81	−0.10	0.04	−0.75	−0.04
GB-Ni$_{117}$Si$_2$$^{(s)}$+Va^{L2}	−1.18	−0.11	−0.33	−1.05	0.02
GB-Ni$_{119}$Si$_2$$^{(i)}$+Va^{L3}	−0.85	−8.98	0.00	−0.84	−3.52

According to this table, it can be said that the binding energy of vacancy to couple of GB+X ($E^b(Va;GB+X)$) is negative for all structures GB-Ni$_{117}$Al$_2$$^{(s)}$+Va^{L2}, GB-Ni$_{117}Si_2$$^{(s)}$+Va^{L2} and GB-Ni$_{119}Si_2$$^{(i)}$+Va^{L2}. This means that the vacancy tends to stay together with the impurity and GB. The highest released binding energy corresponds to the structure GB-Ni$_{117}$Si$_2$$^{(s)}$+Va^{L2} (−1.18 eV) and the lowest released energy is found for the structure GB-Ni$_{117}$Al$_2$$^{(s)}$+Va^{L2} (−0.85 eV).

Considering the second column of Table 8, the binding energy of impurity to couple of lattice defects GB+Va ($E^b(X;GB+Va)$), we can observe negative values for all structures GB-Ni$_{117}$Al$_2$$^{(s)}$+Va^{L2}, GB-Ni$_{117}Si_2$$^{(s)}$+Va^{L2} and GB-Ni$_{119}Si_2$$^{(i)}$+Va^{L2}. The maximum released energy is found for structure GB-Ni$_{119}$Si$_2$$^{(i)}$+Va^{L2} (−8.98 eV). It means that the tendency of interstitial Si to move from the bulk to the GB with a vacancy is very strong. The minimum released value (−0.10 eV) corresponds to the structure GB-Ni$_{117}$Al$_2$$^{(s)}$+Va^{L2}.

When considering the values of vacancy-impurity binding energy at the GB ($E^b(GB+X;GB+Va)$) summarized in the 4th column of Table 8, it can be concluded that the GB-Ni$_{117}$Si$_2$$^{(s)}$+Va^{L2} exhibits the highest released energy (−0.33 eV) which means that these two defects tend to stay together. The structure GB-Ni$_{119}$Si$_2$$^{(i)}$+Va^{L3} exhibits positive value of this quantity (0.04 eV). Hence, the impurity and the vacancy binding are not preferred in this structure. In case of the GB-Ni$_{119}$Si$_2$$^{(i)}$+Va^{L3}, there is neither tendency for binding nor splitting of vacancy and impurity at the GB as the binding energy is zero. Comparing the binding energy $E^b(X;Va)$ of Ni$_{58}$Al$^{(s)}$+Va and Ni$_{58}$Si$^{(s)}$+Va mentioned in Table 2 (being for the most stable structures -0.0451 and −0.1110 eV.X^{-1}Va^{-1}, respectively) with $E^b(GB+X^{(s/i)};GB+Va)$ in GB-Ni$_{117}$Al$_2$$^{(s)}$+Va^{L2} and GB-Ni$_{117}Si_2$$^{(s)}$+Va^{L2} provided in Table 8 (being 0.04 and −0.33 eV.X^{-1}Va^{-1}), we can say that the vacancy and impurity are coupled stronger in case of substitutional Si than in structure with substitutional Al. However, no particular tendency was observed for the binding between a vacancy and an impurity when comparing the structures with and without GB. In case of the Al impurity, the GB causes the weakening of the binding energy (positive value 0.04 eV.X^{-1}Va^{-1}) which results even in the preference of splitting of these two defects. In case of the substitutional Si, the tendency is reverse, i.e. the GB causes the enhancing of the binding energy.

The values of the vacancy-GB binding energy in the presence of X ($E^b(X+Va;X+GB)$) in the fifth column in Table 8 show how the impurity simplifies or complicates the formation of GB+Va couple from the separated GB and Va. In all studied cases, the coupling of Va and GB is favorable as this binding energy is negative. The highest released energy (−1.05 eV.Va^{-1}GB^{-1}) corresponds to the structure GB-Ni$_{117}$Si$_2$$^{(s)}$+Va^{L2}. On the other hand, the lowest binding energy (−0.75 eV.Va^{-1}GB^{-1}) was found for the structure GB-Ni$_{117}$Al$_2$$^{(s)}$+Va^{L2}.

When considering the X-GB binding energy in the presence of Va ($E^b(Va+X;Va+GB)$), (the sixth column in Table 8), we can observe that X can be transferred to GB most easily in the structure GB-Ni$_{119}$Si$_2$$^{(i)}$+Va^{L3} because of the negative binding energy (−3.52 eV.X^{-1}GB^{-1}). On the other hand, the maximum value was obtained for structure GB-Ni$_{117}$Si$_2$$^{(s)}$+Va^{L2} (0.02 eV.X^{-1}GB^{-1}).

It is not necessary to analyze the separate defects only. Their couples can be studied too. The energy of formation of couple defect (X+Va) at the GB from the cell with GB, X in SER state and pure vacancy is, for a substitutional impurity, calculated as follows:

$$E^f(X+Va) = \left(E_{GB-Ni_m X^{(s)}{}_n+Va} + (n+1)E_{1fccNi}\right) - \left(E_{GB-Ni_{m+n+1}} + E_{1SERX} + E_{1Va}\right), \quad (16)$$

and for an interstitial impurity as:

$$E^f(X+Va) = \left(E_{GB-Ni_m X^{(i)}{}_n+Va} + E_{1fccNi}\right) - \left(E_{GB-Ni_{m+1}} + E_{1SERX} + E_{1Va}\right). \quad (17)$$

In the last two Equations (16) and (17), the energy of a vacancy without atoms is considered to be $E_{1Va} = 0$ eV.Va^{-1}. The reason for including this quantity in these equations is purely educational to keep them consistent and complete with respect to the theory. The values of $E^f(X+Va)$ can be interpreted as energies needed/released when the impurity X and Va enter the GB in fcc Ni to form a X+Va couple and the calculated values are listed in the second column of Table 9.

Table 9. The formation energy of the vacancy-impurity couple defect at GB ($E^f(X+Va)$), the binding energy of the triple defect with respect to the total energy of three single defects (GB, X, Va) ($E^b(X;GB;Va)$) and the binding energy of triple defect with respect to the total energy of three couple defects (GB+X, X+Va, Va+GB) ($E^b(X;GB;Va)^*$) for GB-Ni$_{117}$Al$_2$$^{(s)}$+Va^{L2}, GB-Ni$_{117}Si_2$$^{(s)}$+Va^{L2} and GB-Ni$_{119}Si_2$$^{(i)}$+Va^{L3} structure.

Structure	$E^f(X+Va)$ (eV.X^{-1}Va^{-1})	$E^b(X;GB;Va)$ (eV.X^{-1}Va^{-1}GB^{-1})	$E^b(X;GB;Va)^*$ (eV.X^{-1}Va^{-1}GB^{-1})
GB-Ni$_{117}$Al$_2$$^{(s)}$+Va^{L2}	−1.08	−0.92	0.08
GB-Ni$_{117}$Si$_2$$^{(s)}$+Va^{L2}	−1.12	−0.92	−0.22
GB-Ni$_{119}$Si$_2$$^{(i)}$+Va^{L3}	−1.27	−4.35	0.31

In this column, we can observe that all formation energies $E^f(X+Va)$ at the GB acquire quite high negative values. It means that the integration of X+Va couple into the structure with GB is energetically favorable in all three structures. The couple X+Va can incorporate most easily into the structure GB-Ni$_{119}$Si$_2$$^{(i)}$+Va^{L3}. It is because of the negative formation energy of X+Va which equals to −1.27 eV.X^{-1}Va^{-1}. This tendency is lower by 15% for the structure GB-Ni$_{117}$Al$_2$$^{(s)}$+Va^{L2} and by 12% for GB-Ni$_{117}$Si$_2$$^{(s)}$+Va^{L2}.

The binding energy of triple defect (GB;X;Va) with respect to the total energy of three single defects (GB, X and Va) ($E^b(X;GB;Va)$) was calculated according to the following equation:

$$E^b(GB;X;Va) = (E_{GB+X+Va} + E_{bulk}) - (E_{GB} + E_{bulk+X} + E_{bulk+Va}). \tag{18}$$

The results are summarized in Table 9. Because of the negative values of $E^b(GB;X;Va)$, it was concluded, that these three defects (GB, impurity, vacancy) prefer to bind in one structure. The maximum value of $E^b(GB;X;Va)$ (−4.35 eV.X^{-1}Va^{-1}GB^{-1}) corresponds to the structure GB-Ni$_{119}$Si$_2$$^{(i)}$+Va^{L3}, the binding energies of GB-Ni$_{117}$Al$_2$$^{(s)}$+Va^{L2} and GB-Ni$_{117}Si_2$$^{(s)}$+Va^{L2} are rather lower (both −0.92 eV.X^{-1}Va^{-1}GB^{-1}).

Similarly, the binding energy of triple defect (GB;X;Va) with respect to the total energy of couple defects (GB+X, GB+Va and X+Va) ($E^b(X;GB;Va)^*$) was calculated according to Equation (19):

$$E^b(GB;X;Va)^* = (E_{GB+X+Va} + E_{bulk}) - (E_{GB+X} + E_{GB+Va} + E_{bulk+X+Va}). \tag{19}$$

The results are listed in Table 9. According to the values obtained from Equation (19), the triple defects can be considered as unstable with respect to the couple defects in the structure GB-Ni$_{119}$Si$_2$$^{(i)}$+Va^{L3} structure because of their positive binding energy (0.31 eV.X^{-1}Va^{-1}GB^{-1}). On the other hand, the stable triple defect can be found in the structure GB-Ni$_{117}$Al$_2$$^{(s)}$+Va^{L2} where the binding energy reveals negative value.

In principle, the theoretical total energy of studied structures ($E^{theo}_{GB+X^{(s/i)}+Va}$) can be calculated from separated contributions: total energy of bulk Ni and X, total energies of formation of particular defects and total energies of binary and ternary interactions of these defects. For the most complicated structure containing all three defects GB, X and Va, this procedure is described by Equation (20):

$$E^{theo}_{GB+X^{(s/i)}+Va} = E_{nfccNi} + 1E_{1SERX} + E^f_{GB} + E^f_X + E^f_{Va} + E^b_{GB;X} + E^b_{GB;Va} + E^b_{X;Va} + E^b_{GB;X;Va}. \tag{20}$$

The data calculated according to this equation are summarized in Table 10.

Table 10. Theoretical total energies of studied structures $E^{theo}_{GB+X^{(s/i)}+Va}$ and the difference between the total energies calculated ab initio and theoretical total energies (calculated from the particular contributions) E^{diff} for $Ni_{59}Al^{(s)}$, $Ni_{59}Si^{(s)}$, $Ni_{120}Si^{(iO)}$, $Ni_{59}+Va$, $Ni_{58}Al^{(s)}+Va$, $Ni_{58}Si^{(s)}+Va$, $Ni_{59}Si^{(iO)}+Va$, $GB-Ni_{120}$, $GB-Ni_{118}Al_2^{(s)}$, $GB-Ni_{118}Si_2^{(s)}$, $GB-Ni_{120}Si_2^{(i)}$, $GB-Ni_{119}+Va^{L2}$, $GB-Ni_{117}Al_2^{(s)}+Va^{L2}$, $GB-Ni_{117}Si_2^{(s)}+Va^{L2}$ and $GB-Ni_{119}Si_2^{(i)}+Va^{L3}$ structures. In case of the structures with GB, the energies $E^{theo}_{GB+X^{(s/i)}+Va}$ are related to one GB and X. From this reason, these energies are comparable with those of structures without GB.

Structure	$E^{theo}_{GB+X^{(s/i)}+Va}$ (eV.cell^{-1})	E^{diff} (eV.cell^{-1})	Structure	$E^{theo}_{GB+X^{(s/i)}+Va}$ (eV.cell*$^{-1}$)	E^{diff} (eV.cell*$^{-1}$)	E^{diff} (kJ.mol of atoms^{-1})
—	—	—	$GB-Ni_{120}$	−323.654	0.000	0.000
$Ni_{59}Al^{(s)}$	−327.888	0.000	$GB-Ni_{118}Al_2^{(s)}$	−323.596	−0.037	−0.060
$Ni_{59}Si^{(s)}$	−329.598	0.000	$GB-Ni_{118}Si_2^{(s)}$	−324.935	−0.043	−0.070
$Ni_{120}Si^{(iO)}$	−659.820	0.000	$GB-Ni_{120}Si_2^{(i)}$	−330.900	−0.025	−0.039
$Ni_{59}+Va$	−321.155	0.000	$GB-Ni_{119}+Va^{L2}$	−317.601	−0.007	−0.011
$Ni_{58}Al^{(s)}+Va$	−321.054	0.000	$GB-Ni_{117}Al_2^{(s)}+Va^{L2}$	−317.536	−0.014	−0.022
$Ni_{58}Si^{(s)}+Va$	−322.830	0.000	$GB-Ni_{117}Si_2^{(s)}+Va^{L2}$	−319.258	−0.006	−0.010
$Ni_{59}Si^{(iO)}+Va$	−325.222	0.000	$GB-Ni_{119}Si_2^{(i)}+Va^{L3}$	−324.873	−0.011	−0.017

* denotes the cell consisting of one GB, one impurity and one vacancy.

Furthermore, the theoretical energies $E^{theo}_{GB+X^{(s/i)}+Va}$ can be compared with the total energies of particular structures as follows:

$$E^{diff} = E^{calc}_{GB+X^{(s/i)}+Va} - E^{theo}_{GB+X^{(s/i)}+Va}. \tag{21}$$

The results of this comparison are again exhibited in Table 10. In case that all contributions in $E^{theo}_{GB+X^{(s/i)}+Va}$ are calculated properly and no further contribution should be considered, E^{diff} should be zero. In a case of structures with single defect (GB or X or Va) and in a case of structure with two defects (X and Va) the value of E^{theo} and E^{calc} do not differ. In case of structures with all possible combination of defects, the values of E^{diff} are smaller than zero. The lowest value of E^{diff} corresponds to the structure with a GB and substitutional impurity. It means that the fully ab initio calculated values of $E^{calc}_{GB+X^{(s/i)}+Va}$ are more negative than the theoretical ones, $E^{theo}_{GB+X^{(s/i)}+Va}$. This can be caused by the fact that the contributions of particular defects and their combinations were calculated from structures at equilibrium volumes which differ from equilibrium volumes of the final structures. The other explanation for this difference could be that there are possible stabilizing interactions in the directly ab initio calculated values in comparison to the theoretical energy obtained from separated contributions.

To make our research on segregation complete, the further three possible types of segregation should be analyzed taking into the account the presence of Va at the GB or in the bulk or in both of these places. Hence, the segregation energy of impurity from the bulk to the GB with Va was evaluated using the following Equation (22):

$$E^{seg}_{Va^{GB}}(X) = (E_{GB+X+Va} - E_{GB+Va}) - (E_{bulk+X} - E_{bulk}) = E_{Xat(GB+Va)} - E_{Xinbulk}, \tag{22}$$

the segregation energy of impurity from the bulk with Va to the GB was calculated according to the Equation (23):

$$E^{seg}_{Va^{bulk}}(X) = \frac{(E_{GB+X} - E_{GB})}{2} - (E_{bulk+X+Va} - E_{bulk+Va}) = E_{XatGB} - E_{Xin(bulk+Va)}, \tag{23}$$

and, finally, the segregation energy of impurity in the structure with the vacancy both in the bulk and at the GB (impurity-GB binding energy at presence of vacancy) was expressed by Equation (24):

$$E^{seg}_{Va^{bulk};Va^{GB}}(X) = (E_{GB+X+Va} - E_{GB+Va}) - (E_{bulk+X+Va} - E_{bulk+Va})$$
$$= E_{Xat(GB+Va)} - E_{Xin(bulk+Va)}. \quad (24)$$

In these equations, $E_{GB+X+Va}$ and E_{GB+Va} are the energies of the GB+Va structure with and without impurity atom (both containing the same number of Ni atoms) and $E_{bulk+X+Va}$ and $E_{bulk+Va}$ are the total energies of the bulk material containing Va with and without impurity (again, both containing the same number of Ni atoms). The energy difference for bulk Ni material has to be subtracted here to eliminate the energetics of adding an impurity to the bulk material. $E_{XatGB+Va}$ and $E_{Xinbulk+Va}$ are the total energies of the impurity at the GB and in the bulk material, both also containing Va. In the last term of these three equations, the position of impurity in bulk material was always considered as substitutional, because this type of position of impurity is more stable than the interstitial one. After a deeper examination of this issue in Equation (24), it was found that the $E^{seg}_{Va^{bulk};Va^{GB}}(X)$ quantity is, in case of the substitutional impurity, consistent with the X-GB binding energy at presence of Va, which is described by Equation (15). However, in case of the interstitial impurity, the results differ. The reason for this difference is that we used the bulk material with interstitial Si as the reference state in case of Equation (15) and in case of the Equation (24), the bulk material with the more stable substitutional Si was used.

The results obtained using Equations (7) and (22)–(24) are summarized in Table 11.

Table 11. Segregation energies of impurity segregating (a) from the bulk to GB ($E^{seg}(X)$, Equation (7)); (b) from the bulk to GB with Va ($E^{seg}_{Va^{GB}}(X)$, Equation (22)); (c) from the bulk with Va to GB ($E^{seg}_{Va^{bulk}}(X)$, Equation (23)) and (d) from the bulk with Va to GB with Va ($E^{seg}_{Va^{bulk};Va^{GB}}(X)$, Equation (24)) for GB-Ni$_{117}$Al$_2$$^{(s)}$+Va^{L2}, GB-Ni$_{117}Si_2$$^{(s)}$+Va^{L2} and GB-Ni$_{119}Si_2$$^{(i)}$+Va^{L3} structure.

Structure	$E^{seg}(X)$ (eV.X^{-1})	$E^{seg}_{Va^{GB}}(X)$ (eV.X^{-1})	$E^{seg}_{Va^{bulk}}(X)$ (eV.X^{-1})	$E^{seg}_{Va^{bulk};Va^{GB}}(X)$ (eV.X^{-1})
GB-Ni$_{117}$Al$_2$$^{(s)}$+Va^{L2}	−0.125	−0.104	−0.080	−0.042
GB-Ni$_{117}$Si$_2$$^{(s)}$+Va^{L2}	0.241	−0.108	0.352	0.019
GB-Ni$_{119}$Si$_2$$^{(i)}$+Va^{L3}	−3.518	−0.260	−0.128	−0.132

According to the values of segregation energies listed in Table 11, we can conclude that the segregation energies are always the most negative for the structures with interstitial impurity. The segregation energies $E^{seg}_{Va^{bulk}}(X)$ and $E^{seg}_{Va^{bulk};Va^{GB}}(X)$ (the fourth and the fifth column of Table 11) are almost the same (close to −0.13 eV.X^{-1}). These values are much lower than the values listed in the second and third column. This is caused by the presence of the Va in the bulk material which suppress the segregation. The aluminum at the substitutional position also prefers to segregate at the GB because of its negative segregation energy. However, this tendency is lower in comparison with Si$^{(i)}$. The silicon at substitutional position usually prefers to stay in the bulk because of its positive segregation energy except for the case where Va is placed at the GB. In this case Va provides the additional space for the impurity and the segregation energy becomes negative (−0.108 eV.X^{-1}).

It can be concluded that the vacancy makes the segregation less desirable in the case of Al$^{(s)}$ and Si$^{(i)}$ except of the case of $E^{seg}_{Va^{GB}}(Si^{(s)})$, where the vacancy supports the segregation. In the remaining cases, the Si$^{(s)}$ prefers to stay in the bulk.

3.4.3. Magnetism

From a comparison of the distribution of magnetic moments in the structure containing a GB and an impurity with the structure containing a GB, an impurity and a vacancy, it is obvious that the vacancy causes the diversification of magnetic moments of Ni atoms. This means that the difference between the magnetic moment per atom of the structure with and without vacancy ($\Delta\mu_{at}$) is non-zero and it acquires both positive and negative values. The scatter in values of magnetic moments of the Ni atom is up to ± 0.06 μ_B. When the vacancy moves from its position at the GB to another layer the region of scatter of magnetic moments moves in the same direction. This effect was observed in structures with all kinds of impurities. Some examples of the distribution of magnetic moments in the structures GB-Ni_{119}+Va^{L2}, GB-$Ni_{117}Al_2^{(s)}$+Va^{L2}, GB-$Ni_{117}Si_2^{(s)}$+Va^{L2} and GB-$Ni_{119}Si_2^{(i)}$+Va^{L3} are shown in Figure 9.

From Figure 9 it is obvious that the distance between the two adjacent GBs is large enough to achieve a bulk-like behavior of magnetic moments between them, i.e. the magnetic moment of *fcc* bulk Ni of 0.632 μ_B calculated in this work. Comparing Figure 9a with Figure 9b–d, it can be concluded that the structure without impurities GB-Ni_{119}+Va^{L2} (Figure 9a) exhibits a much lower scatter in magnetic moments which occurs only in the nearest vicinity of the vacancy, i.e. at the edges of the figure. On the other hand, the structures with impurities (Figure 9b–d) show a much larger scatter of magnetic moments in the vicinity of impurity. The above-mentioned comparison also shows that the GB and Va cause the increase of magnetic moments, which is demonstrated by the position of all curves in Figure 9a above the value of magnetic moment of *fcc* bulk Ni of 0.632 μ_B calculated in this work. On the contrary, the other structures (Figure 9b–d) contain also atoms with magnetic moments lower than that calculated for *fcc* Ni. In case of the GB-Ni_{119}+Va^{L2} structure (Figure 9a), atoms located at clean GB in different planes but in the same layer have the same magnetic moment (overlapping curves in the middle of figure) which is much higher than that calculated for *fcc* Ni. The maximum of magnetic moment of 0.72 μ_B was reached for atoms in the layer 2. Similar behavior was not observed for structures with impurities which cause the decrease and a large scatter of magnetic moments even at the clean GB.

In case of the structure GB-$Ni_{117}Al_2^{(s)}$+Va^{L2}, the GB with a vacancy enables the Ni atoms to migrate towards the free space of the vacancy and hence to decrease their distance from Al atoms. This change of position causes only a minor decrease of magnetic moments (hundredths of μ_B) and their mild scattering. The differences in magnetic moments between the two planes with impurity are not very large, as it can be seen from Figure 9b—blue line with squares and red with circles. In case of the planes with z = 0.25 and 0.75 (Figure 9b—green line with diamonds), the magnetic moments are even identical. The symmetry of curves with respect to the center of the figure proves that the magnetic moments do not depend dominantly on the position of the vacancy (layer denoted by *) which would cause the asymmetry close to this layer. By other analyses, it was found that the dominating effect is the distance from the impurity, which is different for different atoms in the same layer. This causes the differences of the shape of curves in both the marginal and central part of the figures. In the central part of the figures labelled as GB(Ni,Ni), this effect is visible because of the shift of some lines representing the results of planes with impurities in the middle of the supercell (see the label of Figures 4 and 9). Then the atoms situated in some plane at the GB without impurity GB(Ni,Ni) can be very close to the impurity that is placed in some plane of GB(Ni,X) and can be influenced by them. In case of the planes without the impurity, the value of the magnetic moment at the GB exactly correlates with the distance from the impurity, i.e. the higher distance the higher magnetic moment.

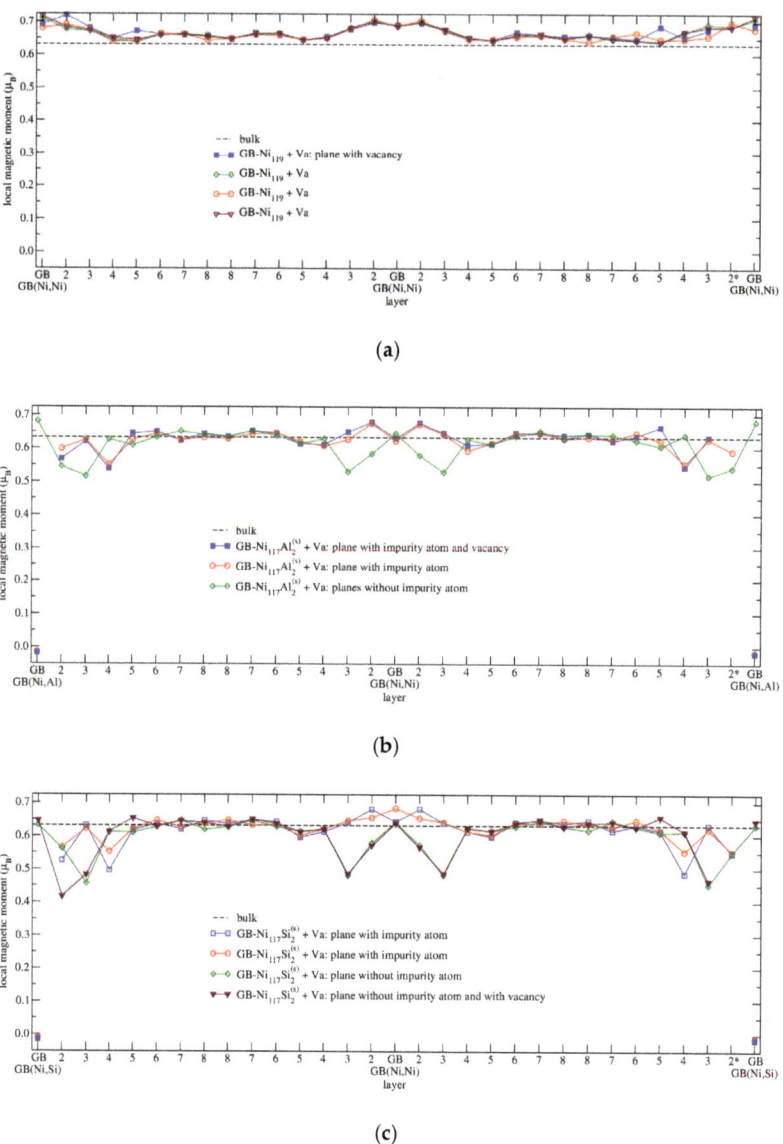

Figure 9. Cont.

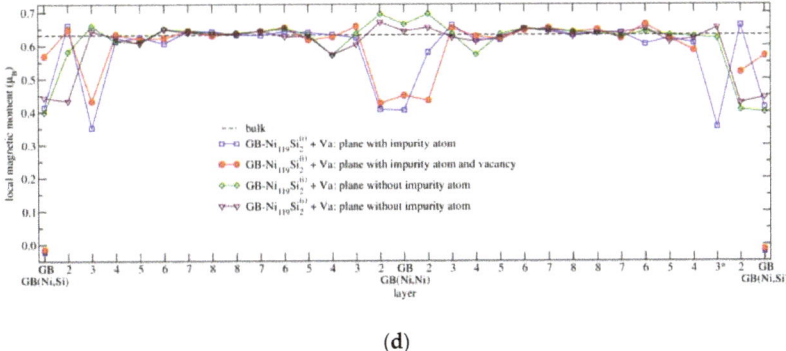

(d)

Figure 9. Magnetic moments of Ni atoms in supercells with a GB, an impurity and a vacancy: (a) GB-Ni$_{119}$+Va^{L2}, (b) GB-Ni$_{117}$Al$_2$$^{(s)}$+Va^{L2}, (c) GB-Ni$_{117}Si_2$$^{(s)}$+Va^{L2}, (d) GB-Ni$_{119}Si_2$$^{(i)}$+Va^{L3}. The superscripts ^{L2} and ^{L3} denote the position of the vacancy in the second and third layer, respectively. * denotes the layer with the vacancy. The full points mark the magnetic moments of impurities. The dashed line shows the magnetic moment of fcc bulk Ni of 0.632 µ$_B$ calculated in this work. For a better demonstration of results in Figures (b–d), the plane with impurity atom is always presented so that the impurity is on the edges of the figures. This was achieved by the shift of $\frac{1}{2}$ of the lattice parameter b of lines representing the results of some planes with impurities (the middle plane of GB-Ni$_{118}$Al$_2$$^{(s)}$ and GB-Ni$_{118}$Si$_2$$^{(s)}$ or the second plane of GB-Ni$_{120}$Si$_2$$^{(i)}$) and one without impurity (the middle plane of GB-Ni$_{120}$Si$_2$$^{(i)}$).

The behavior of magnetic moments in GB-Ni$_{117}$Si$_2$$^{(s)}$+Va^{L2} is similar to that in GB-Ni$_{117}$Al$_2$$^{(s)}$+Va^{L2}. The Ni atoms in the vicinity of the vacancy change their magnetic moments in dependence on the distance from the impurity. The differences between the two planes with impurities (Figure 9c—blue curve with squares and red with circles) are not very large. They reveal the same trends. The same behavior is observed for planes without impurity (Figure 9c—green curve with diamonds and violet with triangles). The couples of planes even reveal similar trends in the vicinity of the vacancy (the right part of the figure denoted by *). The only case when they differ is the opposite part of the GB with respect to the position of the vacancy (the left-hand side of the figure), where the Ni atoms in the plane of the vacancy migrate towards the vacancy and at the same time towards the impurity, which causes the decrease of the magnetic moment on the violet curve with triangles in the 2nd layer and bigger scatter in magnetic moments of other atoms as their positions are also slightly influenced.

Comparing the effects of substitutional impurities (Figure 9b,c), it can be concluded that Si causes a larger decrease of magnetic moments of the neighboring Ni atoms than Al. This corresponds to the larger structural changes in the configuration with substitutional Si which is obvious from the left-hand part of the Figure 8b,c.

In case of the structure GB-Ni$_{119}$Si$_2$$^{(i)}$+Va^{L3} (Figure 9d), we can find again two couples of curves with similar variations corresponding to planes with (the red curve with circles and blue with squares) and without (the green curve with diamonds and violet with triangles) the impurity. The atomic positions in this structure have also changed significantly due to the relaxation. As the silicon atom in the vicinity of the vacancy left its position at the grain boundary, the layout of magnetic moments became asymmetric with respect to the GB. Nevertheless, the distance of the Ni atoms from the impurity plays the dominant role also here. At the GB without a vacancy, the magnetic moment of Ni atoms increases from 0.35 to 0.45 µ$_B$ with the increasing distance of Ni atom from the Si impurity from 2.18 to 2.46 Å. For atoms in the distance of 2.93 Å and further, the magnetic moments reach the values of 0.61 µ$_B$ and higher converging to the bulk value. At the GB with the vacancy, the magnetic moments reveal the same trend as at the GB without vacancy. They change from 0.41 to 0.58 µ$_B$ for the distances from 2.26 to 2.80 Å. The Ni atoms placed further from Si reach the magnetic moments of 0.61 μ_B and higher. The Ni atoms located simultaneously at the GB and in the vicinity of the Si atom possess low magnetic moments despite of the fact that they are placed in the region with a larger free volume causing in general an increase of the magnetic moment. This demonstrates the dominating effect of the impurity on the magnetic moments. This effect could also be proved by the fact that the Ni atoms placed at different GBs (with and without Va) but having the same distance from impurity exhibit also similar magnetic moments. For example, the magnetic moments of 0.41 and 0.52 μ_B are reached at the distances of 2.26 and 2.52 Å from Si atom in the layer with a vacancy. For comparison, at the GB without a vacancy, the magnetic moments of 0.41 and 0.45 μ_B are reached at the distances of 2.28 and 2.46 Å from the Si atom. Magnetic moments of interstitial Si atoms are identical within the computational error. The asymmetrical behavior of the blue line with squares in the center of Figure 9d is caused by the presence of the shifted impurity in the neighboring plane. This impurity gets to the vicinity of the Ni atom in the 2nd plane on the right-hand side of central part of the blue line with squares in this figure, which causes the decrease of the magnetic moment of Ni.

The curves in Figure 9d reveal a larger scatter at the GB with vacancy as its free space enables the movement of atoms and, consequently, the change of the distances between Ni atoms and interstitial Si.

It can be concluded that in case of all structures, the distance between the Ni atom and an impurity has the dominating effect on the magnetic moments of Ni. The effect of the vacancy is due to the mediated migration of atoms.

Furthermore, Figure 9b–d also provide the information about the magnetic moments of the impurities which are −0.017, −0.013 and −0.019 μ_B for $Al^{(s)}$, $Si^{(s)}$ and $Si^{(i)}$, respectively. These values are comparable with those found in previously studied materials/configurations in this work. The magnetic moments of impurities in the structure with a GB and without a vacancy are −0.016, −0.011 and −0.021 μ_B for $Al^{(s)}$, $Si^{(s)}$ and $Si^{(i)}$. In the bulk material with a vacancy, the magnetic moment of $Al^{(s)}$, $Si^{(s)}$ and $Si^{(i)}$ amounts to −0.019, −0.016 and −0.014 μ_B, respectively, and for the bulk without a vacancy, it is −0.018, −0.014 and −0.021 μ_B, respectively. Except for the bulk structure with a vacancy, silicon at the interstitial position has always the lowest magnetic moment and silicon at the substitutional position has always the highest magnetic moment.

In the experimental study [71], a GB configuration close to the twist (001) GB was examined. Those authors found magnetic moment enhancement up to about 100%. On the other hand, our values of the enhancement of magnetic moment are rather small. They amount, in maximum, to about 9% (with respect to the bulk). On the other hand, the magnetic moment of Ni atom situated at the Σ5(210) GB was experimentally determined using a transmission electron microscopy/electron energy loss spectroscopy method [72] to be 0.63 μ_B. This is essentially the same as the magnetic moment of bulk Ni atom (0.632 μ_B) calculated in this work. Some lattice defects at the GBs, for example vacancies studied here, might slightly enhance this phenomenon although we still were not able to reach the enhancement reported in [71]. However, other theoretical studies [2,49,57,73,74] also demonstrate that the enhancement of magnetic moments at various GBs in Ni is rather small, if not negative [75] (unfortunately, GBs with other defects than segregated impurities have been studied only rarely up to now). Thus, it is not excluded that such a large enhancement reported in [71] may be a spurious effect. Another questionable issue is the fact that the enhancement of magnetic moment found in [71] extends up to a nearly 100 Å distance from the GB plane whereas most theoretical calculations locate it in the region of about several Å. Thus, further experimental as well as theoretical studies are desirable to shed light on these problems.

There are also some issues connected with non-zero temperature. As the atomistic simulations [76,77] demonstrate, with increasing temperature vacancies become unstable; they delocalize and may lead to instability. Consequently, effective disappearance of the free volume connected with vacancies at temperatures of about 600 K at most may occur. Of course, GBs act as vacancy sinks and the effect described above is in full accordance with this fact.

As it is mentioned in [78] the optimized model of a tensile test on clean GB in *fcc* Ni predicted a structural transformation from the Σ5(210) GB to a Σ11(311) GB. In addition, the ab initio simulations of hydrostatic and uniaxial tensile tests on *fcc* crystals of Ni predicted instabilities of an elastic nature in Ni crystal [79]. Further, in [80] the effect of applied stress on vacancy segregation near the Σ5(210) GB in Ni has been reported with the help of a methodology similar to that in the present paper. Here, we have analyzed more vacancy positions – not only those at the GB plane. The authors of the study [80] suggested that vacancies can be absorbed by the GB under a compressive stress with just an opposite effect for a tensile stress applied perpendicular to the GB. Those predictions, however, are not in agreement with known thermodynamic models. In addition to that, the effect of temperature (vibrations) on the behavior of vacancies was not included. This, of course, could influence the results of those calculations.

4. Conclusions

A detailed and systematic first-principles study of substitutional $Al^{(s)}$ and both substitutional $Si^{(s)}$ and interstitial $Si^{(i)}$ impurities in the ferromagnetic *fcc* bulk nickel and at the Σ5(210) GB is presented, including various concentrations and positions of impurities not previously studied. This research was complemented by cases where vacancies have also occurred. In all instances, the structural, energetic and magnetic aspects were investigated.

In the bulk material, the addition of impurities caused an increase of the volume per atom in comparison with elemental Ni. The most significant increase was observed in the structures with the tetrahedral (iT) or octahedral (iO) interstitial Si atom. These configurations nevertheless possess a positive energy of formation related to standard states of pure elements, which makes them unstable. Larger structures with interstitial impurities are more stable than the smaller ones. The structure $Ni_{120}Si^{(iO)}$ reveals a higher stability than $Ni_{120}Si^{(iT)}$ which reflects the fact that the octahedral site is larger than the tetrahedral one. Introducing the substitutional Al or Si into the material led to structure stabilization. The impurities reduce the magnetic moments of their neighboring Ni atoms and they themselves obtain a small negative induced magnetic moment.

Adding vacancies to the bulk material caused a decrease of the cell volume but an increase of the volume per atom in comparison with the bulk structures without a vacancy. The addition of a vacancy always induced an increase of the total energy of the system or the vacancy disappearance during the recombination with the interstitial $Si^{(i)}$ atom. In case of $Ni_{59}Al^{(s)}$+Va and $Ni_{59}Si^{(s)}$+Va structures, the minimum-energy configurations are those with the vacancy and impurity being the nearest neighbors. In structures with interstitial silicon, the vacancy is stable only if it is far enough from the silicon atom. The preference of the structures with substitutional impurities is sustained even in the presence of the Va. The presence of a vacancy causes an increase of magnetic moments of neighboring Ni atoms by 0.02–0.04 μ_B. However, this effect is not as strong as the "impurity-caused-decreasing" effect.

The structures containing a GB and an impurity reveal a scattering in the interlayer distances. The largest distance is found between the GB and the second layer and the lowest one occurs between the second and the third layer. For the next interlayer distances, the difference from calculated data [2,49] was not so large and from the sixth layer on, the interlayer distances closely oscillated around the calculated bulk value. The excess free volume gains the values in the range between 0.11 and 0.34 $Å^3 \cdot Å^{-2}$. In all studied cases, the GBs are unstable with respect to the bulk material in the sequence of GB-$Ni_{120}Si_2^{(i)}$ < GB-Ni_{60} = GB-Ni_{120} < GB-$Ni_{118}Al_2^{(s)}$ < GB-$Ni_{118}Si_2^{(s)}$. The segregation energy of substitutional $Al^{(s)}$ atom and interstitial $Si^{(i)}$ atom is negative which means that both impurities tend to be placed in the neighborhood of GB rather than to stay in the bulk material. The substitutional $Si^{(s)}$ atom prefers to be located in the bulk material. However, this arrangement is less stable than the structure with interstitial $Si^{(i)}$. For the structures where the impurity is located at grain boundaries, the "magnetic-moment-decreasing" effect of impurity is suppressing the "magnetic-moment-increasing" effect of the grain boundary. The largest decrease in magnetic moment near to the impurity can be observed in the GB-$Ni_{120}Si_2^{(i)}$ structure.

In case of the structures with vacancies and GBs, the introduction of a vacancy causes changes in positions of atoms, which tend to occupy the free space provided by a vacancy. However, the vacancies in these structures are preserved. For GB-Ni$_{119}$+Va, GB-Ni$_{117}$Al$_2^{(s)}$+Va and GB-Ni$_{117}$Si$_2^{(s)}$+Va (GB-Ni$_{119}$Si$_2^{(i)}$+Va), the configuration with vacancy in layer 2 (layer 3) reveals the highest decrease of the volume and the lowest but positive both energy of vacancy formation and the energy of formation with respect to the standard element reference states. This means that the largest decrease in volume corresponds to the most easily but still unfavorably formed vacancy and to the most stable arrangement of structure with vacancy in a given position. However, as all configurations of structures possess the positive values of the energy of formation they must be considered as unstable at 0 K. Nevertheless, these structures can be stabilized at higher temperatures. In such a case, we expect the Si$^{(i)}$ to be more stable than Si$^{(s)}$. One of the contributions to this instability is the vacancy formation energy, which ranges from 0.25 to 0.62 eV.Va^{-1}.

Analyzing the binding energies between the defects, it was found that the binding energy $E^b(X;Va)$ (where X stands for segregated Al or Si) reaches the negative values, which can be interpreted as the tendency to the binding between vacancy and impurity. The $E^b(GB;Va)$ is −0.849 eV for the most stable configuration GB-Ni$_{119}$+Va^{L2}. The binding energies between one defect and the remaining couple of defects, $E^b(Va;GB+X)$ and $E^b(X;GB+Va)$, were analyzed too. It was found that the vacancy tends to stay together with the impurity and GB. The highest released binding energy $E^b(Va;GB+X)$ corresponds to the structure GB-Ni$_{117}$Si$_2^{(s)}$+Va^{L2} (−1.18 eV) and the other values are also reasonable (more than −0.8 eV). Additionally, the impurities prefer to stay in the neighborhood of the couple GB+Va which is related to the negative binding energy $E^b(X;GB+Va)$. In particular, the tendency of interstitial Si to move from the bulk to a GB with a vacancy is very strong (−8.98 eV). The effects of how two defects interact in the presence of the third one were also investigated. In all studied cases, the interaction is associated with negative (the defects tend to stay together) or negligible energy effects (their interaction neither preferred nor rejected). Considering the binding energies of the triple defect from single defects $E^b(X;GB;Va)$ or couple of defects $E^b(X;GB;Va)^*$, it was again concluded that the defects tend to stay together except for the case of combining the couple of defects in structure GB-Ni$_{119}$Si$_2^{(i)}$+Va^{L3}.

The theoretical total energy $E^{theo}_{GB+X^{(s/i)}+Va}$ of studied structures calculated from separated contributions was also evaluated. The lowest value (−330.900 and −324.873 eV.cell^{-1}) corresponds to the GB-Ni$_{120}$Si$_2^{(i)}$ and GB-Ni$_{119}$Si$_2^{(i)}$+Va^{L3} structure. The difference between the directly calculated and the theoretical energy value E^{diff} was negative for all studied structures with GB and impurity. It indicates that for these structures the directly calculated energy could contain some stabilizing interactions in comparison to the theoretical energy obtained from separated contributions. However, this can be also caused by the fact that the contributions of particular defects and their combinations were calculated from structures at equilibrium volumes which differ from equilibrium volumes of the final structures.

The presence of a vacancy at GB induced changes in the behavior of impurities with respect to segregation—all studied impurities prefer to segregate. When the vacancy is present in both structures (at the GB and in the bulk material) or in the bulk material, the segregation becomes less advantageous and in case of the Si$^{(s)}$ even unfavorable. The strongest tendency to segregation is observed in structures with substitutional impurities and without vacancies.

The GBs and vacancies cause an increase of magnetic moments, nevertheless the impurities, having induced magnetic moments of the opposite direction than Ni atoms, exhibit the dominating decreasing effect on the magnetism of Ni atoms.

Let us note that the segregation energy itself (calculated at 0 K) cannot completely describe the solute segregation but very well characterizes the tendency of the solute to segregate at a grain boundary [1]. Nevertheless, as it follows from Langmuir–McLean segregation isotherm, the segregation energy is an important controlling parameter of the temperature dependence of grain boundary segregation and this constitutes the main point of our paper. The problem of entropy is substantial as it modifies the solute concentration at the boundary [81]. Despite the crucial importance of entropy, any information about the segregation energy—as a part of the total driving force for segregation—is also valuable for experimentalists.

Supplementary Materials: The following are available online at http://www.mdpi.com/2079-4991/10/4/691/s1. Table S1: Data depicted in Figure 4: the equilibrium interlayer distances D for structures GB-Ni$_{60}$, GB-Ni$_{118}$Al$_2^{(s)}$, GB-Ni$_{118}$Si$_2^{(s)}$ and GB-Ni$_{120}$Si$_2^{(i)}$ obtained from the automatic relaxation. Here, the interlayer distance of type 2/3 stands for the distance between the 2nd and 3rd layer under the assumption that the 1st layer is the layer of GB. Further, Ni,Ni and X,Ni stand for the clean plane and the plane with impurity X = Al, Si, respectively.

Author Contributions: Conceptualization, M.Š. and J.P.; methodology, M.V., J.P. and M.M.; calculations and formal analysis, M.M. and M.V.; resources, M.Š.; writing—original draft preparation, M.M.; writing—review and editing, M.M., M.V., J.P. and M.Š.; visualization, M.V. and M.M.; supervision, M.V., J.P. and M.Š.; project administration, M.Š. All authors participated in the discussions of the results. All authors have read and agreed to the published version of the manuscript.

Funding: This research was supported by the Grant Agency of the Czech Republic (Project No. GA16-24711S), by the Ministry of Education, Youth and Sports of the Czech Republic (CEITEC 2020-Project No. LQ1601), and by the Academy of Sciences of the Czech Republic (Institutional Project No. RVO:68081723).

Acknowledgments: Computational resources were provided by the Ministry of Education, Youth and Sports of the Czech Republic under the Projects CESNET (LM2015042), CERIT-Scientific Cloud (LM2015085) and IT4Innovations National Supercomputing Center (LM2015070) within the program Projects of Large Research, Development and Innovations Infrastructures.

Conflicts of Interest: The authors declare no conflict of interest.

References

1. Lejček, P. Grain Boundary Segregation in Metals. In *Springer Series in Materials Science*; Springer: Berlin/Heidelberg, Germany, 2010; Volume 136.
2. Všianská, M.; Šob, M. The effect of segregated sp-impurities on grain-boundary and surface structure, magnetism and embrittlement in nickel. *Prog. Mater. Sci.* **2011**, *56*, 817. [CrossRef]
3. Lejček, P.; Šob, M.; Paidar, V. Interfacial segregation and grain boundary embrittlement: An overview and critical assessment of experimental data and calculated results. *Prog. Mater. Sci.* **2017**, *87*, 83. [CrossRef]
4. Všianská, M.; Vémolová, H.; Šob, M. Segregation of sp-impurities at grain boundaries and surfaces: Comparison of fcc cobalt and nickel. *Model. Simul. Mater. Sci. Eng.* **2017**, *25*, 085004. [CrossRef]
5. Razumovskiy, V.I.; Lozovoi, A.Y.; Razumovskii, I.M. First-principles-aided design of a new Ni-base superalloy: Influence of transition metal alloying elements on grain boundary and bulk cohesion. *Acta Mater.* **2015**, *82*, 369. [CrossRef]
6. Liu, W.; Han, H.; Ren, C.; Yin, H.; Zou, Y.; Huai, P.; Xu, H. Effects of rare-earth on the cohesion of Ni Σ5(012) grain boundary from first-principles calculations. *Comput. Mater. Sci.* **2015**, *96*, 374. [CrossRef]
7. Wenguan, L.; Yuan, Q.; Dongxun, Z.; Youshi, Z.; Xingbo, H.; Xinxin, C.; Huiqin, Y.; Guo, Y.; Guanghua, W.; Shengwei, W.; et al. A theoretical study of the effects of sp-elements on hydrogen in nickel-based alloys. *Comput. Mater. Sci.* **2017**, *128*, 37.
8. Bentria, E.T.; Lefkaier, I.K.; Benghia, A.; Bentria, B.; Kanoun, M.B.; Goumri-Said, S. Toward a better understanding of the enhancing/embrittling effects of impurities in Nickel grain boundaries. *Sci. Rep.* **2019**, *9*, 1424. [CrossRef]
9. Divinski, S.V.; Edelhoff, H. Diffusion and segregation of silver in copper Σ5(310) grain boundary. *Phys. Rev. B* **2012**, *85*, 144104. [CrossRef]
10. Muller, D.A.; Subramanian, S.; Batson, P.E.; Silcox, J.; Sass, S.L. Structure, chemistry and bonding at grain boundaries in Ni$_3$Al—I. The role of boron in ductilizing grain boundaries. *Acta Mater.* **1996**, *44*, 1637. [CrossRef]
11. Aksoy, D.; Dingreville, R.; Spearot, D.E. Embedded-atom method potential parameterized for sulfur-induced embrittlement of nickel. *Model. Simul. Mater. Sci. Eng.* **2019**, *27*, 085016. [CrossRef]
12. Lejček, P.; Všianská, M.; Šob, M. Recent trends and open questions in grain boundary segregation. *J. Mater. Res.* **2018**, *33*, 2647. [CrossRef]
13. Prakash, S. Structural alloys for power plants. In *Operational Challenges and High-Temperature Materials*; Woodhead Pub.: Waltham, MA, USA, 2014.

14. Wang, M.; Deng, Q.; Du, J.; Tian, Y.; Zhu, J. The Effect of aluminum on microstructure and mechanical properties of ATI 718Plus alloy. *Mater. Trans.* **2015**, *56*, 635. [CrossRef]
15. Zhu, H.Q.; Guo, S.R.; Guan, H.R.; Zhu, V.X.; Hu, Z.Q.; Murata, V.; Morinaga, M. The effect of silicon on the microstructure and segregation of directionally solidified IN738 superalloy. *Mater. High Temp.* **2016**, *12*, 285. [CrossRef]
16. Lefkaier, I.K.; Bentria, E.T. The Effect of Impurities in Nickel Grain Boundary: Density Functional Theory Study. In *Study of Grain Boundary Character*; Tanski, T., Borek, W., Eds.; IntechOpen: Rijeka, Croatia, 2017; Volume 1, p. 1.
17. Smith, R.W.; Geng, W.T.; Geller, C.B.; Wu, R.; Freeman, A.J. The effect of Li, He and Ca on grain boundary cohesive strength in Ni. *Scr. Mater.* **2000**, *43*, 957. [CrossRef]
18. Brodetskii, I.L.; Kharchevnikov, V.P.; Belov, B.F.; Trotsan, A.I. Effect of calcium on grain boundary embrittlement of structural steel strengthened with carbonitrides. *Met. Sci. Heat Treat.* **1995**, *37*, 200. [CrossRef]
19. Floreen, S.; Westbrook, J.H. Grain boundary segregation and the grain size dependence of strength of nickel-sulfur alloys. *Acta Metall.* **1969**, *17*, 1175. [CrossRef]
20. Kronberg, M.L.; Wilson, F.H. Secondary recrystallization in copper. *Trans. AIME* **1949**, *185*, 501. [CrossRef]
21. Wang, G.J.; Sutton, A.P.; Vítek, V. A computer simulation study of <001> and <111> tilt boundaries: The multiplicity of structures. *Acta Metall.* **1984**, *32*, 1093. [CrossRef]
22. Wetzel, J.T.; Machlin, E.S. On calculated energies of segregation, grain boundary energies and lattice energy functions. *Scr. Metall.* **1983**, *17*, 555. [CrossRef]
23. Chen, P.; Srolovitz, D.J.; Voter, A.F. Computer simulation on surfaces and [001] symmetric tilt grain boundaries in Ni, Al, and Ni$_3$Al. *J. Mater. Res.* **1989**, *4*, 62. [CrossRef]
24. Kresse, G.; Hafner, J. Ab initio molecular dynamics for open-shell transition metals. *Phys. Rev. B* **1993**, *48*, 13115. [CrossRef]
25. Kresse, G.; Furthmüller, J. Efficient iterative schemes for ab initio total-energy calculations using a plane-wave basis set. *Phys. Rev. B* **1996**, *54*, 11169. [CrossRef]
26. Kresse, G.; Furthmüller, J. Efficiency of ab-initio total energy calculations for metals and semiconductors using a plane-wave basis set. *Comput. Mater. Sci.* **1996**, *6*, 15. [CrossRef]
27. Blöchl, P.E. Projector augmented-wave method. *Phys. Rev. B* **1994**, *50*, 17953. [CrossRef] [PubMed]
28. Kresse, G.; Joubert, D. From ultrasoft pseudopotentials to the projector augmented-wave method. *Phys. Rev. B* **1999**, *59*, 1758. [CrossRef]
29. Perdew, J.P.; Burke, K.; Ernzerhof, M. Generalized gradient approximation made simple. *Phys. Rev. Lett.* **1996**, *77*, 3865. [CrossRef]
30. Haglund, J.; Guillermet, F.; Grimvall, G.; Korling, M. Theory of bonding in transition-metal carbides and nitrides. *Phys. Rev. B* **1993**, *48*, 11685. [CrossRef]
31. Kohlhaas, R.; Donner, P.; Schmitz-Pranghe, N.; Angew, Z. The temperature-dependence of the lattice parameters of iron, cobalt, and nickel in the high temperature range. *Physics* **1967**, *23*, 245.
32. Všianská, M. Electronic Structure and Properties of Grain Boundaries in Nickel. Ph.D. Thesis, Masaryk University, Brno, Czech Republic, 2013.
33. Saal, J.E.; Kirklin, S.; Aykol, M.; Meredig, B.; Wolverton, C. The open quantum materials database (OQMD): Materials design and discovery with high-throughput density functional theory. *J. Miner. Met. Mater. Soc.* **2013**, *65*, 1501. [CrossRef]
34. Kirklin, S.; Saal, J.E.; Meredig, B.; Thompson, A.; Doak, J.W.; Aykol, M.; Rühl, S.; Wolverton, C. The open quantum materials database (OQMD): Assessing the accuracy of DFT formation energies. *Comput. Mater.* **2015**, *1*, 15010. [CrossRef]
35. Connétable, D.; Andrieu, E.; Monceau, D. First-principles nickel database: Energetics of impurities and defects. *Comput. Mater. Sci.* **2015**, *101*, 77. [CrossRef]
36. Kandaskalov, D.; Monceau, D.; Mijoule, C.; Connétable, D. First-principles study of sulfur multi-absorption in nickel and its segregation to the Ni(100) and Ni(111) surfaces. *Surf. Sci.* **2013**, *617*, 15. [CrossRef]
37. Korhonen, T.; Puska, M.J.; Nieminen, R.M. Vacancy-formation energies for fcc and bcc transition metals. *Phys. Rev. B* **1995**, *51*, 9526. [CrossRef] [PubMed]
38. Wolff, J.; Franz, M.; Kluin, J.E.; Schmid, D. Vacancy formation in nickel and α-nickel-carbon alloy. *Acta Mater.* **1997**, *45*, 4759. [CrossRef]

39. Connétable, D.; Ter-Ovanessian, B.; Andrieu, E. Diffusion and segregation of niobium in fcc-nickel. *J. Phys. Condens. Matter* **2012**, *24*, 095010. [CrossRef]
40. Nazarov, R.; Hickel, T.; Neugebauer, J. Ab initio study of H-vacancy interactions in fcc metals: Implications for the formation of superabundant vacancies. *Phys. Rev. B* **2014**, *89*, 144108. [CrossRef]
41. Tanguy, D.; Wang, Y.; Connétable, D. Stability of vacancy-hydrogen clusters in nickel from first-principles calculations. *Acta Mater.* **2014**, *78*, 135. [CrossRef]
42. Subashiev, A.V.; Nee, H.H. Hydrogen trapping at divacancies and impurity-vacancy complexes in nickel: First principles study. *J. Nucl. Mater.* **2017**, *487*, 135. [CrossRef]
43. Kostromin, B.F.; Plishkin, Y.M.; Podchinyonov, I.E.; Trakhtenberg, I.S. Detecting the relation between the diffusion parameters and the point defect microscopic characteristics by a computer simulation method. *Fiz. Met. Metalloved.* **1983**, *55*, 450.
44. Ackland, G.I.; Tichy, G.; Vitek, V.; Finnis, M.W. Simple N-body potentials for the noble metals and nickel. *Philos. Mag. A* **1987**, *56*, 735. [CrossRef]
45. Krause, U.; Kuska, J.P.; Wedell, R. Monovacancy formation energies in cubic crystals. *Phys. Status Solidi* **1989**, *151*, 479. [CrossRef]
46. Rosato, V.; Guillope, M.; Legrand, B. Thermodynamical and structural properties of f.c.c. transition metals using a simple tight-binding model. *Philos. Mag. A* **1989**, *59*, 321. [CrossRef]
47. Ghorai, A. Calculation of some defect parameters in F.C.C. Metals. *Phys. Status Solidi* **1991**, *167*, 551. [CrossRef]
48. Černý, M.; Šesták, P.; Řehák, P.; Všianská, M.; Šob, M. Atomistic approaches to cleavage of interfaces. *Model. Simul. Mater. Sci. Eng.* **2019**, *27*, 035007. [CrossRef]
49. Čák, M.; Šob, M.; Hafner, J. First-principles study of magnetism at grain boundaries in iron and nickel. *Phys. Rev. B* **2008**, *78*, 054418. [CrossRef]
50. Yamaguchi, M.; Shiga, M.; Kaburaki, H. Energetics of segregation and embrittling potency for non-transition elements in the Ni Σ5 (012) symmetrical tilt grain boundary: A first-principles study. *J. Phys. Condens. Matter* **2004**, *16*, 3933. [CrossRef]
51. Shvindlerman, L.S.; Gottstein, G. Cornerstones of grain structure evolution and stability: Vacancies, boundaries, triple junctions. *J. Mater. Sci.* **2005**, *40*, 819. [CrossRef]
52. Shvindlerman, L.S.; Gottstein, G.; Ivanov, V.A.; Molodov, D.A.; Kolesnikov, D.; Lojkowski, W. Grain boundary excess free volume—direct thermodynamic measurement. *J. Mater. Sci.* **2006**, *41*, 7725. [CrossRef]
53. Berthier, F.; Legrand, B.; Tréglia, G. New structures and atomistic analysis of the polymorphism for the Σ5(210) [001] Tilt Boundary. *Interface Sci.* **2000**, *8*, 55. [CrossRef]
54. Rajagopalan, M.; Tschopp, M.A.; Solanki, K.N. Grain boundary segregation of interstitial and substitutional impurity atoms in alpha-iron. *J. Miner. Met. Mater. Soc.* **2014**, *66*, 129. [CrossRef]
55. Yamaguchi, M.; Shiga, M.; Kaburaki, H. Grain boundary decohesion by impurity segregation in a nickel-sulfur system. *Science* **2005**, *307*, 393. [CrossRef] [PubMed]
56. Všianská, M.; Šob, M. Magnetically dead layers at sp-impurity-decorated grain boundaries and surfaces in nickel. *Phys. Rev. B* **2011**, *84*, 014418. [CrossRef]
57. Turek, I.; Kudrnovský, J.; Šob, M.; Drchal, V.; Weinberger, P. Ferromagnetism of imperfect ultrathin Ru and Rh films on a Ag (001) substrate. *Phys. Rev. Lett.* **1995**, *74*, 2551. [CrossRef] [PubMed]
58. Kisker, H.; Kronmüller, H.; Schaefer, H.E.; Suzuki, T. Magnetism and microstructure of nanocrystalline nickel. *J. Appl. Phys.* **1996**, *79*, 5143. [CrossRef]
59. Sulițanu, B.; Brînză, F. Structure-properties relationships in electrodeposited Ni-W thin films with columnar nanocrystallites. *J. Optoelectron. Adv. Mater.* **2003**, *5*, 421.
60. Tanimoto, H.; Pasquini, L.; Prümmer, R.; Kronmüller, H.; Schaefer, H.E. Self-diffusion and magnetic properties in explosion densities nanocrystalline Fe. *Scr. Mater.* **2000**, *42*, 961. [CrossRef]
61. Kecskes, L.; Qiu, X.; Lin, R.; Graeter, J.; Guo, S.M.; Wang, J. Combustion synthesis reaction behavior of cold-rolled Ni/Al and Ti/Al multilayers. In *Army Research Laboratory Report ARL-TR-5507*; U.S. Army Research Research Laboratory: Aberdeen Proving Ground, MD, USA, 2011.
62. Nash, P.; Nash, A. The Ni–Si (Nickel-Silicon) system. *Bull. Alloy Phase Diagr.* **1987**, *8*, 6. [CrossRef]
63. Lejček, P.; Šob, M.; Paidar, V.; Vitek, V. Why calculated energies of grain boundary segregation are unreliable when segregant solubility is low. *Scr. Mater.* **2013**, *68*, 547. [CrossRef]

64. Farkas, D. Atomistic theory and computer simulation of grain boundary structure and diffusion. *J. Phys. Condens. Matter* **2000**, *12*, 497. [CrossRef]
65. Sørensen, M.R.; Mishin, Y.; Voter, A.F. Diffusion mechanisms in Cu grain boundaries. *Phys. Rev. B* **2000**, *62*, 3658. [CrossRef]
66. Brokman, A.; Bristove, P.D.; Ballufi, R.W. Computer simulation study of the structure of vacancies in grain boundaries. *J. Appl. Phys.* **1981**, *52*, 6116. [CrossRef]
67. Gillan, M.J. Calculation of the vacancy formation energy in aluminium. *J. Phys. Condens. Matter* **1989**, *1*, 689. [CrossRef]
68. Mattsson, T.R.; Mattsson, A.E. Calculating the vacancy formation energy in metals: Pt, Pd, and Mo. *Phys. Rev. B* **2002**, *66*, 214110. [CrossRef]
69. Zhang, P.; Zou, T.; Zhao, J.; Zheng, P.; Chen, J. Effect of helium and vacancies in a vanadium grain boundary by first-principles. *Nucl. Instrum. Methods Phys. Res. Sect. B* **2015**, *352*, 121. [CrossRef]
70. Janot, C.; George, B.; Delcroix, P. Point defects in vanadium investigated by Mossbauer spectroscopy and positron annihilation. *J. Phys. F Met. Phys.* **1982**, *12*, 47. [CrossRef]
71. Fitzsimmons, M.R.; Röll, A.; Burkel, E.; Sickafus, K.E.; Nastasi, M.A.; Smith, G.S.; Pynn, R. The magnetization of a grain boundary in nickel. *Nanostructured Mater.* **1995**, *6*, 539. [CrossRef]
72. Hirayama, K.; Ii, S.; Tsurekawa, S. Transmission electron microscopy/electron energy loss spectroscopy measurements and ab initio calculation of local magnetic moments at nickel grain boundaries. *Sci. Technol. Adv. Mater.* **2014**, *15*, 015005. [CrossRef]
73. Crampin, S.; Vvedensky, D.D.; MacLaren, J.M.; Eberhart, M.E. Electronic structure near (210) tilt boundaries in nickel. *Phys. Rev. B* **1989**, *40*, 3413. [CrossRef]
74. Geng, W.T.; Freeman, A.J.; Wu, R.; Geller, C.B.; Raynolds, J.E. Embrittling and strengthening effects of hydrogen, boron, and phosphorus on a 5 nickel grain boundary. *Phys. Rev. B* **1999**, *60*, 7149. [CrossRef]
75. Szpunar, B.; Erb, U.; Palumbo, G.; Aust, K.T.; Lewis, L.J. Magnetism in complex atomic structures: Grain boundaries in nickel. *Phys. Rev. B* **1996**, *53*, 5547. [CrossRef]
76. Kuriplach, J.; Melikhova, O.; Hou, M.; Van Petegem, S.; Zhurkin, E.; Šob, M. Positron annihilation in vacancies at grain boundaries in metals. *Appl. Surf. Sci.* **2008**, *255*, 128. [CrossRef]
77. Kuriplach, J.; Melikhova, O.; Hou, M.; Van Petegem, S.; Zhurkin, E.; Šob, M. Positron annihilation at grain boundaries in metals. *Phys. Status Solidi C* **2007**, *4*, 3461. [CrossRef]
78. Černý, M.; Šesták, P.; Řehák, P.; Všianská, M.; Šob, M. Ab initio tensile tests of grain boundaries in the fcc crystals of Ni and Co with segregated sp-impurities. *Mater. Sci. Eng. A Struct. Mater. Prop. Microstruct. Process.* **2016**, *669*, 218. [CrossRef]
79. Řehák, P.; Černý, M.; Šob, M. Mechanical stability of Ni and Ir under hydrostatic and uniaxial loading. *Model. Simul. Mater. Sci. Eng.* **2015**, *23*, 055010. [CrossRef]
80. Xiao, W.; Liu, C.S.; Tian, Z.X.; Geng, W.T. Effect of applied stress on vacancy segregation near the grain boundary in nickel. *J. Appl. Phys.* **2008**, *104*, 053519. [CrossRef]
81. Lejček, P.; Hofmann, S.; Paidar, V. The significance of entropy in grain boundary segregation. *Materials* **2019**, *12*, 492. [CrossRef]

© 2020 by the authors. Licensee MDPI, Basel, Switzerland. This article is an open access article distributed under the terms and conditions of the Creative Commons Attribution (CC BY) license (http://creativecommons.org/licenses/by/4.0/).

Communication

Generalized Stacking Fault Energy of Al-Doped CrMnFeCoNi High-Entropy Alloy

Xun Sun [1,2], Hualei Zhang [2,*], Wei Li [1], Xiangdong Ding [2], Yunzhi Wang [3] and Levente Vitos [1,4,5]

1. Applied Materials Physics, Department of Materials Science and Engineering, Royal Institute of Technology, SE-100 44 Stockholm, Sweden; xunsun@kth.se (X.S.); wei2@kth.se (W.L.); levente@kth.se (L.V.)
2. Frontier Institute of Science and Technology, State Key Laboratory for Mechanical Behavior of Materials, Xi'an Jiaotong University, Xi'an 710049, China; dingxd@mail.xjtu.edu.cn
3. Department of Materials Science and Engineering, The Ohio State University, 2041 College Road, Columbus, OH 43210, USA; wang.363@osu.edu
4. Division of Materials Theory, Department of Physics and Materials Science, Uppsala University, P.O. Box 516, SE-75120 Uppsala, Sweden
5. Research Institute for Solid State Physics and Optics, Wigner Research Center for Physics, P.O. Box 49, H-1525 Budapest, Hungary
* Correspondence: hualei@xjtu.edu.cn

Received: 12 December 2019; Accepted: 25 December 2019; Published: 26 December 2019

Abstract: Using first-principles methods, we investigate the effect of Al on the generalized stacking fault energy of face-centered cubic (fcc) CrMnFeCoNi high-entropy alloy as a function of temperature. Upon Al addition or temperature increase, the intrinsic and extrinsic stacking fault energies increase, whereas the unstable stacking fault and unstable twinning fault energies decrease monotonously. The thermodynamic expression for the intrinsic stacking fault energy in combination with the theoretical Gibbs energy difference between the hexagonal close packed (hcp) and fcc lattices allows one to determine the so-called hcp-fcc interfacial energy. The results show that the interfacial energy is small and only weakly dependent on temperature and Al content. Two parameters are adopted to measure the nano-twinning ability of the present high-entropy alloys (HEAs). Both measures indicate that the twinability decreases with increasing temperature or Al content. The present study provides systematic theoretical plasticity parameters for modeling and designing high entropy alloys with specific mechanical properties.

Keywords: high-entropy alloys; generalized stacking fault energy; first-principles; interfacial energy

---

1. Introduction

High-entropy alloys (HEAs) have attracted significant attention in recent years [1–7]. Excellent combination of strength-ductility properties is one of the great advantages of the face-cubic centered (fcc) HEAs [1,2], which is usually attributed to the deformation twins [1,8].

In conventional alloys, it is often a challenge to improve the strength and ductility at the same time. Usually higher strength is achieved by sacrificing ductility and vice versa. On the other hand, the deformation twinning mechanism can be used to overcome the strength-ductility trade-off. The deformation twins are created by the dislocation gliding in the slip systems under external stress. The newly created twin boundaries hinder the dislocation motion, resulting in an increased work hardening rate ("dynamic Hall-Petch effect"). At the same time, twinning maintains the elongation of alloys during work hardening by delaying the onset of plastic instability by necking [9].

The generalized stacking fault energy (GSFE) plays an important role in understanding the deformation mechanism of fcc alloys [10–12]. There are four important parameters of GSFE corresponding to the first four extrema on the energy versus slip vector curve: the unstable stacking fault energy (γ_{usf}), the intrinsic stacking fault energy (γ_{isf}), the unstable twin fault energy (γ_{utf}), and the extrinsic stacking fault energy (γ_{esf}). γ_{isf} is the most widely used parameter to predict the twinning ability, twinning stress [10], and phase stability [13]. Classical theories generally predict that the critical twinning stress (τ_{crit}) is proportional to the intrinsic stacking fault energy γ_{isf} [10], i.e., $\tau_{crit} \cdot b_{112} \sim \gamma_{isf}$, where b_{112} is the Burgers vector of partial dislocation. A lower γ_{isf} suggests that deformation twins are easier to form, and the twinning stress is relatively low. For example, the twinning easily happens in Cu but not in Al, which is attributed to the lower γ_{isf} of Cu (45 mJ/m^2) than that of Al (122 mJ/m^2) [14]. The measured γ_{isf} of many medium-entropy alloys (MEAs) and HEAs are as low as (or even lower than) those obtained for the twinning-induced plasticity (TWIP) steels [10], which is consistent with the large amount of nano-twins observed in these systems. γ_{isf} of CrMnFeCoNi HEA is 25–35 mJ/m^2 [9] and 22–31 mJ/m^2 [15] measured at room temperature by different works. The deformation twins are easily found at cryogenic temperature in CrMnFeCoNi HEA [1], indicating that γ_{isf} decreases with lowering temperature. γ_{isf} of MEAs and HEAs were studied by theoretical methods as well. Both CrCoNi and CrMnFeCoNi have negative theoretical γ_{isf} [16–18], which is ascribed to the metastable character of these alloys [19].

2. Methodology

The ab initio calculations were performed using the exact muffin-tin orbitals (EMTO) method [20]. The Perdew–Burke–Ernzerhof (PBE) [21] exchange-correlation functional was adopted to perform the self-consistent and total energy calculations. The chemical and magnetic disorders were treated within the coherent-potential approximation (CPA) [22]. The paramagnetic (PM) state of Al$_y$(CrMnFeCoNi)$_{100-y}$ was modeled within the disordered local magnetic moment approach [23]. The EMTO-CPA method successfully described the lattice constants [24] and the elastic moduli [5] of Al-doped CrMnFeCoNi HEAs in our previous works.

According to the Mahajan–Chin model [25], the nucleation and propagation of deformation twins in fcc systems by shearing successive {111} planes along the <112> direction [26], as shown in Figure 1. The GSFE was calculated by adopting a 9-layers supercells with and without one fault per unit cell [26]. Due to the periodic boundary condition used, the number of atomic layers needs to be large enough to prevent the influence of the interaction between the two adjacent stacking faults. The 9-layers supercell is proved to be accurate enough for the GSFEs [18]. This approach has been successfully applied in pure metals, binary alloys, and HEAs in previous studies [27,28]. The GSFE was calculated as $\gamma_{GSFE} = (F^{fault} - F^0)/A$, where F^{fault} and F^0 are the free energies of supercell with and without the fault, respectively, and A is the area. The free energy is approximated as $F = E - TS_{mag}$, where E is the total energy and T is the temperature. Within the mean-field approximation, the magnetic entropy is $S_{mag} = k_B \sum_{i=1}^{6} c_i \ln(1 + \mu_i)$, where k_B is the Boltzmann constant, c_i is the concentration, and μ_i the local magnetic moment of the ith alloying element, respectively. The total energy E at each temperature and Al concentration was calculated at the corresponding lattice constant. We started from the experimental lattice constants of fcc Al$_y$(CrMnFeCoNi)$_{100-y}$ (y = 0, 2, 4, 6, 8) alloys at room temperature [29]. Then we used the coefficient of thermal expansion [30] of fcc CrMnFeCoNi alloy to evaluate the lattice constants of Al$_y$(CrMnFeCoNi)$_{100-y}$ alloys as a function of temperature, i.e., we assumed that Al addition has a negligible influence on the thermal expansion coefficient. This assumption is supported by the fact that the Debye temperature of Al-doped fcc CrMnFeCoNi varies little with the amount of Al [5]. Namely, the Debye temperature changes from 525 to 490 K [5] as the Al concentration increases from 0% to 8%. According to the quasi-harmonic Debye model [31], the corresponding change in the thermal expansion coefficient is less than 10% at 300 K, which leads to less than 0.05% uncertainty in the room-temperature lattice parameters.

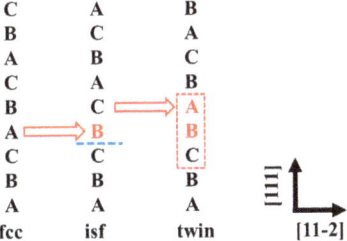

Figure 1. Schematic of planar fault path on {111} planes. Perfect face-cubic centered (fcc) bulk forms an intrinsic stacking fault by shearing one b_p along <112> direction. Extrinsic stacking fault (twin) formed by shearing one b_p along <112> direction in the adjacent layer of intrinsic stacking fault.

3. Results and Discussion

In Figure 2, we present the GSFE of paramagnetic fcc CrMnFeCoNi at room temperature. For γ_{usf}, γ_{isf}, γ_{utf}, and γ_{esf} our predictions give 285, −6, 281, and 3 mJ/m^2, respectively. We observe that γ_{isf} is negative, and γ_{esf} is also very small. On the other hand, the energy barriers γ_{usf} and γ_{utf} are relatively large as compared to the energy barriers obtained for pure metals with low γ_{isf} [27], such as Cu, Au, and Ag. We find that the present results of GSFE satisfy with a good accuracy the universal scaling law [32], i.e., $\gamma_{usf} \simeq \gamma_{utf} - \frac{1}{2}\gamma_{isf}$. It is interesting that although γ_{isf} is negative, the universal scaling law remains valid, which suggests that our results are reasonable. Similar negative γ_{isf} was reported in previous works. For instance, Huang et al. [18] used a similar approach as the one adopted here and obtained −7 mJ/m^2 for γ_{isf} of PM fcc CrMnFeCoNi at room temperature. On the other hand, the present γ_{isf} at 300 K is smaller than the former EMTO result (21 mJ/m^2) reported by Huang et al. [28]. The relatively large deviation between the two sets of data should be ascribed to the fact that Huang et al. [28] employed the local density appropriation (LDA) instead of the PBE functional adopted here and used an experimental lattice parameter of 3.6 Å compared to 3.59 Å [29] considered here. Furthermore, here we neglect the positive strain contribution to γ_{isf}, which was considered by Huang et al. [28].

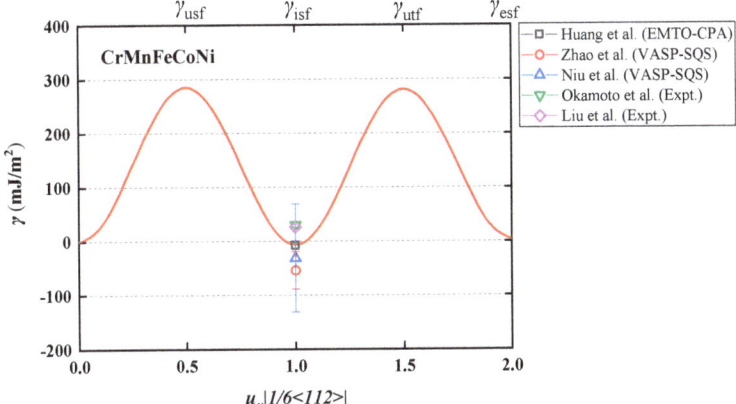

Figure 2. Theoretical generalized stacking fault energy (γ, in mJ/m^2) of paramagnetic (PM) fcc CrMnFeCoNi alloy calculated at room temperature. The theoretical (exact muffin-tin orbitals-coherent-potential approximation (EMTO-CPA), Vienna ab initio simulation package-special quasi-random structure (VASP-SQS)) [18,33,34] and experimental (Expt.) [9,15] γ_{isf} are plotted for comparison.

At 0 K and for the ferromagnetic (collinear) state, calculations based on the Vienna ab initio simulation package (VASP) combined with the special quasi-random structure (SQS) approach for the intrinsic stacking fault energy of CrMnFeCoNi gave average values of −54 [33] and −31 mJ/m² [34] with scatter of about ±35 and ±100 mJ/m², respectively. The deviations between the present result and the above VASP values [33,34] can be attributed to the different magnetic states and different alloy theories adopted in those calculations.

In Figure 3, we present the calculated GSFE for the PM fcc $Al_y(CrMnFeCoNi)_{100-y}$ (y = 0, 2, 4, 6, 8) alloys as a function of temperature and composition. In the considered temperature and Al concentration range, both γ_{isf} and γ_{esf} increase monotonously with increasing temperature and Al content. At the same time, γ_{usf} and γ_{utf} decrease with increasing temperature and Al addition. It is found that temperature has a large effect on the GSFE. The values of γ_{isf} and γ_{esf} for the Al-free alloy are negative when the temperature is below 400 and 200 K, respectively, as shown in Figure 3. We notice that the present temperature dependence of γ_{isf} follows closely the one predicted by Huang et al. [28] in spite of the methodological differences discussed above.

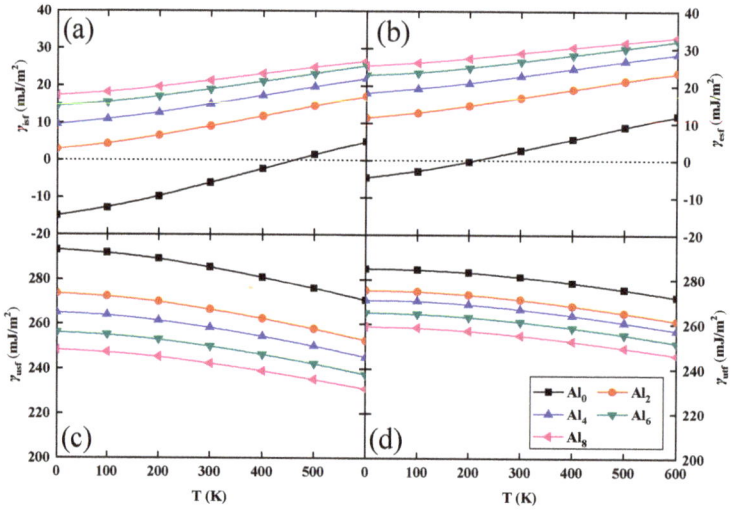

Figure 3. Theoretical generalized stacking fault energy (γ, in mJ/m²) of PM fcc $Al_y(CrMnFeCoNi)_{100-y}$ (y = 0, 2, 4, 6, 8) alloys as a function of temperature and composition. (**a**) The intrinsic stacking fault energy (γ_{isf}, in mJ/m²); (**b**) The extrinsic stacking fault energy (γ_{esf}, in mJ/m²); (**c**) The unstable stacking fault energy (γ_{usf}, in mJ/m²); (**d**) The unstable twinning fault energy (γ_{utf}, in mJ/m²).

For elemental metals and homogeneous solid solutions, the intrinsic stacking fault energy can be approximated by the energy difference between the hexagonal close packed (hcp) and fcc lattices, viz. $\gamma_{isf} = \frac{2(F^{hcp}-F^{fcc})}{A} + 2\sigma$ [13,35], where F_{hcp} and F_{fcc} are the free energies per atom for the hcp and fcc phase, respectively. The last term is the interfacial contribution describing the transition zone between the fcc matrix and the hcp embryo. The interfacial energy σ was estimated to be of the order of 7 mJ/m² [36] for the present Al-free CrMnFeCoNi HEA. In Figure 4a, we compared γ_{isf} and the stacking fault energy γ_0 obtained merely from the structural energy difference, viz. $\gamma_0 = \frac{2(F^{hcp}-F^{fcc})}{A}$. We find γ_0 is very close to γ_{isf} for all Al concentrations and temperature. Thus, the free energy difference between hcp and fcc can reflect the value of γ_{isf}. The small difference between γ_{isf} and γ_0 is equal to the double of the interfacial energy σ. In Figure 4b, we also plotted the interfacial energy $\sigma = (\gamma_{isf} - \gamma_0)/2$ as a function of temperature. We find that σ slightly decreases with increasing temperature and Al content, but it remains in the range of 4–7 mJ/m². Thus, at least for the present alloy family, the composition and temperature dependence of σ is rather weak and could safely be omitted. Similar observation was made by Dong et al. using calculations based on floating spin and longitudinal spin fluctuations schemes [37].

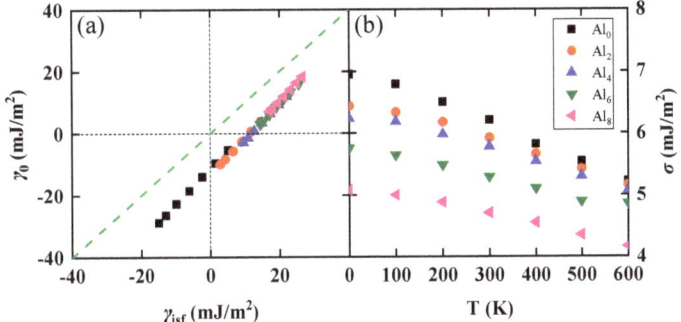

Figure 4. (a) Comparison between γ_{isf} and γ_0 (in mJ/m^2) of PM fcc Al$_y$(CrMnFeCoNi)$_{100-y}$ ($y = 0$, 2, 4, 6, 8) alloys. Symbols correspond to temperatures between 0 and 600 K (from left to right) with increments of 100 K. (b) The interfacial energy σ (in mJ/m^2) is plotted as a function of temperature and composition.

Previous theoretical study on the PM CrMnFeCoNi system [38] found that the hcp structure is more stable than the fcc structure below 370 K, which means that the negative values of γ_{isf} and γ_{esf} shown in Figure 3 are reasonable. Similar to our findings, first-principles calculations [19] discovered that the negative γ_{isf} in fcc CrCoNi and CrFeCoNi alloys originate from the thermodynamic stability of the hcp phase at low temperatures. Recently, Zhang et al. [39] confirmed that the hcp phase is more stable thermodynamically than the fcc one at relatively lower temperatures, agreeing well with the theoretical results [38].

Despite the fact that low γ_{isf} generally improves the twinning ability of alloys, the combined effects of all energy parameters determining the GSFE should be considered when studying the twinning affinity. That is because the intrinsic energy parameters in Figure 3 exhibit complex temperature and alloying trends. To describe the twinning ability in the Al$_y$CrMnFeCoNi system, we adopt two twinning ability parameters. The parameter T_{tw} proposed by Asaro et al. [40] is defined as

$$T_{tw} = \sqrt{(3\gamma_{usf} - 2\gamma_{isf})/\gamma_{utf}} \tag{1}$$

when $T_{tw} > 1$ a twin is more favorable than the dislocation slip and vice versa. We plot T_{tw} as a function of temperature and Al content in Figure 5a. We find that all alloys considered here have good twinning ability within the entire temperature range. All T_{tw} decrease with increasing temperature and Al content, suggesting that the Al addition and increasing temperature decrease the ability of twinning. However, the twinning parameter remains far above 1, meaning that the present alloys remain prone to twinning even when the Al level comes close to the solubility limit within the fcc phase and temperature increases up to 600 K. A second twinning parameter was introduced by Jo et al. [41] as

$$r_d = \gamma_{isf}/(\gamma_{usf} - \gamma_{isf}) \tag{2}$$

In terms of this parameter, one can distinguish four regimes corresponding to different deformation mechanisms. Namely, for $r_d < -0.5$ we have stacking fault only, for $-0.5 < r_d < 0$ both stacking fault and full slip can be realized, for $0 < r_d < 2$ full slip is combined with twinning, and for $r_d > 2$ we have full slip only. Within the range of $0 < r_d < 2$, $r_d = 0$ corresponds to the maximum twinning ability. As shown in Figure 5b, all r_d values are much lower than 2, indicating a strong twinning ability. We find that r_d increases with increasing Al content and temperature, meaning that the twinning ability decreases, which is fully consistent with the prediction from T_{tw}. There is one exceptional case that the negative r_d in Al-free CrMnFeCoNi HEA below 400 K predict that only stacking fault and full slip will happen. However, plenty of deformation twins are found in the CrMnFeCoNi HEA [1,42]. The current twinning parameter cannot explain this phenomenon. Further theory is needed to resolve this question. Byun formula would suggest that the critical stress for twinning also increases and thus it is unclear whether twinning can indeed be realized at elevated temperatures. That depends crucially on the alloy preparation, micro-structure, and strain rate. Describing these effects is a very complex problem and calls for further advanced models built among others on the presently disclosed intrinsic energy parameters.

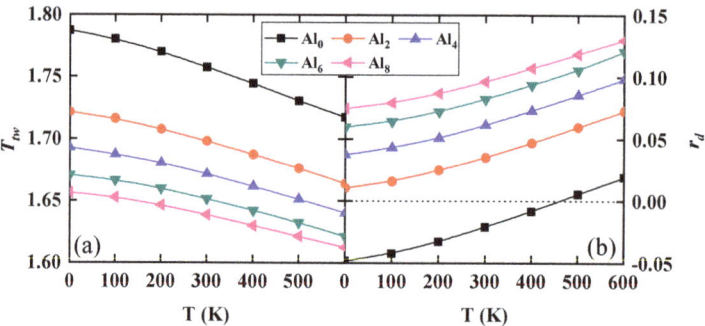

Figure 5. The effect of Al on the twinning ability of PM fcc $Al_y(CrMnFeCoNi)_{100-y}$ (y = 0, 2, 4, 6, 8) alloys as a function of temperature and composition. Shown are (**a**) T_{tw} and (**b**) r_d as a function of temperature and composition.

4. Conclusions

Using first-principles alloy theory formulated within the EMTO method, we have calculated the GSFE of paramagnetic fcc $Al_y(CrMnFeCoNi)_{100-y}$ (y = 0, 2, 4, 6, 8) alloys as a function of temperature. The present theoretical results show that the GSFE can be tuned by adding Al or changing the temperature. In particular, the intrinsic and extrinsic stacking fault energies increase, whereas the unstable stacking and twinning fault energies decrease with increasing temperature and Al doping. The thermodynamic phase stability can reflect γ_{isf} accurately due to the fact that γ_0 is very close to γ_{isf}. The interfacial energy σ slightly decreases with increasing temperature and Al content, but the change within the present composition-temperature interval always remains below ~30% compared to its mean value. Furthermore, from two parameters for twinning ability, it is predicted that Al addition and temperature increase cause a small decrease of the ability of twinning, but the alloys still remain prone to twinning even at the largest temperature and the highest Al-level considered here. The present theoretical data is expected to serve as input for modeling and design of new HEAs with desired mechanical properties.

Author Contributions: X.S., H.Z. and L.V initiated the study. X.S. and W.L. performed the calculations. X.S. and L.V. analyzed the results. X.S., H.Z. and L.V. wrote the manuscript and all the authors comment on the manuscript. All authors have read and agree to the published version of the manuscript. All authors have read and agreed to the published version of the manuscript.

Funding: This research was funded by The National Natural Science Foundation of China (No. 51871175), the National Key Research and Development Program of China (No. 2016YFB0701302), the Swedish Research Council (VR), the Swedish Foundation for Strategic Research (SSF), the Swedish Foundation for International Cooperation in Research and Higher Education (STINT), the Carl Tryggers Foundations, the Swedish Governmental Agency for Innovation Systems (VINNOVA), the China Scholarship Council, and the Hungarian Scientific Research Fund (research project OTKA 128229).

Acknowledgments: The computations were performed on resources provided by the Swedish National Infrastructure for Computing (SNIC) at Linköping, and the "H2" High Performance Cluster at Xi'an, China. Y. W. acknowledges the financial support of NSF under Grant DMR-1534826.

Conflicts of Interest: The authors declare no conflict of interest.

References

1. Gludovatz, B.; Hohenwarter, A.; Catoor, D.; Chang, E.H.; George, E.P.; Ritchie, R.O. A fracture-resistant high-entropy alloy for cryogenic applications. *Science* **2014**, *345*, 1153–1158. [CrossRef] [PubMed]
2. Li, Z.; Pradeep, K.G.; Deng, Y.; Raabe, D.; Tasan, C.C. Metastable high-entropy dual-phase alloys overcome the strength-ductility trade-off. *Nature* **2016**, *534*, 227–230. [CrossRef] [PubMed]
3. Jiang, S.; Wang, H.; Wu, Y.; Liu, X.; Chen, H.; Yao, M.; Gault, B.; Ponge, D.; Raabe, D.; Hirata, A.; et al. Ultrastrong steel via minimal lattice misfit and high-density nanoprecipitation. *Nature* **2017**, *544*, 460–464. [CrossRef]
4. Lei, Z.; Liu, X.; Wu, Y.; Wang, H.; Jiang, S.; Wang, S.; Hui, X.; Wu, Y.; Gault, B.; Kontis, P.; et al. Enhanced strength and ductility in a high-entropy alloy via ordered oxygen complexes. *Nature* **2018**, *563*, 546–550. [CrossRef] [PubMed]
5. Zhang, H.; Sun, X.; Lu, S.; Dong, Z.; Ding, X.; Wang, Y.; Vitos, L. Elastic properties of $Al_xCrMnFeCoNi$ ($0 \leq x \leq 5$) high-entropy alloys from ab initio theory. *Acta Mater.* **2018**, *155*, 12–22. [CrossRef]
6. Qin, G.; Chen, R.; Liaw, P.K.; Gao, Y.; Li, X.; Zheng, H.; Wang, L.; Su, Y.; Guo, J.; Fu, H. A novel face-centered-cubic high-entropy alloy strengthened by nanoscale precipitates. *Scr. Mater.* **2019**, *172*, 51–55. [CrossRef]
7. Ding, Q.; Zhang, Y.; Chen, X.; Fu, X.; Chen, D.; Chen, S.; Gu, L.; Wei, F.; Bei, H.; Gao, Y.; et al. Tuning element distribution, structure and properties by composition in high-entropy alloys. *Nature* **2019**, *574*, 223–227. [CrossRef]
8. Wei, D.; Li, X.; Jiang, J.; Heng, W.; Koizumi, Y.; Choi, W.-M.; Lee, B.-J.; Kim, H.S.; Kato, H.; Chiba, A. Novel Co-rich high performance twinning-induced plasticity (TWIP) and transformation-induced plasticity (TRIP) high-entropy alloys. *Scr. Mater.* **2019**, *165*, 39–43. [CrossRef]
9. Okamoto, N.L.; Fujimoto, S.; Kambara, Y.; Kawamura, M.; Chen, Z.M.; Matsunoshita, H.; Tanaka, K.; Inui, H.; George, E.P. Size effect, critical resolved shear stress, stacking fault energy, and solid solution strengthening in the CrMnFeCoNi high-entropy alloy. *Sci. Rep.* **2016**, *6*, 35863. [CrossRef]
10. De Cooman, B.C.; Estrin, Y.; Kim, S.K. Twinning-induced plasticity (TWIP) steels. *Acta Mater.* **2018**, *142*, 283–362. [CrossRef]
11. Wang, Y.; Liu, B.; Yan, K.; Wang, M.; Kabra, S.; Chiu, Y.-L.; Dye, D.; Lee, P.D.; Liu, Y.; Cai, B. Probing deformation mechanisms of a FeCoCrNi high-entropy alloy at 293 and 77 K using in situ neutron diffraction. *Acta Mater.* **2018**, *154*, 79–89. [CrossRef]
12. Xu, X.D.; Liu, P.; Tang, Z.; Hirata, A.; Song, S.X.; Nieh, T.G.; Liaw, P.K.; Liu, C.T.; Chen, M.W. Transmission electron microscopy characterization of dislocation structure in a face-centered cubic high-entropy alloy Al0.1CoCrFeNi. *Acta Mater.* **2018**, *144*, 107–115. [CrossRef]
13. Li, R.; Lu, S.; Kim, D.; Schonecker, S.; Zhao, J.; Kwon, S.K.; Vitos, L. Stacking fault energy of face-centered cubic metals: Thermodynamic and ab initio approaches. *J. Phys. Condens. Matter* **2016**, *28*, 395001. [CrossRef] [PubMed]
14. Zhu, Y.T.; Liao, X.Z.; Srinivasan, S.G.; Lavernia, E.J. Nucleation of deformation twins in nanocrystalline face-centered-cubic metals processed by severe plastic deformation. *J. Appl. Phys.* **2005**, *98*. [CrossRef]
15. Liu, S.F.; Wu, Y.; Wang, H.T.; He, J.Y.; Liu, J.B.; Chen, C.X.; Liu, X.J.; Wang, H.; Lu, Z.P. Stacking fault energy of face-centered-cubic high entropy alloys. *Intermetallics* **2017**, *93*, 269–737. [CrossRef]
16. Ding, J.; Yu, Q.; Asta, M.; Ritchie, R.O. Tunable stacking fault energies by tailoring local chemical order in CrCoNi medium-entropy alloys. *Proc. Natl. Acad. Sci. USA* **2018**, *115*, 8919–8924. [CrossRef]
17. Zhang, Z.; Sheng, H.; Wang, Z.; Gludovatz, B.; Zhang, Z.; George, E.P.; Yu, Q.; Mao, S.X.; Ritchie, R.O. Dislocation mechanisms and 3D twin architectures generate exceptional strength-ductility-toughness combination in CrCoNi medium-entropy alloy. *Nat. Commun.* **2017**, *8*, 14390. [CrossRef]
18. Huang, H.; Li, X.; Dong, Z.; Li, W.; Huang, S.; Meng, D.; Lai, X.; Liu, T.; Zhu, S.; Vitos, L. Critical stress for twinning nucleation in CrCoNi-based medium and high entropy alloys. *Acta Mater.* **2018**, *149*, 388–396. [CrossRef]
19. Zhang, Y.H.; Zhuang, Y.; Hu, A.; Kai, J.J.; Liu, C.T. The origin of negative stacking fault energies and nano-twin formation in face-centered cubic high entropy alloys. *Scr. Mater.* **2017**, *130*, 96–99. [CrossRef]

20. Vitos, L. Total-energy method based on the exact muffin-tin orbitals theory. *Phys. Rev. B* **2001**, *64*. [CrossRef]
21. Perdew, J.P.; Burke, K.; Ernzerhof, M. Generalized gradient approximation made simple. *Phys. Rev. Lett.* **1996**, *77*, 3865–3868. [CrossRef] [PubMed]
22. Vitos, L. The EMTO Method and Applications. In *Computational Quantum Mechanicals for Materials Engineers*; Springer: London, UK, 2007.
23. Gyorffy, B.L.; Pindor, A.J.; Staunton, J.; Stocks, G.M.; Winter, H. A first-principles theory of ferromagnetic phase transitions in metals. *J. Phys. F Met. Phys.* **1985**, *15*, 1337. [CrossRef]
24. Sun, X.; Zhang, H.; Lu, S.; Ding, X.; Wang, Y.; Vitos, L. Phase selection rule for Al-doped CrMnFeCoNi high-entropy alloys from first-principles. *Acta Mater.* **2017**, *140*, 366–374. [CrossRef]
25. Mahajan, S.; Chin, G.Y. Formation of deformation twins in f.c.c. crystals. *Acta Metall.* **1973**, *21*, 1353–1363. [CrossRef]
26. Kibey, S.; Liu, J.B.; Johnson, D.D.; Sehitoglu, H. Predicting twinning stress in fcc metals: Linking twin-energy pathways to twin nucleation. *Acta Mater.* **2007**, *55*, 6843–6851. [CrossRef]
27. Li, W.; Lu, S.; Hu, Q.M.; Kwon, S.K.; Johansson, B.; Vitos, L. Generalized stacking fault energies of alloys. *J. Phys. Condens. Matter* **2014**, *26*, 265005. [CrossRef]
28. Huang, S.; Li, W.; Lu, S.; Tian, F.; Shen, J.; Holmström, E.; Vitos, L. Temperature dependent stacking fault energy of FeCrCoNiMn high entropy alloy. *Scr. Mater.* **2015**, *108*, 44–47. [CrossRef]
29. He, J.Y.; Liu, W.H.; Wang, H.; Wu, Y.; Liu, X.J.; Nieh, T.G.; Lu, Z.P. Effects of Al addition on structural evolution and tensile properties of the FeCoNiCrMn high-entropy alloy system. *Acta Mater.* **2014**, *62*, 105–113. [CrossRef]
30. Laplanche, G.; Gadaud, P.; Horst, O.; Otto, F.; Eggeler, G.; George, E.P. Temperature dependencies of the elastic moduli and thermal expansion coefficient of an equiatomic, single-phase CoCrFeMnNi high-entropy alloy. *J. Alloys Compd.* **2015**, *623*, 348–353. [CrossRef]
31. Blanco, M.A.; Francisco, E.; Luaña, V. GIBBS: Isothermal-isobaric thermodynamics of solids from energy curves using a quasi-harmonic Debye model. *Comput. Phys. Commun.* **2004**, *158*, 57–72. [CrossRef]
32. Jin, Z.H.; Dunham, S.T.; Gleiter, H.; Hahn, H.; Gumbsch, P. A universal scaling of planar fault energy barriers in face-centered cubic metals. *Scripta Mater.* **2011**, *64*, 605–608. [CrossRef]
33. Zhao, S.; Stocks, G.M.; Zhang, Y. Stacking fault energies of face-centered cubic concentrated solid solution alloys. *Acta Mater.* **2017**, *134*, 334–345. [CrossRef]
34. Niu, C.; LaRosa, C.R.; Miao, J.; Mills, M.J.; Ghazisaeidi, M. Magnetically-driven phase transformation strengthening in high entropy alloys. *Nat. Commun.* **2018**, *9*, 1363. [CrossRef] [PubMed]
35. Olson, G.B.; Cohen, M. A general mechanism of martensitic nucleation: Part I. General concepts and the FCC → HCP transformation. *Metall. Trans. A* **1976**, *7*, 1897–1904. [CrossRef]
36. Huang, S.; Huang, H.; Li, W.; Kim, D.; Lu, S.; Li, X.; Holmstrom, E.; Kwon, S.K.; Vitos, L. Twinning in metastable high-entropy alloys. *Nat. Commun.* **2018**, *9*, 2381. [CrossRef]
37. Dong, Z.; Schonecker, S.; Li, W.; Chen, D.; Vitos, L. Thermal spin fluctuations in CoCrFeMnNi high entropy alloy. *Sci. Rep.* **2018**, *8*, 12211. [CrossRef]
38. Ma, D.; Grabowski, B.; Körmann, F.; Neugebauer, J.; Raabe, D. Ab initio thermodynamics of the CoCrFeMnNi high entropy alloy: Importance of entropy contributions beyond the configurational one. *Acta Mater.* **2015**, *100*, 90–97. [CrossRef]
39. Zhang, F.; Wu, Y.; Lou, H.; Zeng, Z.; Prakapenka, V.B.; Greenberg, E.; Ren, Y.; Yan, J.; Okasinski, J.S.; Liu, X.; et al. Polymorphism in a high-entropy alloy. *Nat. Commun.* **2017**, *8*, 15687. [CrossRef]
40. Asaro, R.J.; Suresh, S. Mechanistic models for the activation volume and rate sensitivity in metals with nanocrystalline grains and nano-scale twins. *Acta Mater.* **2005**, *53*, 3369–3382. [CrossRef]
41. Jo, M.; Koo, Y.M.; Lee, B.J.; Johansson, B.; Vitos, L.; Kwon, S.K. Theory for plasticity of face-centered cubic metals. *Proc. Natl. Acad. Sci. USA* **2014**, *111*, 6560–6565. [CrossRef]
42. Kireeva, I.V.; Chumlyakov, Y.I.; Pobedennaya, Z.V.; Kuksgausen, I.V.; Karaman, I. Orientation dependence of twinning in single crystalline CoCrFeMnNi high-entropy alloy. *Mater. Sci. Eng. A* **2017**, *705*, 176–181. [CrossRef]

© 2019 by the authors. Licensee MDPI, Basel, Switzerland. This article is an open access article distributed under the terms and conditions of the Creative Commons Attribution (CC BY) license (http://creativecommons.org/licenses/by/4.0/).

Communication

Mixed-Solvent Polarity-Assisted Phase Transition of Cesium Lead Halide Perovskite Nanocrystals with Improved Stability at Room Temperature

Rui Yun, Li Luo *, Jingqi He, Jiaxi Wang, Xiaofen Li, Weiren Zhao, Zhaogang Nie and Zhiping Lin

School of Physics and Optoelectronic Engineering, Guangdong University of Technology, Guangzhou 510006, China; 13570980324@163.com (R.Y.); 13690103732@163.com (J.H.); 18635208773@163.com (J.W.); selfiemua@163.com (X.L.); zwrab@163.com (W.Z.); zgniegdut@163.com (Z.N.); zhipinglphy@gdut.edu.cn (Z.L.)
* Correspondence: luoli@gdut.edu.cn

Received: 16 September 2019; Accepted: 28 October 2019; Published: 30 October 2019

Abstract: Cesium lead halide perovskite nanocrystals (NCs) have attracted enormous interest in light-emitting diode, photodetector and low-threshold lasing application in terms of their unique optical and electrical performance. However, little attention has been paid to other structures associated with $CsPbBr_3$, such as $CsPb_2Br_5$. Herein, we realize a facile method to prepare dual-phase NCs with improved stability against polar solvents by replacing conventional oleylamine with cetyltrimethyl ammonium bromide (CTAB) in the reprecipitation process. The growth of NCs can be regulated with different ratios of toluene and ethanol depending on solvent polarity, which not only obtains NCs with different sizes and morphologies, but also controls phase transition between orthorhombic $CsPbBr_3$ and tetragonal $CsPb_2Br_5$. The photoluminescence (PL) and defect density calculated exhibit considerable solvent polarity dependence, which is ascribed to solvent polarity affecting the ability of CTAB to passivate surface defects and improve stoichiometry in the system. This new synthetic method of perovskite material will be helpful for further studies in the field of lighting and detectors.

Keywords: $CsPbBr_3$; $CsPb_2Br_5$; solvent polarity; CTAB; phase transition

---

1. Introduction

Cesium lead halide perovskite ($CsPbX_3$ X = Cl, Br, I) nanocrystals (NCs) have attracted enormous attention, having emerged as promising materials in the field of displays, lighting, lasing and photodetection [1,2]. A large number of studies on the thin-film, micro-structure and single crystals of these materials have been devoted to technological explorations for diverse applications [3,4], based on their outstanding optoelectronic performance, including ultrahigh photoluminescence quantum yield, narrow-band emission, flexible wavelength, high charge carrier mobilities and facile synthetic process [5,6]. In addition, other phases of cesium lead halide perovskite derivatives such as hexagonal Cs_4PbBr_6 and tetragonal $CsPb_2Br_5$ are observed in the form of quantum dots and so on [7,8], which possess structure-dependent physical properties and greatly expands the potential applications in sensing, catalysis, electro-chemistry and optoelectronics [9,10].

Recently, some reports indicated that $CsPb_2Br_5$ played an important role in improving the emission lifetime and stability of $CsPbBr_3$ and enhancing solar cell efficiency [11]. Even so, studies on controllable syntheses of $CsPb_2Br_5$ and the mechanism of photoluminescence are still not abundant, especially in the area of theoretical simulation that cannot match the experimental results [12]. Some studies insisted on the pure $CsPb_2Br_5$ microplates with a bandgap of 2.44 eV exhibited lasing emission under both one- and two-photon excitation [13]. Jiang's group indicated that the strong green emission is originated from coexisting phase CsPbBr$_3$ rather than CsPb$_2$Br$_5$ and concluded that CsPb$_2$Br$_5$ is an indirect bandgap semiconductor with a bandgap of 3.1 eV and has high nonradiative Auger recombination, indicating that no luminescence will be generated from the CsPb$_2$Br$_5$ [12]. Some investigations have reported that the emissive CsPb$_2$Br$_5$ is associated to the sub-bandgap defects such as Br vacancy or Pb and Cs vacancies [14–16], while other researchers have proposed that the lead bromide complex in CsPb$_2$Br$_5$ is the reason for the luminescence [17,18]. Therefore, the research of the luminescence mechanism on CsPb$_2$Br$_5$ cannot be neglected.

Cesium lead halide perovskite NCs can rapidly nucleate and grow during synthesis, which is assigned to its low formation energy and fast crystallization rate [19]. What's more, the crystalline phases are extremely sensitive to the ratios of the elements in the precursors, the post-processing and the film-formation [20]. Therefore, several strategies have been developed to control the composition, morphology and size of NCs in this rapid reaction by changing the experimental parameters. Jiang et al. found a phase transition from orthorhombic CsPbBr$_3$ to tetragonal CsPb$_2$Br$_5$ and a shape evolution from octagonal to square by controlling the reaction time [12]. Deng et al. prepared uniform CsPb$_2$Br$_5$ nanowires and nanosheets with superior stability and high yield by mediating the ligands at room temperature [21]. Sun et al. synthesized dual-phase CsPbBr$_3$-CsPb$_2$Br$_5$ composites at lower temperature and employed them as an emitting layer in LEDs, which exhibits a distinct improvement of about 21- and 18-fold in CE and EQE compared with reported CsPbBr$_3$ LEDs [11]. However, little work has been devoted to the investigation of solvent polarity. The major role of the solvent is affecting the charge transfer rates between NCs and the surface bonding structure [22]. This is the first report where a systematic study that the morphology, phase structure and PL of NCs have a regular change under variable polar conditions. Such a study would trigger a deeper consideration of solvent effects and provide a new direction for improving optoelectronic device performance.

In this work, we demonstrate a new approach to synthesize NCs with cetyltrimethyl ammonium bromide (CTAB) instead of oleylamine under variable polar conditions. The role of CTAB in the system is discussed and NCs synthesized with it show enhanced stability against ethanol. Subsequently, we have revealed the phase structure of NCs transit from orthorhombic CsPbBr$_3$ to tetragonal CsPb$_2$Br$_5$ with the solvent polarity increase and solvent polarity dependence of PL and defect density. Furthermore, the possible mechanism for NCs by combining ligand CTAB with mixed-solvent polarity is investigated. This would provide new guidance to modify the reprecipitation method.

2. Materials and Methods

2.1. Materials

PbBr$_2$ (Macklin, 99.0%, Shanghai, China), CsBr (Macklin, 99.5%), PbI$_2$ (Macklin, 98.0%), CsI (Macklin, 99.9%), PbCl$_2$ (Macklin, 99.5%), CsCl (Macklin, AR), cetyltrimethyl ammonium bromide (CTAB, Macklin, 99.0%), oleic acid (OA, Aladdin, AR, Shanghai, China), oleylamine (OAm, Aladdin, AR), dimethylformamide, (DMF, Aladdin, 99.8%), ethanol and toluene were purchased from Guangzhou Chemical Reagent Co. Ltd (Guangzhou, China) and were used directly without further purification.

2.2. Methods

2.2.1. Synthesis of CsPbBr$_3$/CsPb$_2$Br$_5$ NCs

In the typical synthesis of CsPbBr$_3$/CsPb$_2$Br$_5$ QDs, PbBr$_2$ (0.2 mmol) and CsBr (0.2 mmol) were dissolved in DMF (5 mL) at room temperature, OA (0.3 mL) and CTAB (0.05 g) were added to the constantly vigorous stirred DMF solution for 1h. Finally, 1 mL of mixed precursor solution was mixed in a new beaker with toluene (20 mL) under vigorous stirring for 20 min. The as-synthesized QDs were dispersed in toluene for further characterization.

2.2.2. Synthesis of CsPbBr$_3$/CsPb$_2$Br$_5$ NCs with Ethanol

Similar as the above CsPbBr$_3$/CsPb$_2$Br$_5$ NCs approach, 1 mL of mixed precursor solution was injected into a new beaker with a mixture (20 mL) of ethanol (E) and toluene (T) (E:T = 0.2, 0.3, 0.4, 0.5, 0.6, 0.7) under vigorous stirring for 20 min.

2.2.3. Synthesis of CsPbX$_3$/CsPb$_2$X$_5$ NCs

Similar as the above CsPbBr$_3$/CsPb$_2$Br$_5$ NCs approach, CsPb(Cl/Br)$_3$/CsPb$_2$(Cl/Br)$_5$ NCs were prepared by PbCl$_2$ (0.07 mmol), CsCl (0.07 mmol), PbBr$_2$ (0.10 mmol), CsBr (0.10 mmol), OA (0.30 mL) and CTAB (0.02 g); CsPb(Br/I)$_3$/CsPb$_2$(Br/I)$_5$ NCs were prepared by PbBr$_2$ (0.10 mmol), CsBr (0.10 mmol), PbI$_2$ (0.10 mmol), CsI (0.10 mmol), OA (0.30 mL) and CTAB (0.02 g).

2.2.4. Synthesis of CsPbBr$_3$/CsPb$_2$Br$_5$ NCs with OAm

Similar as the above CsPbBr$_3$/CsPb$_2$Br$_5$ NCs approach, the CTAB were replaced with OAm (0.5 mL). 1 mL of mixed precursor solution was mixed in a new beaker with a mixture (20 mL) of ethanol (E) and toluene (T) (E:T = 0, 0.2, 0.3, 0.4, 0.5, 0.6, 0.7) under vigorous stirring for 20 min.

2.3. Characterization

Photoluminescence (PL) spectra were acquired on a fluorescence spectrophotometer (F-7000, Hitachi, Tokyo, Japan). Ultraviolet and visible absorption (UV-vis) spectra were measured with a UV-3600 plus spectrophotometer (Shimadzu, Kyoto, Japan). The absolute PLQYs were obtained on a integration sphere (mod. 2100, Otsuka Electronics, Tokyo, Japan). Fluorescence lifetimes were gained using a FM-4P time-corrected single-photon-counting (TCSPC) system (Horiba, Kyoto, Japan) at an excitation wavelength of 325 nm. FTIR spectra were measured on a Nicolet instrument (iS5, Madison, WI, United States) in the region of 3200–900 cm^{-1}. X-ray diffractometry (XRD) patterns were collected with a D8 Advanced X-ray diffractometer (Bruker, Karlsruhe, Germany) using Cu Kα radiation (wavelength 1.55406 Å). Transmission electron microscopy (TEM) was performed on a JEM electron microscope (2100F, Tokyo, Japan) operating at 200 kV and energy dispersive spectra (EDS) were obtained with EDAX Genesis XM2 spectrometry. X-ray photoelectron spectroscopy (XPS) were recorded with an Escalab 250Xi X-ray photoelectron spectrometer (Thermo Fisher, Waltham, MA, United States) in the 3900–750 eV region.

3. Results

It is well known that surface ligands have a major impact on the shape, size and composition of NCs, and the size and morphology can be correlated with the performance of NCs in optics and electricity due to changes in band structure [23]. CTAB is one of the most common surfactants in the synthesis of gold nanorods, which is attributed to its electrostatic interaction with NCs [24,25]. Moreover, it's generally accepted that the negative exciton trapping effect of Br vacancies (V$_{Br}$) generated before nucleation cannot compensate for missing Br ions due to the fast nucleation rate, leading to a large amount of V$_{Br}$ and some researchers have suggested the reduced V$_{Br}$ density by passivation would lead to a higher QY [26,27]. The CTAB-modified NCs exhibit enhanced stability against polar solvents due to avoiding the ligand loss and low stability caused by the interligand proton transfer between oleylamine (OLA) and oleic acid (OA) [28]. Therefore, we explore a new synthetic method to trigger a deeper consideration of the nucleation and growth mechanism of NCs.

Furthermore, the effect of the solvent environment on the dispersibility, stability and photoelectric properties of the NCs has been widely investigated [29]. The strategy for the synthesis of CsPbBr$_3$/CsPb$_2$Br$_5$ NCs are carried out by tuning solvent polarity, that is to change the faction of ethanol and toluene in the system. The main reason why ethanol was chosen as a solvent polarity regulator is that CTAB has a higher solubility in ethanol, leading to the concentration of Br only slightly changing [30]. The crystallized phases with different morphology are obtained by mixing a precursor in good solvent (N,N-dimethylformamide, DMF) into a poor mixture under ambient conditions at room temperature. The details for synthesis can be found in the Methods section and Table S1.

Figure 1 shows the transmission electron microscopy (TEM) images and high-resolution TEM (HRTEM) images of the representative samples a and d, which were synthesized with a poor mixture at V_E: V_T = 0 and 0.4, respectively. It is observed that the NCs of 8–22 nm are uniform and monodispersed in sample a (Figure 1a,b), while Figure 1c,d show spherical nanoparticles (NPs) of 3–4 nm uniformly embedded in NCs with a similar size distribution (Figure S2b) and the yellow circles indicate the position of the embedded NPs. With an increase of the solvent polarity, a significant difference in shape is presented in Figure S3, from which we can see that the NCs around 13.5 nm are sharply reduced to be spherical quantum dots of 1-4 nm after adding ethanol (Figure 1b). The size of the NCs grows dramatically to about 24.3 nm (Figure S3d) and then reduce slightly to about 14.7 nm (Figure S2a). Subsequently, the embedded NPs vanish and the dispersion of NCs becomes worse after slight adjustment of the ratio between E and T (Figure S3g). The length of NCs continues to increase and reaches a maximum of approximately 120 nm (Figure S3j). Finally, agglomerated NCs of about 30 nm in size are obtained when V_E:V_T = 0.7. (Figure S3m). This reveals that the solvent polarity has an important role in the final morphology of products [31] and implies changes in the structural phase. It is worth noting that the overall size of NCs shows a gradual increase during the process and the dispersion becomes worse as the solvent polarity increases, which could be related to the decrease of ligand efficiency caused by excessive ethanol [32].

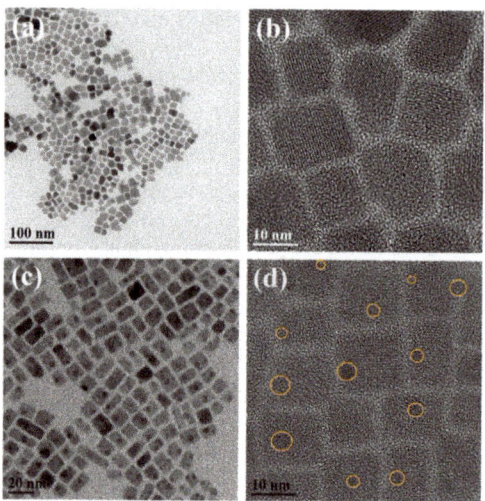

Figure 1. (a) Typical TEM overview image of sample a; (b) HRTEM image of selected lager sample a; (c) Typical TEM overview image of sample d; (d) HRTEM image of selected lager sample d.

The HRTEM images reveal the $CsPb_2Br_5$ structure with the lattice fringe spacing of 0.3 nm and 0.42 nm as shown in Figure S1a. Furthermore, the selected area electron diffraction (SAED) patterns of sample a, where the lattice fringe spacings of 0.59 nm, 0.24 nm and 0.21 nm correspond to the (2,2,2), (1,3,2) and (4,0,0) crystal planes of $CsPbBr_3$ and the lattice fringe spacing of 0.43 nm and 0.30 nm are associated with the (2,0,0) and (2,2,0) crystal planes of $CsPb_2Br_5$, respectively. These results suggest that $CsPbBr_3$ and $CsPb_2Br_5$ phases coexist in these regions. What is more, the NPs (Figure S2a) with the lattice fringe spacing of 0.228 nm assigned to $CsPbBr_3$ and the NCs corresponds to $CsPb_2Br_5$. It is confirmed that the $CsPbBr_3$ NPs are uniformly embedded in the $CsPb_2Br_5$ NCs.

In order to investigate the effect of solvent on the composition of NCs, XRD measurements for the above seven samples were performed as shown in Figure 2, where the diffraction patterns from

NCs are matched well with the main diffraction peaks at 15.21, 21.64, 30.70° corresponding to (110), (200), (220) plane of the orthorhombic CsPbBr$_3$ (JCPDS No. 01-072-7929) in the yellow area and 11.67, 23.39, 35.44, 37.90, 47.86° corresponding to (002), (210), (312), (313) and (420) plane of the tetragonal CsPb$_2$Br$_5$ (JCPDS No. 00-025-0211) in the blue area, respectively. It is easy to see that the peaks in the blue region, especially at 15.21° and 30.70°, are gradually weakened until $V_E:V_T = 0.6$ and then almost disappear. However, the sharp and intense peaks in 11.67° are obviously enhanced when $V_E:V_T > 0.2$. Furthermore, the percentage of CsPbBr$_3$ and CsPb$_2$Br$_5$ in a mixture can be roughly estimated by the ratio of their strongest XRD peaks. It's obvious that the percentage of CsPbBr$_3$ decreased from about 66% to 7% corresponding to samples a-g, while the content of CsPb$_2$Br$_5$ increased from around 34% to nearly pure phase. The above observations demonstrate that the solvent polarity controls the molar ratio of CsPbBr$_3$/CsPb$_2$Br$_5$ in the composite NCs and ethanol has a positive effect on phase transition.

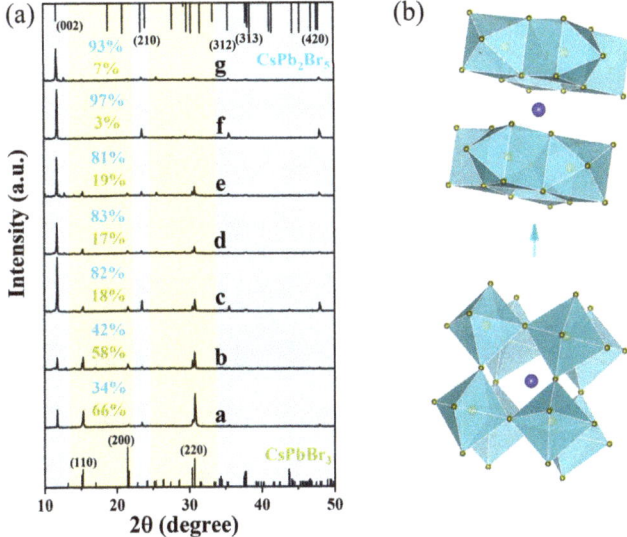

Figure 2. (a) X-ray diffraction (XRD) patterns of sample a-g; (b) Schematic representation of mixed-solvent polarity assisted the transition of orthorhombic CsPbBr$_3$ to tetragonal CsPb$_2$Br$_5$.

The absorption spectra (Figure 3a) were measured to gain more insight into the degree of electronic disorder in the crystals, which is attributed to the fact that the absorption edge is known as the Urbach tail. The Urbach energy (E_U) reflects the cumulative effect of impurities, defects and electron-phonon interactions on NCs, which could be obtained by fitting the Urbach tails in the logarithmic absorption spectra according to the Urbach's rule [33]:

$$\alpha(E) = \alpha_0 \exp\left[\frac{(E - E_0)}{E_U}\right] \quad (1)$$

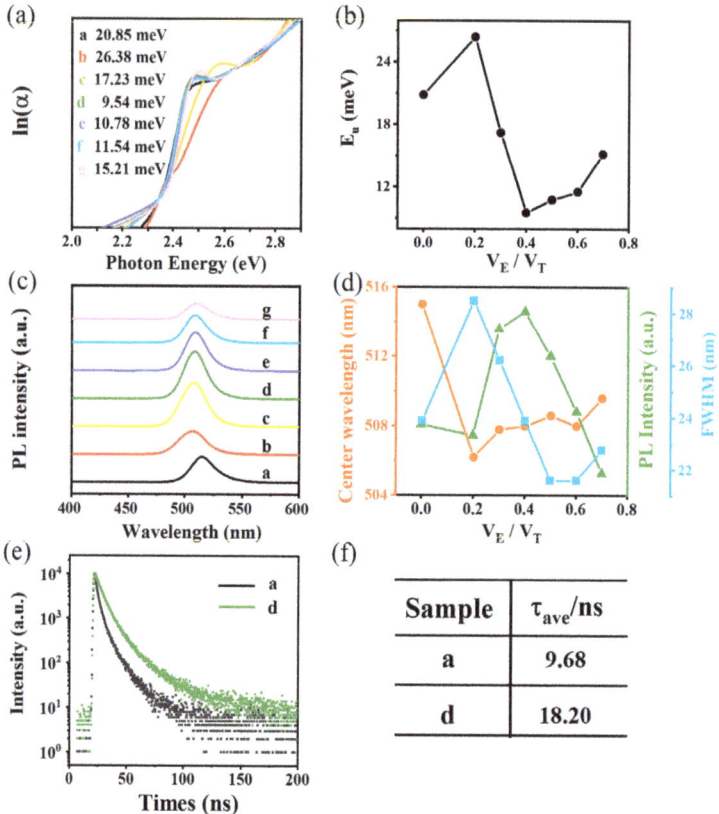

Figure 3. (a) Logarithmic absorption coefficient of sample a-g as a function of photon energy; (b) Relationship between VE/VT and Urbach energy; (c) Emission spectra of samples a-g excited by 365 nm; (d) Relationship between VE/VT and Central wavelength, PL intensity and FWHM; (e) PL decay for samples a and d with a 325 nm pulse laser; (f) The fitted average lifetime of samples a and d.

We obtain $E_U = k_B T/\sigma(T)$, where k_B is the Boltzmann constant, T is the absolute temperature and σ is the steepness parameter [34]. Figure 3b displays the E_U value of NCs synthesized in a mixture, which was obtained by fitting curves. It's observed that the E_U gradually reduces to a minimum (9.54 meV) and the maximum value reached is approximately 26.38 meV, which reveals the solvent polarity dependence upon the E_U. The NCs with lower E_U means that they possess a lower degree of structurual disorder and/or defect density than other NCs [35]. This indicates that the solvent polarity is a key factor to control internal defect density or structural disorder of NCs during the reprecipitation process. Furthermore, some investigations proved that the Br$^-$ concentration in the octahedron can be characterized by the red-shift of the absorption spectra [36]. Therefore, we can conclude that solvent polarity may be useful to control the CTAB passivation effect on NCs.

Figure 3c shows photoluminescence (PL) spectra of the $CsPbBr_3/CsPb_2Br_5$ NCs synthesized in different mixtures with a strong green emission, which varies slightly in PL intensity, central wavelength and full width at half-maximum (FWHM). The specific relationship between V_E/V_T and three parameters are shown in Figure 3d. It can be seen that the PL intensity increases first and reaches its maximum when the ratio between $CsPb_2Br_5$ and $CsPbBr_3$ phase is around 4.6 (Figure 2), and then the decrease of PL intensity when this ratio excess 4.6. The significant improvement in PL intensity of samples c-e is associated with the suitable volume fraction ratio of both structures [37]. Figure S4 shows the NCs synthesized with oleic acid and oleylamine quench after adding ethanol, which demonstrates that the NCs synthesized with CTAB have solvent-resistant ability as described in previous reports. Another solvent effect on NCs is the Stokes' shift of 8 nm, and the small Stokes' shift originates from band edge radiative recombination [38]. It's worth noting that the FWHM of the emission peak in sample a-g is between 21 nm to 29 nm, which roughly agrees with the narrow size distribution of the NCs (Figure S5).

The PL decays and lifetime obtained by triexponential decay functions are shown in Figure 3e. The triexponential functions (Equations (S1) and (S2)) and specific data obtained are recorded in Table S2. It's observed that the sample d has a longer lifetime (18.20 ns) than sample a (9.68 ns). Some reports have mentioned that the lifetime is decreased with the increase in hydrogen solvent polarity and inferred that solvent polarity plays an important role in changing the NCs trap states [39].

To further investigate the composition and phase transitions process, the film formed by samples a and d on the glass were characterized by X-ray photoelectron spectroscopy (XPS). All XPS spectra were calibrated with C 1s peak at 284.6 eV. Figure 4 shows the XPS survey spectra and high-resolution XPS spectra of sample a and d at Cs, Pb, Br. It can be seen that the peaks of Cs 3d, Pb 4f and Br 3d all are shifted to lower binding energy (BE) after adding ethanol. Pb^{2+} into NCs is in two chemical environments. The BE curves of Pb $4f_{5/2}$ and Pb $4f_{7/2}$ located at approximate 143 eV and 138 eV. It is noteworthy that the peaks marked as pink and orange after fitting are ascribed to the surface Pb ions and their areas occupied are smaller compared to sample a, implying the V_{Br} defects in sample d being reduced under higher polar condition [40]. Similarly, Br in NCs also exists in two chemical environments, and the BE curves of Br $3d_{3/2}$ and Br $3d_{5/2}$ appearing at approximately 68.5 eV and 67 eV are assigned to Pb-Br and Cs-Br. The significant differences of phases with different proportions are ascribed to their bond and structure [41]. Furthermore, the element ratio of Cs to Pb obtained by XPS (Table S3) is about 1:1 (sample a) and 1:2 (sample d), which is in good agreement with EDS results (Figure S6), while the excessive Br is originated from CTAB.

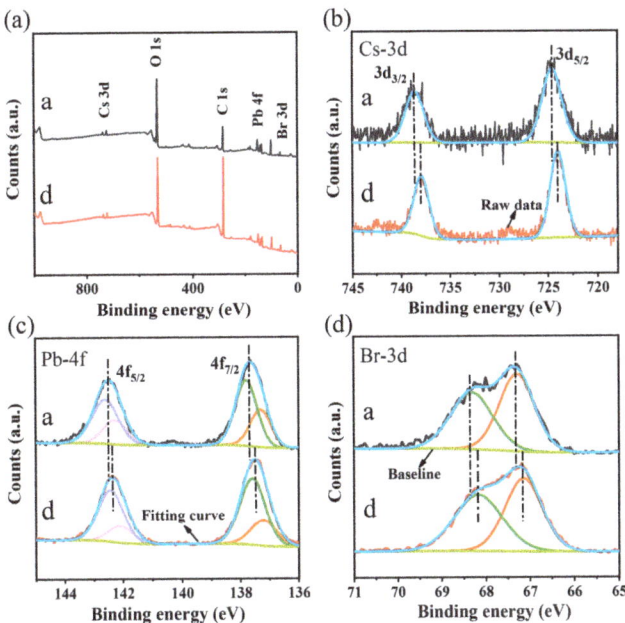

Figure 4. (a) XPS survey spectra of sample a film; (b) high-resolution XPS spectra of sample a and d at Cs; (c) high-resolution XPS spectra of sample a and d at Pb; (d) high-resolution XPS spectra of sample a and d at Br.

Moreover, the Fourier transform infrared (FTIR) spectra show the surface groups of NCs synthesized in different polar condition, as shown in Figure 5a. The peaks located at 2980 cm^{-1} and 2895 cm^{-1} are due to ν(C-H) in the -CH$_2$ group [29]. The intense peaks at 1679, 1394 cm^{-1} and 1265 cm^{-1} are assigned to ν(C=O). It's worth noting that the C=O bond is obviously shifted to a lower frequency with the increase of the solvent polarity, as shown in Figure 5b, which could be attributed to one dimensional structures formed by the coordination between one DMF molecule and Pb [42]. The peaks appearing at 1095 cm^{-1} and 1053 cm^{-1} are originated from C-N stretching vibrations in CTAB molecules [43]. Figure 5b shows highly magnified FTIR spectra in the 1000–1200 cm^{-1} region, in which the 1053 cm^{-1} peak of sample a isn't obvious, while the 1053 cm^{-1} peak exists in samples with ethanol. These four samples have similar peaks positions and no significant change in transmittance value. It is shown that the change in PL intensity is mainly associated with its defect density, rather than the charge transfer rates between NCs and the surface bonds.

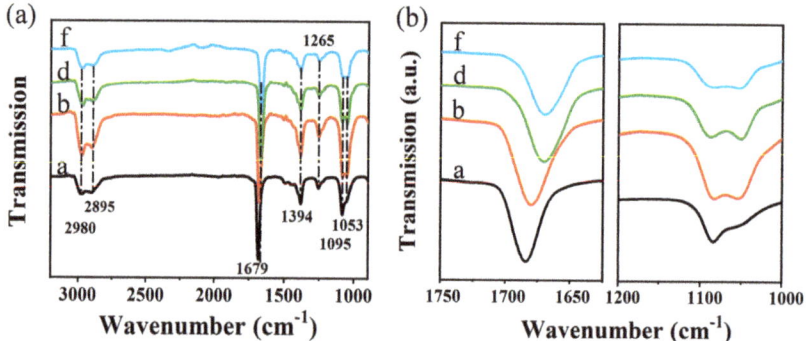

Figure 5. (a) Fourier transform infrared (FTIR) spectra of samples a, b, d, f; (b) Highly magnified FTIR spectra of samples a, b, d, f ranging from 1750 to 1625 cm^{-1} and 1200 to 1000 cm^{-1}.

The sharp emission peak of NCs can be tuned from 458 nm to 600 nm by changing the halogen ratio and UV/Vis spectra exhibit intense absorption, as shown in Figure 6a. Besides, the PL spectra of CsPbBr$_3$/CsPb$_2$Br$_5$ NCs synthesized with ethanol, isopropyl alcohol, cyclohexane, hexane, ether, ethyl acetate, methanol, acetone and toluene in a ratio of 0.4 were measured and are presented in Figure 6b. The PL of initially NCs is quenched after the addition of polar or non-polar solvent, which could be attributed to several reasons: the introduction of some functional groups causes a decrease in carrier mobility [44] and even the crystal structure is destroyed due to the nature of the ionic lattice and highly dynamic ligands process [45].

Sehrawat et al. indicated that the variation in PL properties could be demonstrated by geminate recombination and an associated variation in Onsager length related to the dielectric constant [46]. However, a significant improvement in the PL of NCs formed in toluene and ethanol (Table S4) is due to CTAB dissolved in ethanol will ionize into CTA$^+$ and Br$^-$ [30] and the higher Br$^-$ concentration in the system could improve the internal defect of NCs and stoichiometry. Hyun et al. proposed that solvent molecules can affect the charge transfer process by intervening the dielectric layer or the rearrangement of solvent molecules on the surface of NCs [47]. Majima et al. revealed that a strong solvent-polarity dependence on the electron-transfer process [48]. Therefore, it can be inferred that the solvent effects on NCs can't be neglected from the variable performance of NCs synthesized under different polar conditions.

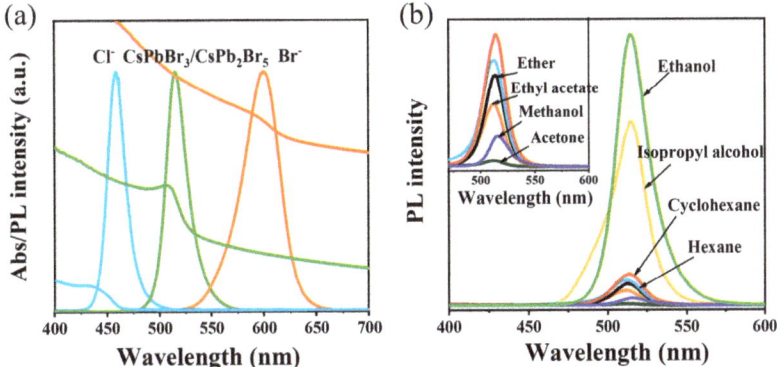

Figure 6. (a) UV-vis absorption and PL spectra of NCs synthesized with pure toluene; (b) PL spectra of NCs synthesized with different organic solvents: toluene = 0.4.

4. Conclusions

In conclusion, the solvent polarity-assisted transition from dual-phase to $CsPb_2Br_5$ phase offers a technique to alter the morphology of NCs. Such a phase transition could be related to two reasons: the degree of CTAB dissolution and growth of NCs under different polar conditions. The obtained NCs show enhanced stability and solvent polarity dependence of PL intensity, which could be assigned to the fact that CTAB molecules are highly soluble in ethanol and the produced Br^- can effectively passivate defects and improve the stoichiometry in the system. These guesses can be proved by defect density calculated in absorption spectra. Therefore, we can conclude that solvent polarity affects the ability of CTAB to passivate surface defects and it's a key factor for the final performance of the resulting NCs. This work provides new insights for deeper understanding in the field of perovskite NCs.

Supplementary Materials: The following are available online at http://www.mdpi.com/2079-4991/9/11/1537/s1.

Author Contributions: R.Y.: prepared samples, XRD, TEM, PL spectra, UV visible absorption spectra, XPS and FTIR tests and wrote manuscript, J.H. and J.W.: organized the data and plotted in cooperation with X.L., L.L.: designed this experiment and revised this manuscript, W.Z.: performed XPS analysis in cooperation with Z.N., Z.L.: performed FTIR analysis.

Funding: This work was supported by the National Nature Science Foundation of China [grant numbers 11574058]; the Major Program for Cooperative Innovation of Production, Education & Research of Guangzhou City [201704030106]; and the Special Fund for Application, Science and Technology Planning Projects of Guangdong Province of China [2017B010127002]. We acknowledge the technical help of Prof. Zhe Chuan Feng and Mr. Jiabin Wang at Guangxi University and Miss. Lingling Bai and the center of forecasting and analysis at Guangdong University of Technology.

Conflicts of Interest: The authors declare no conflict of interest.

References

1. Kim, Y.H.; Cho, H.; Heo, J.H.; Kim, T.S.; Lee, T.W. Multicolored organic/inorganic hybrid perovskite light-emitting diodes. *Adv. Mater.* **2015**, *27*, 1248–1254. [CrossRef] [PubMed]
2. Zhang, F.; Chen, C.; Wu, X.; Hu, X.; Huang, H.; Han, J.; Zou, B.; Dong, Y. Brightly luminescent and colortunable colloidal $CH_3NH_3PbX_3$ (X = Br, I, Cl) quantum dots: Potential alternatives for display technology. *ACS Nano* **2015**, *9*, 4533–4542. [CrossRef] [PubMed]
3. Grätzel, M. The light and shade of perovskite solar cells. *Nat. Mater.* **2014**, *13*, 838–842. [CrossRef] [PubMed]
4. Cha, J.H.; Han, J.H.; Yin, W.; Park, C.; Jung, D.Y. Photoresponse of $CsPbBr_3$ and Cs_4PbBr_6 perovskite single crystals. *J. Phys. Chem. Lett.* **2017**, *8*, 565–570. [CrossRef]

5. Protesescu, L.; Yakunin, S.; Bodnarchuk, M.I.; Krieg, F.; Caputo, R.; Hendon, C.H.; Yang, R.X.; Walsh, A.; Kovalenko, M.V. Nanocrystals of cesium lead halide perovskites (CsPbX$_3$, X = Cl, Br and I): Novel optoelectronic materials showing bright emission with wide color gamut. *Nano Lett.* **2015**, *15*, 3692–3696. [CrossRef]
6. Li, X.; Wu, Y.; Zhang, S.; Cai, B.; Gu, Y.; Song, J.; Zeng, H. CsPbX3 Quantum Dots for Lighting and displays: Room temperature synthesis, photoluminescence superiorities, underlying origins and white light-emitting diodes. *Adv. Funct. Mater.* **2016**, *26*, 2435–2445. [CrossRef]
7. Xuan, T.; Lou, S.; Huang, J.; Cao, L.; Yang, X.; Li, H.; Wang, J. Monodisperse and brightly luminescent CsPbBr$_3$/Cs$_4$PbBr$_6$ perovskite composite nanocrystals. *Nanoscale* **2018**, *10*, 9840–9844. [CrossRef]
8. Tong, G.; Li, H.; Li, D.; Zhu, Z.; Xu, E.; Li, G.; Yu, L.; Xu, J.; Jiang, Y. Dual-phase CsPbBr$_3$–CsPb$_2$Br$_5$ perovskite thin films via vapor deposition for high-performance rigid and flexible photodetectors. *Small* **2017**, *14*, 1702523. [CrossRef]
9. Chia, X.; Eng, A.Y.S.; Ambrosi, A.; Tan, S.M.; Pumera, M. Electrochemistry of nanostructured layered transition-metal dichalcogenides. *Chem. Rev.* **2015**, *115*, 11941–11966. [CrossRef]
10. Duan, X.; Wang, C.; Pan, A.; Yu, R.; Duan, X. Two-dimensional transition metal dichalcogenides as atomically thin semiconductors: Opportunities and challenges. *Chem. Soc. Rev.* **2015**, *44*, 8859–8876. [CrossRef]
11. Zhang, X.; Xu, B.; Zhang, J.; Gao, Y.; Zheng, Y.; Wang, K.; Sun, X.W. All-inorganic perovskite nanocrystals for high-effciency light emitting diodes: Dual-phase CsPbBr$_3$-CsPb$_2$Br$_5$ composites. *Adv. Funct. Mater.* **2016**, *26*, 4595–4600. [CrossRef]
12. Li, G.; Wang, H.; Zhu, Z.; Chang, Y.; Zhang, T.; Song, Z.; Jiang, Y. Shape and phase evolution from CsPbBr$_3$ perovskite nanocubes to tetragonal CsPb$_2$Br$_5$ nanosheets with an indirect bandgap. *Chem. Commun.* **2016**, *52*, 11296–11299. [CrossRef] [PubMed]
13. Tang, X.; Hu, Z.; Yuan, W.; Hu, W.; Shao, H.; Han, D.; Zheng, J.; Hao, J.; Zang, Z.; Du, J.; et al. Perovskite CsPb$_2$Br$_5$ microplate laser with enhanced stability and tunable properties. *Adv. Opt. Mater.* **2017**, *5*, 1600788. [CrossRef]
14. Iyikanat, F.; Sari, E.; Sahin, H. Thinning CsPb$_2$Br$_5$ perovskite down to monolayers: Cs-dependent stability. *Phys. Rev. B* **2017**, *96*, 155442. [CrossRef]
15. Dursun, I.; Bastiani, M.D.; Turedi, B.; Alamer, B.; Shkurenko, A.; Yin, J.; El-Zohry, A.M.; Gereige, I.; AlSaggaf, A.; Mohammed, O.F.; et al. CsPb$_2$Br$_5$ single crystals: Synthesis and characterization. *ChemSusChem* **2017**, *10*, 3746. [CrossRef]
16. Yin, J.; Yang, H.; Song, K.; El-Zohry, A.M.; Han, Y.; Bakr, O.M.; Bredas, J.L.; Mohammed, O.F. Point defects and green emission in zero-dimensional perovskites. *J. Phys. Chem. Lett.* **2018**, *9*, 5490–5495. [CrossRef]
17. Wang, K.H.; Wu, L.; Li, L.; Yao, H.B.; Qian, H.S.; Yu, S.H. Large-scale synthesis of highly luminescent perovskite-related CsPb$_2$Br$_5$ nanoplatelets and their fast anion exchange. *Angew. Chem.* **2016**, *128*, 1–6.
18. Balakrishnan, S.K.; Kamat, P.V. Ligand assisted transformation of cubic CsPbBr$_3$ nanocrystals into two-dimensional CsPb$_2$Br$_5$ nanosheets. *Chem. Mater.* **2018**, *30*, 74–78. [CrossRef]
19. Huang, H.; Raith, J.; Kershaw, S.V.; Kalytchuk, S.; Tomanec, O.; Jing, L.; Susha, A.S.; Zboril, R.; Rogach, A.L. Growth mechanism of strongly emitting CH$_3$NH$_3$PbBr$_3$ perovskite nanocrystals with a tunable bandgap. *Nat. Commun.* **2017**, *8*, 996. [CrossRef]
20. Palazon, F.; Urso, C.; Trizio, L.D.; Akkerman, Q.; Marras, S.; Locardi, F.; Nelli, I.; Ferretti, M.; Prato, M.; Manna, L. Postsynthesis transformation of insulating Cs$_4$PbBr$_6$ nanocrystals into bright perovskite CsPbBr$_3$ through physical and chemical extraction of CsBr. *ACS Energy Lett.* **2017**, *2*, 2445–2448. [CrossRef]
21. Ruan, L.; Shen, W.; Wang, A.; Xiang, A.; Deng, Z. Alkyl-thiol ligand-induced shape- and crystalline phase-controlled synthesis of stable perovskite-related CsPb$_2$Br$_5$ nanocrystals at room temperature. *J. Phys. Chem. Lett.* **2017**, *8*, 3853–3860. [CrossRef] [PubMed]
22. Qiu, T.; Wu, L.X.; Kong, F.; Ma, H.B.; Chu, P.K. Solvent effect on light-emitting property of Si nanocrystals. *Phys. Lett. A* **2005**, *334*, 447–452. [CrossRef]
23. Grim, J.Q.; Manna, L.; Moreels, I. A sustainable future for photonic colloidal nanocrystals. *Chem. Soc. Rev.* **2015**, *44*, 5897–5914. [CrossRef] [PubMed]
24. Ye, X.; Jin, L.; Caglayan, H.; Chen, J.; Xing, G.; Zheng, C.; Nguyen, V.D.; Kang, Y.; Engheta, N.; Kagan, C.R.; et al. Improved size-tunable synthesis of monodisperse gold nanorods through the use of aromatic additives. *ACS Nano* **2012**, *6*, 2804–2817. [CrossRef]

25. Ferhan, A.R.; Guo, L.; Kim, D.H. Influence of ionic strength and surfactant concentration on electrostatic surfacial assembly of cetyltrimethylammonium bromide-capped gold nanorods on fully immersed glass. *Langmuir* **2010**, *26*, 12433–12442. [CrossRef]
26. Wu, Y.; Wei, C.; Li, X.; Li, Y.; Qiu, S.; Shen, W.; Cai, B.; Sun, Z.; Yang, D.; Deng, Z.; et al. In situ passivation of $PbBr_6^{4-}$ octahedra toward blue luminescent $CsPbBr_3$ nanoplatelets with near 100% absolute quantum yield. *ACS Energy Lett.* **2018**, *3*, 2030–2037. [CrossRef]
27. Pan, J.; Quan, L.N.; Zhao, Y.; Peng, W.; Murali, B.; Sarmah, S.P.; Yuan, M.; Sinatra, L.; Alyami, N.M.; Liu, J.; et al. Highly efficient perovskite-quantum-dot light-emitting diodes by surface engineering. *Adv. Mater.* **2016**, *28*, 8718–8725. [CrossRef]
28. Cai, Y.; Wang, L.; Zhou, T.; Zheng, P.; Li, Y.; Xie, R.J. Improved stability of $CsPbBr_3$ perovskite quantum dots achieved by suppressing interligand proton transfer and applying a polystyrene coating. *Nanoscale* **2018**, *10*, 21441–21450. [CrossRef]
29. Mei, J.; Wang, F.; Wang, Y.; Tian, C.; Liu, H.; Zhao, D. Energy transfer assisted solvent effects on $CsPbBr_3$ quantum dots. *J. Mater. Chem. C* **2017**, *5*, 11076–11082. [CrossRef]
30. Wang, Y.X.; Sun, J.; Fan, X.Y.; Yu, X. A CTAB-assisted hydrothermal and solvothermal synthesis of ZnO nanopowders. *Ceram. Int.* **2017**, *37*, 3431–3436. [CrossRef]
31. Zhu, L.; Xie, Y.; Zheng, X.; Liu, X.; Zhou, G. Fabrication of novel urchin-like architecture and snowflake-like pattern CuS. *J. Cryst. Growth* **2004**, *260*, 494–499. [CrossRef]
32. Meng, S.; Zhang, J.; Wu, C.; Zhang, Y.; Xiao, Q.; Lu, G. Dissipative particle dynamics simulations of surfactant CTAB in ethanol/water mixture. *Mol. Simul.* **2014**, *40*, 1052–1058. [CrossRef]
33. Wasim, S.M.; Marín, G.; Rincón, C.; Bocaranda, P.; Pérez, G.S. Urbach's tail in the absorption spectra of the ordered vacancy compound $CuGa_3Se_5$. *J. Appl. Phys.* **1998**, *84*, 5823–5825. [CrossRef]
34. Rai, R.C. Analysis of the urbach tails in absorption spectra of undoped ZnO thin films. *J. Appl. Phys.* **2013**, *113*, 153508. [CrossRef]
35. Liu, F.; Zhang, Y.; Ding, C.; Kobayashi, S.; Izuishi, T.; Nakazawa, N.; Toyoda, T.; Ohta, T.; Hayase, S.; Minemoto, T.; et al. Highly luminescent phase-stable $CsPbI_3$ perovskite quantum dots achieving near 100% absolute photoluminescence quantum yield. *ACS Nano* **2017**, *11*, 10373–10383. [CrossRef] [PubMed]
36. Yang, D.; Li, X.; Wu, Y.; Wei, C.; Qin, Z.; Zhang, C.; Sun, Z.; Li, Y.; Wang, Y.; Zeng, H. Surface halogen compensation for robust performance enhancements of $CsPbX_3$ perovskite quantum dots. *Adv. Opt. Mater.* **2019**, *7*, 1900276. [CrossRef]
37. Tan, T.; Li, R.; Xu, H.; Qin, Y.; Song, T.; Sun, B. Ultrastable and reversible fluorescent perovskite films used for flexible instantaneous display. *Adv. Funct. Mater.* **2019**, *29*, 1900730. [CrossRef]
38. Liang, Z.; Zhao, S.; Xu, Z.; Qiao, B.; Song, P.; Gao, D.; Xu, X. Shape-controlled synthesis of all-inorganic $CsPbBr_3$ perovskite nanocrystals with bright blue emission. *ACS Appl. Mater. Interfaces* **2016**, *8*, 28824–28830. [CrossRef]
39. Kumar, P.; Bohidar, H.B. Observation of fluorescence from non-functionalized carbon nanoparticles and its solvent dependent spectroscopy. *J. Lumin.* **2013**, *141*, 155–161. [CrossRef]
40. Woo, J.Y.; Kim, Y.; Bae, J.; Kim, T.G.; Kim, J.W.; Lee, D.C.; Jeong, S. Highly stable cesium lead halide perovskite nanocrystals through in situ lead halide inorganic passivation. *Chem. Mater.* **2017**, *29*, 7088–7092. [CrossRef]
41. Li, M.; Zhang, X.; Dong, T.; Wang, P.; Postolek, K.M.; Yang, P. Evolution of morphology, phase composition and photoluminescence of cesium lead bromine nanocrystals with temperature and precursors. *J. Phys. Chem. C* **2018**, *122*, 28968–28976. [CrossRef]
42. Wang, H.; Zhang, X.; Wu, Q.; Cao, F.; Yang, D.; Shang, Y.; Ning, Z.; Zhang, W.; Zheng, W.; Yan, Y.; et al. Trifluoroacetate induced small-grained $CsPbBr_3$ perovskite films result in efficient and stable lightemitting devices. *Nat. Commun.* **2019**, *10*, 665. [CrossRef] [PubMed]
43. Ghosh, S.; Acharyya, S.S.; Kumar, M.; Bal, R. One-pot preparation of nanocrystalline $Ag–WO_3$ catalyst for the selective oxidation of styrene. *RSC Adv.* **2015**, *5*, 37610. [CrossRef]
44. Alas, M.O.; Genc, R. An investigation into the role of macromolecules of different polarity as passivating agent on the physical, chemical and structural properties of fluorescent carbon nanodots. *J. Nanopart. Res.* **2017**, *19*, 185. [CrossRef]
45. Ruan, L.; Shen, W.; Wang, A.; Zhou, Q.; Zhang, H.; Deng, Z. Stable and conductive lead halide perovskites facilitated by X-type ligands. *Nanoscale* **2017**, *9*, 7252. [CrossRef]

46. Sehrawat, K.; Mehra, R.M. Modification and photoluminescence spectra of porous silicon by changing the surrounding dielectric environment. *Indian. J. Pure Appl. Phys.* **2004**, *42*, 419–422.
47. Hyun, B.R.; Bartnik, A.C.; Lee, J.K.; Imoto, H.; Sun, L.; Choi, O.J.; Chujo, Y.; Hanrath, T.; Ober, C.K.; Wise, F.W. Role of solvent dielectric properties on charge transfer from PbS nanocrystals to molecules. *Nano Lett.* **2010**, *10*, 318–323. [CrossRef]
48. Cui, S.C.; Tachikawa, T.; Fujitsuka, M.; Majima, T. Solvent-polarity dependence of electron-transfer kinetics in a Cdse/ZnS quantum dot-pyromellitimide conjugate. *J. Phys. Chem. C* **2010**, *114*, 1217–1225. [CrossRef]

© 2019 by the authors. Licensee MDPI, Basel, Switzerland. This article is an open access article distributed under the terms and conditions of the Creative Commons Attribution (CC BY) license (http://creativecommons.org/licenses/by/4.0/).

Article

Structural Evolution of AlN Nanoclusters and the Elemental Chemisorption Characteristics: Atomistic Insight

Xi Nie [1,†], Zhao Qian [1,*,†], Wenzheng Du [1], Zhansheng Lu [2], Hu Li [3], Rajeev Ahuja [4] and Xiangfa Liu [1]

1. Key Laboratory for Liquid-Solid Structural Evolution and Processing of Materials (Ministry of Education), School of Materials Science and Engineering, Shandong University, Jinan 250061, China; 17861412028@139.com (X.N.); 13127150302@163.com (W.D.); xfliu@sdu.edu.cn (X.L.)
2. School of Materials Science and Engineering, Henan Normal University, Xinxiang 453007, China; zslu@henannu.edu.cn
3. School of Electrical and Electronic Engineering, University of Manchester, Manchester M139PL, UK; Hu.Li@manchester.ac.uk
4. Condensed Matter Theory Group, Department of Physics and Astronomy, Ångström Laboratory, Uppsala University, 75120 Uppsala, Sweden; rajeev.ahuja@physics.uu.se
* Correspondence: qianzhao@sdu.edu.cn
† These authors contributed equally to this work.

Received: 1 September 2019; Accepted: 2 October 2019; Published: 4 October 2019

Abstract: A theoretical insight into the structural evolution of AlN atomic clusters and the chemisorption of several common alloying elements on a large cluster has been performed in the framework of state-of-the-art density functional theory calculations. We report the findings that the longitudinal growth takes precedence during the early stage of structural evolution of small AlN clusters, when the longitudinal dimension becomes stable, the AlN cluster proceeds with cross-growth and blossoms into the large-size $Al_{60}N_{60}$. Upon the growth of clusters, the structures tend to become well-knit gradually. As for the evolution of electronic structures of AlN clusters through the HSE06 calculations, the density of states curves become more and more nondiscrete with the atomic structures evolving from small to large size and tend to resemble that of the Wurtzite AlN. The chemisorption characteristics of the large $Al_{60}N_{60}$ cluster towards different elements such as Al, N, Fe and Cu are also theoretically unveiled, in which it is interestingly found that the N and Cu atoms are likely to be adsorbed similarly at the growth edge position of the $Al_{60}N_{60}$ cluster and the density of states curves of these two chemisorption systems near the Fermi level also show some interesting similarities.

Keywords: AlN; low-dimensional material; atomic cluster; electronic structure; HSE06 hybrid functional

1. Introduction

Atomic clusters play important roles in the nucleation of solid phases. It is worthwhile to perform fundamental research on clusters to reveal their structures and properties at the atomic and electronic levels. Li et al. systematically investigated the structural evolution of gold–germanium bimetallic clusters and the nonlinear optical properties, chemical properties of a series of alkali-metals-adsorbed gold–germanium bimetallic clusters, in which it was found that the atomic structure of gold–germanium bimetallic clusters with adsorbed alkali metals did not change significantly and alkali metals tended to adhere to the surface or edge of clusters [1]. Yan et al. studied the structural evolution of as clusters using the first principles method [2]. Die et al. investigated the structural and magnetic properties of Cu4M clusters and found that in the most stable Cu4M clusters, the positions of M atoms are the most coordinated [3]. In recent years, the light-element aluminum-based clusters have received extensive attention and have been studied [4–6], especially the aluminum–pnictogen system. Aluminum nitride (AlN) is one example, which has many desirable properties, such as high thermal conductivity, high temperature resistance, impact resistance and a low expansion coefficient [7]. AlN can be prepared by the reaction of aluminum salt with ammonia, chemical vapor deposition, etc. [8–10]. In vacuum, researchers have used magnetron reactive sputtering technology to make the sputtered aluminum react with nitrogen to prepare a new AlN nanofilm and Kishimoto et al. [11] studied AlN film growth on sapphire in an experiment. In addition, in the magnetron reactive sputtering experiment, a series of Al_nN_m clusters were also observed [12]. Meanwhile, some researchers used the ab initio methods to study the aluminum–nitrogen system [13–16]. Saeedi et al. performed density functional theory to calculate the electronic properties of octahedral Al_nN_n cages and Al_nP_n cages to discuss the isotropic chemical shielding parameters of Al_nN_n cages or Al_nP_n cages in different electrostatic environments [17]. BelBruno designed the structures of Al_nN_n (n = 2–4) clusters using density functional theory and compared them with the carbon and boron nitride clusters [18]. Furthermore, some scholars have studied the hydrogen storage and gas detecting properties of AlN clusters [19,20]. However, there are few reports on the electronic structures' evolution and growth of AlN clusters employing the hybrid functionals, although the growth of aluminum nitride has been reported in experiments concerning the preparation of AlN thin films, blocks or single crystals [21–24].

In this article, we have theoretically investigated the structural evolution concerning the growth of AlN clusters from the perspective of atomic and electronic structures using the HSE06 hybrid functional, along with the characteristics of the bond lengths and energetics of AlN clusters. In order to understand the chemisorption of common alloying elements such as Fe, Cu, Al, N on AlN clusters, we also simulated the interaction between these elements and the $Al_{60}N_{60}$ large clusters to unveil their chemisorption characteristics. In the previous experimental study, we found that copper can be observed on the AlN_p/Al interface by adding copper powder to AlN_p-reinforced Al composites. In addition, when the Fe power is added to the AlN_p-reinforced Al composites, it is not aggregated at the interface but dispersedly distributed in the composite material [25]. This work is thus proposed to help understand the experimental result and expose structures and chemisorption properties of AlN atomic clusters, as well as providing a deep theoretical guidance for experimentalists.

2. Methods

In this study, the calculations of total energies, atomic forces, and structure optimizations have been performed using the generalized gradient approximation (GGA) [26] in the form of Perdew–Burke–Ernzerhof (PBE) based on the density functional theory (DFT) [27,28]. The projector augmented wave (PAW) [29] method has been employed in the Vienna Ab initio Simulation Package [30]. Considering that the experimental exposed surfaces of AlN are usually made up of (10$\bar{1}$0) and (000$\bar{1}$) [31], we have established the models in the supercell of Wurtzite AlN through cutting out the smallest structural unit of the hexagonal prism (Al_6N_6 cluster). In order to avoid interaction between clusters in the x, y and z directions, all the models are separated by the vacuum space of 20 Å. The K-points mesh of 1×1×1 has been used within the Monkhorst–Pack scheme in the stage of geometry optimizations and electronic structure calculations. The plane-wave energy cut-off of 520 eV has been employed for all the structure optimizations. We used a conjugate gradient algorithm to perform the structure optimizations and relax all ionic positions until the force on each ion is lower than 0.02 eV/Å. In order to calculate the electronic structure of the clusters and their chemisorption systems more accurately, we have considered the effects of nonlocal exchange and used the screened hybrid functional of Heyd, Scuseria, and Ernzerhof (HSE06) [32]. In HSE06, only the local part of the exact exchange energy is treated by Hartree-Fork theory, while the remaining part is treated by DFT. In our paper, the screening parameter μ was set to 0.2, conforming to the HSE06 functional.

3. Results and Discussion

3.1. Structural Evolution of AlN Clusters

During the structural evolution and growth of AlN clusters, the five most representative clusters are selected. The smallest cluster unit is Al_6N_6. When the Al_6N_6 cluster first grows longitudinally into a stable two-layer hexagonal prism structure, the representative cluster is Al_9N_9. For the $Al_{15}N_{15}$ cluster, it is formed when the longitudinal dimension of the cluster becomes stable. The $Al_{15}N_{15}$ cluster then evolves to $Al_{30}N_{30}$ in the transverse direction after the cluster size is stabilized in the longitudinal direction. The $Al_{60}N_{60}$ is a large cluster structure evolved from $Al_{30}N_{30}$ with continued transverse growth. The structural evolution process of AlN clusters in the atomic scale is shown in Figure 1. The small AlN unit cluster prefers growing along the longitudinal direction in the beginning, after which, the larger cluster would continue growing along the transverse direction when the longitudinal growth reaches stability. For the $Al_{15}N_{15}$ cluster, each aluminum or nitrogen atom of the hexatomic ring connects with a nitrogen or aluminum atom to lead to the formation of the $Al_{30}N_{30}$ cluster. The dangling aluminum and nitrogen atoms in the $Al_{30}N_{30}$ cluster would further absorb atoms, forming complete hexatomic rings and growing into a large aluminum nitride cluster ($Al_{60}N_{60}$ cluster).

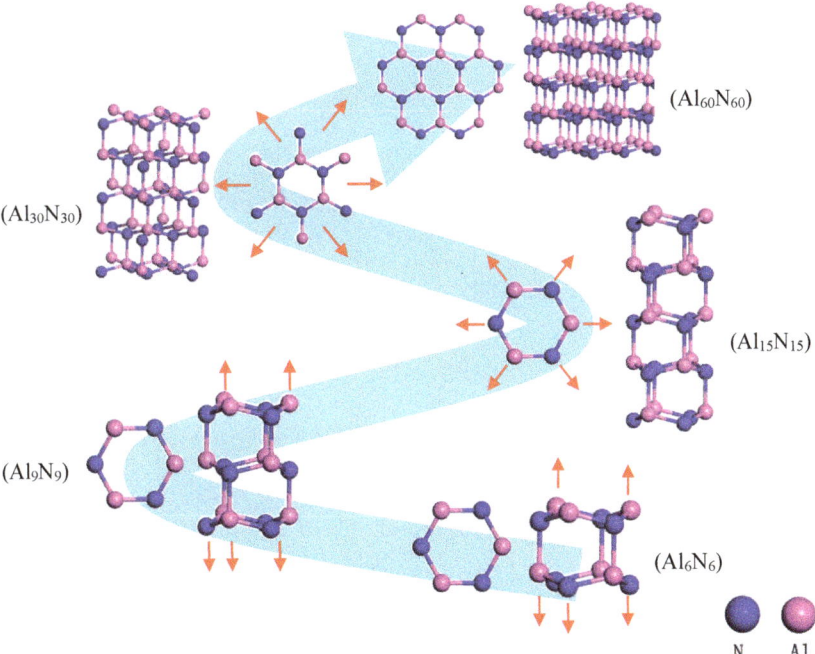

Figure 1. Outline of the stepwise formation (both the top view and the side view) of the $Al_{60}N_{60}$ large clusters. The red arrows stand for the growth direction of each respective cluster, and the big blue arrow stands for the overall evolution of clusters from small to large.

In addition to the evolution of AlN clusters in atomic structures, we have also investigated the evolution of AlN clusters in electronic structures. In order to obtain the more accurate electronic structure, we have employed the HSE06 hybrid functional in calculations. The electronic structures of the Al_6N_6, Al_9N_9, $Al_{15}N_{15}$, $Al_{30}N_{30}$ and $Al_{60}N_{60}$ clusters are compared with the Wurtzite AlN crystal, which can be shown in Figure 2. The Wurtzite AlN crystal has the continuous curve of density of states. From the figure, it is obvious that the small AlN clusters, such as Al_6N_6, Al_9N_9 and $Al_{15}N_{15}$, have the discrete density of states curves. With the evolution and growth of AlN clusters, the density of states curves tend to be continuous and become more similar to the DOS curve of the Wurtzite AlN crystal gradually. For the larger $Al_{30}N_{30}$ and $Al_{60}N_{60}$ atomic clusters, the density of states curves show four and two deep levels, respectively, in the band gap through the HSE06 calculations. Near the Fermi level, in accordance with the increase of AlN clusters, the HOMO–LUMO gap is also gradually increased.

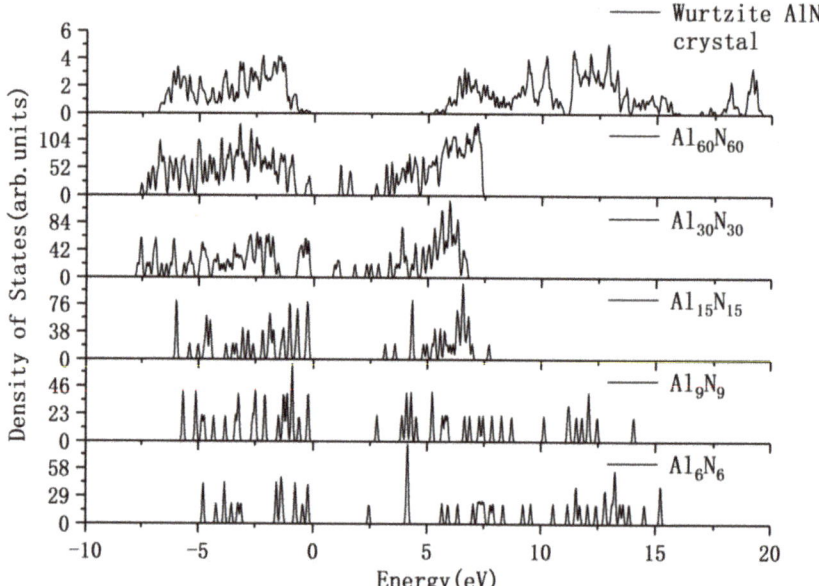

Figure 2. The electronic density of states of the Al_6N_6, Al_9N_9, $Al_{15}N_{15}$, $Al_{30}N_{30}$, $Al_{60}N_{60}$ atomic clusters and the WurtziteAlN crystal through the HSE06 calculations (the Fermi level is set at zero).

In order to further investigate the properties of AlN clusters, the average bond length, cohesive energy and total energy changes during the structural evolution of atomic clusters have been analyzed, which can be seen in Figure 3. Figure 3a shows the average Al-N bond lengths of $(AlN)_n$ clusters: with the growth and evolution of clusters, the average bond length gradually increases; when the cluster reaches a certain size, the bond length attains the maximum value. After that, the average bond length begins to decrease and the large cluster tends to shrink. Figure 3b provides the cohesive energies of $(AlN)_n$ clusters, which are calculated using the previous method [33,34] based on the formula as follows:

$$E_{coh} = -[E(AlN)_n - nE(Al) - nE(N)]/2n \qquad (1)$$

where $E(AlN)_n$ is the total energy of the investigated $(AlN)_n$ (n = 6,9,12,15,30,35,55,60) cluster system, while E_{Al} and E_N stand for the total energies of a single aluminum atom and nitrogen atom, respectively. When the AlN clusters grow in the early stage, the cohesive energies tend to increase. While, the cohesive energy decreases rapidly when n = 30, which is related to the structure of the $Al_{30}N_{30}$ cluster. Each atom in the hexatomic ring of the $Al_{30}N_{30}$ cluster is connected with a hetero-atom, which dangles outside and results in a sudden decrease in cohesive energy. After those dangling bonds are saturated by growth, the cohesive energy of the cluster would increase again. Figure 3c shows us the total energies of the investigated clusters, from which it can be seen that with the growth and evolution of the clusters, the total energy increases almost linearly.

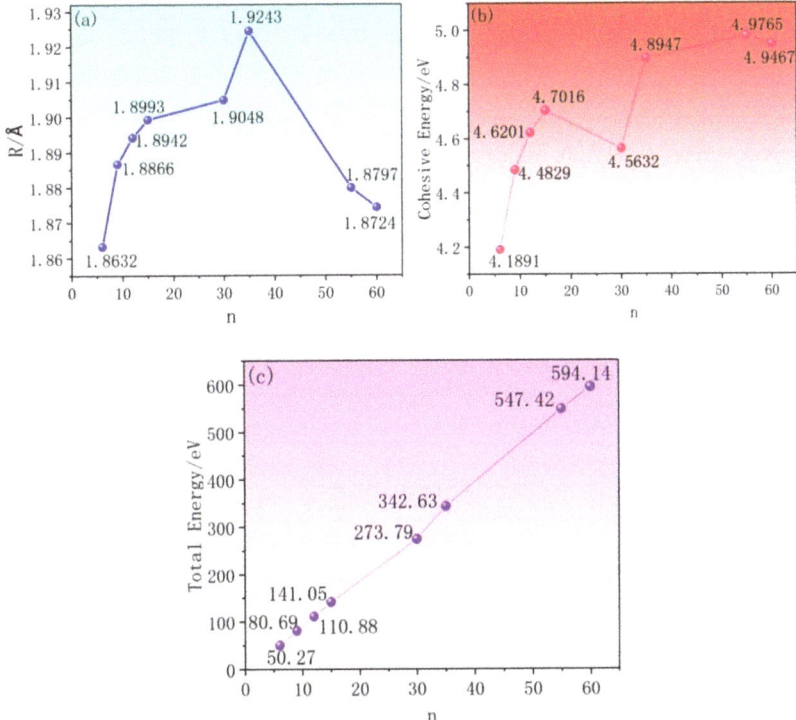

Figure 3. (a) The average bond lengths, (b) the cohesive energies and (c) the total energies of the $(AlN)_n$ (n = 6,9,12,15,30,35,55,60) clusters.

3.2. Chemisorption of Al, N, Fe, Cu Atoms on the $Al_{60}N_{60}$ Large Cluster

As is known, the large clusters especially, in short- or medium-range order could play important roles in nucleation of the corresponding solid phase in materials science. Take the in-situ AlN-reinforced Al alloys for example, Cu is a common strengthening element in aluminum alloys, which can significantly increase strength by precipitation hardening; Fe is also an important element in rapidly cooled aluminum alloys. For chemisorptions of Al and N atoms, it is useful to understand the evolution and growth of AlN clusters. It is essential to study the interactions between the AlN cluster and the alloying elements, which can help to understand the distribution of those alloying elements in solidified materials. In this work, the fundamental interactions between several common alloying elements (Al, N, Fe, Cu) and the $Al_{60}N_{60}$ large cluster are investigated considering different chemisorption sites. There are three different positions on the $Al_{60}N_{60}$ large cluster: top, edge and side. Due to the existence of equivalent sites, two different sites are studied for the top position: one above the center of the six rings and the other above the Al-N bond. There are four different sites for the edge position: near the N atom with four bonds, near the Al atom with three bonds, near the N atom with three bonds and near the Al atom with two bonds. There are two different sites for the side position: one is near the recessed Al atom, and the other is near the Al-N bond. The adsorption energy E_{ad} is calculated as follows:

$$E_{ad} = E_{Al60N60+M} - E_{Al60N60} - E_M \quad (2)$$

where $E_{Al60N60+M}$ is the total energy of the system with M element (Al, N, Fe or Cu) adsorbed on the $Al_{60}N_{60}$ cluster, while $E_{Al60N60}$ and E_M stand for the total energies of the $Al_{60}N_{60}$ cluster and the isolated M (Al, N, Fe or Cu) atom, respectively. We considered the chemisorption of Al, N, Fe and Cu at different positions (top, side and edge) of the $Al_{60}N_{60}$ large cluster. A total of eight chemisorption sites for every elemental sorption are studied. After calculations, we have chosen the respective site with the largest adsorption energy for each position (i.e., the adsorption energies described below are the largest for the corresponding positions) in our research.

The chemisorption characteristics towards the Al atom are shown in Figure 4. When the Al atom is adsorbed at the side of the $Al_{60}N_{60}$ large cluster, the E_{ad} is the maximum of 10.4775 eV, almost three times those at the top or edge positions. Thus, the aluminum is likely to adsorb and accumulate at the side of the $Al_{60}N_{60}$ cluster. Figure 5 shows the chemisorption of nitrogen. When the nitrogen atom is adsorbed at the edge position, the E_{ad} is the maximum of 3.0128 eV. Thus, the nitrogen is more likely to be adsorbed at the edge of the growth frontier of AlN. In the Al melt, the growth atmosphere of AlN is rich in aluminum atoms and poor in nitrogen atoms, so the growth of AlN mainly depends on the deposition/adsorption of nitrogen atoms on AlN. As shown in Figure 5, the nitrogen atoms are likely to adsorb and accumulate at the edge site instead of at the top or side sites of the cluster, which makes AlN tend to grow incliningly instead of longitudinally or transversely. This theoretical finding may provide some guidance for the growth of AlN.

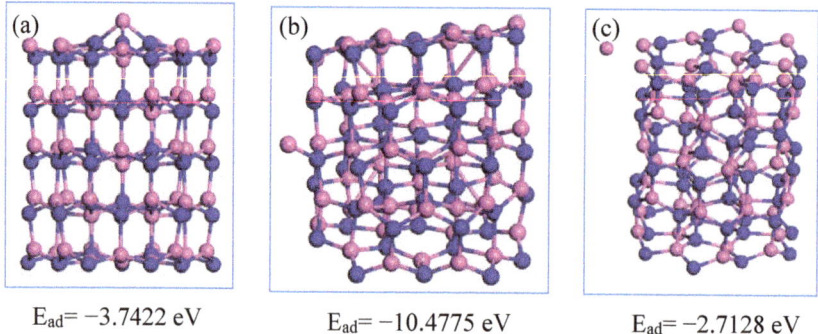

Figure 4. The chemisorption of Al at different sites of the $Al_{60}N_{60}$ cluster. The Al and N atoms are shown in pink and blue. The Al atom is located at the top, side and edge position of the $Al_{60}N_{60}$ cluster shown in (**a**–**c**), respectively.

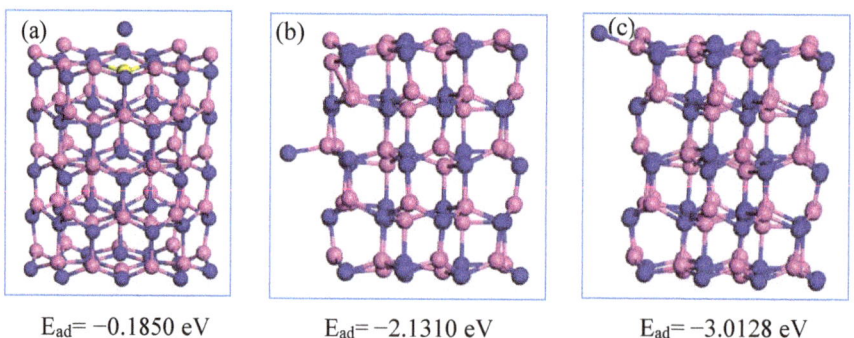

Figure 5. The chemisorption of N element at different sites of the $Al_{60}N_{60}$ cluster. The Al and N atoms are shown in pink and blue. The N atom is located at the top, side and edge position of the $Al_{60}N_{60}$ cluster shown in (**a**–**c**), respectively.

The chemisorption of Cu atoms on the $Al_{60}N_{60}$ large cluster at different positions is shown in Figure 6. In the $Al_{60}N_{60}$ cluster, compared with the top and side sites, the copper atom has the highest adsorption energy at the edge site. It can be seen that copper is more likely to be adsorbed at the edge site of the cluster, which is similar to the stacking mode of nitrogen atoms. The chemisorption of iron atoms at different positions of the $Al_{60}N_{60}$ cluster is shown in Figure 7. Compared with the top and edge positions, the E_{ad} of Fe element at the side is the largest. It is shown that the Fe atom is easier to accumulate at the side of the $Al_{60}N_{60}$ cluster.

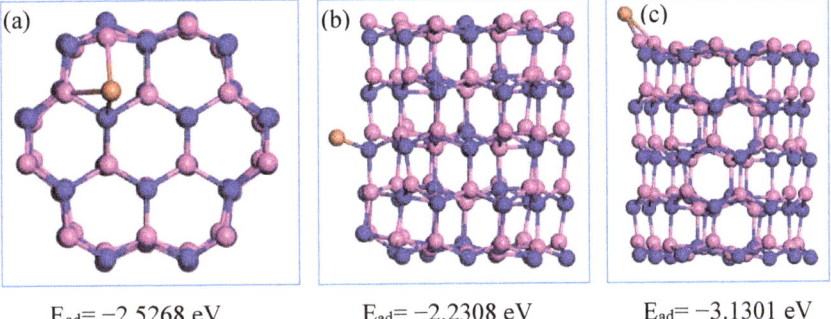

Figure 6. The chemisorption of Cu at different sites of the $Al_{60}N_{60}$ cluster. The Al, N and Cu atoms are shown in pink, blue and orange. The Cu atom is located at the top, side and edge position of $Al_{60}N_{60}$ cluster shown in (**a**) (top view), (**b**,**c**), respectively.

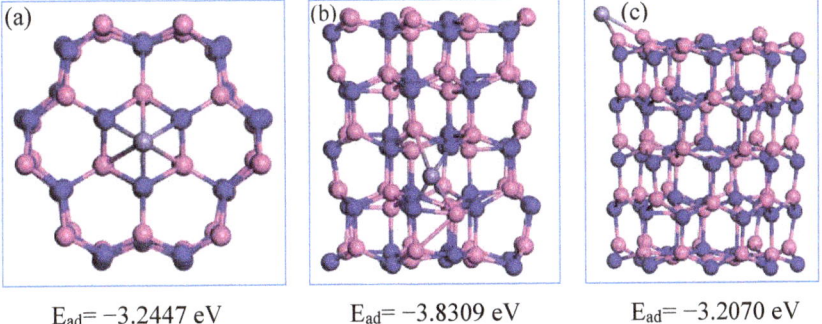

Figure 7. The chemisorption of Fe at different sites of the $Al_{60}N_{60}$ cluster. The Al, N and Fe atoms are shown in pink, blue and gray. The Fe atom is located at the top, side and edge position of $Al_{60}N_{60}$ cluster shown in (**a**) (top view), (**b**,**c**), respectively.

In order to discuss the effects of Al, N, Fe and Cu atoms on the electronic properties of the $Al_{60}N_{60}$ cluster, we have also investigated the electronic structures based on the configurations corresponding to the respective chemisorption systems with the largest adsorption energies of Al, N, Fe and Cu atoms on the cluster through the HSE06 calculations (shown in Figure 8). After Al or Fe atomic chemisorption, the electronic band gap of the system decreases, which illustrates that the $Al_{60}N_{60}$ large cluster is sensitive to the atomic chemisorption. What is more, the electronic density of states curves of the N-chemisorption and Cu-chemisorption systems near the Fermi level show some interesting similarities. This corresponds with the above absorption site similarity of the two systems. After adsorbing N or Cu atoms, a small number of deep levels are produced between the HOMO and LUMO.

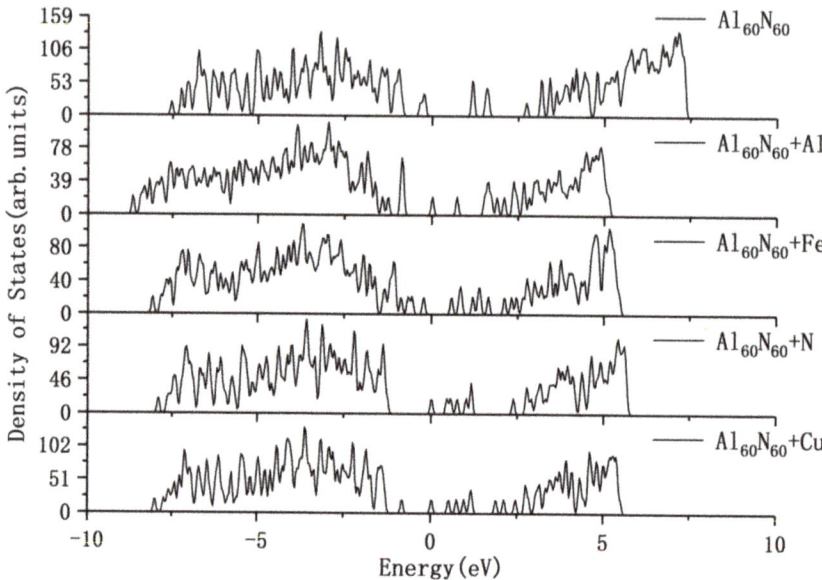

Figure 8. The electronic density of states of the pristine $Al_{60}N_{60}$ cluster and the cluster adsorbing Al, N, Cu and Fe atoms, respectively, through the HSE06 calculations (Fermi energy is set to zero).

4. Summary and Outlook

In summary, the AlN clusters prefer to grow longitudinally in the early stage and then evolve to the large cluster transversely by using a first principles method. During the structural evolution process, the cohesive energy generally increases and the large-size AlN clusters tend to shrink. The elemental chemisorption studies show that the copper and nitrogen atoms have similar chemisorption characteristics on the $Al_{60}N_{60}$ large cluster. It is also found that the electronic density of states curves of the N-chemisorption and Cu-chemisorption systems near the Fermi level show some interesting similarities through HSE06 calculations. This work is proposed to provide valuable theoretical clues and ab initio analyses for the experimentalists in the field and help to deepen understanding of the broader aluminum–pnictogen system at the basic atomic and electronic levels.

Author Contributions: Conceptualization, Z.Q.; methodology, Z.Q. and X.N.; formal analysis, Z.Q. and X.N.; data curation, Z.Q., X.N., W.D., Z.L., H.L. and X.L.; writing—original draft preparation, Z.Q. and X.N.; writing—review and editing, Z.Q. and R.A.; project administration, Z.Q.; funding acquisition, Z.Q.

Funding: We would like to thank the supports from the Natural Science Foundation of China (51801113) and the Natural Science Foundation of Shandong Province (ZR2018MEM001). The Young Scholars Program of Shandong University (YSPSDU), the Natural Science Foundation of China (51731007) and the Key Foundation of Shandong Province (ZR2016QZ005) are also thanked. R.A. thanks the Swedish Research Council (VR) and Swedish supercomputer facility (SNIC) for support.

Acknowledgments: Some scientific calculations in this work have been done on the HPC Cloud Platform of Shandong University.

Conflicts of Interest: The authors declare no conflict of interest.

References

1. Li, X.; Li, S.; Ren, H.; Yang, J.; Tang, Y. Effect of alkali metal atoms doping on structure and nonlinear optical properties of the gold-germanium bimetallic clisters. *Nanomaterials* **2017**, *7*, 184. [CrossRef]
2. Yan, J.; Xia, J.; Zhang, Q.; Zhang, B.; Wang, B. First-principles studieds on the structural and electronic properties of As clusters. *Materials* **2018**, *11*, 1596. [CrossRef]

3. Die, D.; Zheng, B.; Kuang, X.-y.; Zhao, Z.-Q.; Guo, J.-J.; Du, Q. Exploration of the structural, electronic and tunable magnetic properties of Cu4M (M=Sc-Ni) clusters. *Materials* **2017**, *10*, 946. [CrossRef]
4. Rao, B.K.; Jena, P. Evolution of the electronic structure and properties of neutral and charged aluminum clusters: A comprehensive analysis. *J. Chem. Phys.* **1999**, *111*, 1890–1904. [CrossRef]
5. Deshpande, M.D.; Kanhere, D.G.; Vasiliev, I.; Martin, R.M. Ab initio absorption spectra of Aln (n = 2–13) clusters. *Phys. Rev. B* **2003**, *68*, 035428. [CrossRef]
6. Guo, L. Density functional study of structural and electronic properties of AlnAs (1 ≤ n ≤ 15) clusters. *J. Alloys. Compd.* **2012**, *527*, 197–203. [CrossRef]
7. Ruiz, E.; Alvarez, S.; Alemany, P. Electronic structure and properties of AlN. *Phys. Rev. B* **1994**, *49*, 7115. [CrossRef]
8. Kim, H.J.; Egashira, Y.; Komiyama, H. Temperature dependence of the sticking probability and molecular size of the film growth species in an atmospheric chemical vapor deposition process to form AlN from AlCl3 and NH3. *Appl. Phys. Lett.* **1991**, *59*, 2521–2523. [CrossRef]
9. Jiang, Z.; Interrante, L.V. N, N'-Bis (triethylaluminum) ethylenediamine -and N, N'-Bis (trimethylaluminum) ethylenediamine-derived organometallic precursors to aluminum nitride: Syntheses, structures, and pyrolyses. *Chem. Mater.* **1990**, *2*, 439–446. [CrossRef]
10. Sauls, F.C.; Interrante, L.V.; Jiang, Z.P. Me3Al·NH3 formation and pyrolytic methane loss: thermodynamics, kinetics, and mechanism. *Inorg. Chem.* **1990**, *29*, 2989–2996. [CrossRef]
11. Kishimoto, K.; Funato, M.; Kawakami, Y. Effects of Al and N2 flow sequences on the interface formation of AlN on sapphire By EVPE. *Crystals* **2017**, *7*, 123. [CrossRef]
12. Búc, D.; Hotový, I.; Haščík, S.; Červeň, I. Reactive unbalanced magnetron sputtering of AlN thin films. *Vacuum* **1998**, *50*, 121–123. [CrossRef]
13. Beheshtian, J.; Peyghan, A.A.; Bagheri, Z. Quantum chemical study of fluorinated AlN nano-cage. *Appl. Surf. Sci.* **2012**, *259*, 631–636. [CrossRef]
14. Costales, A.; Blanco, M.A.; Francisco, E.; Pendas, A.M.; Pandey, R. First principles study of neutral and anionic (medium-size) aluminum nitride clusters: Al_nN_n, n=7–16. *J. Phys. Chem. B* **2006**, *110*, 4092–4098. [CrossRef]
15. Wu, H.S.; Zhang, F.Q.; Xu, X.H.; Zhang, C.J.; Jiao, H. Geometric and energetic aspects of aluminum nitride cages. *J. Phys. Chem. A* **2003**, *107*, 204–209. [CrossRef]
16. Kandalam, A.K.; Pandey, R.; Blanco, M.A.; Costales, A.; Recio, J.M.; Newsam, J.M. First principles study of polyatomic clusters of AlN, GaN, and InN. 1. structure, stability, vibrations, and ionization. *J. Phys. Chem. B* **2000**, *104*, 4361–4367. [CrossRef]
17. Saeedi, M.; Anafcheh, M.; Ghafouri, R.; Hadipour, N.L. A computational investigation of the electronic properties of Octahedral Al_nN_n and Al_nP_n cages (n = 12, 16, 28, 36, and 48). *Struct. Chem.* **2013**, *24*, 681–689. [CrossRef]
18. BelBruno, J.J. The structure of AlnNn (n = 2–4) clusters: a DFT study. *Chem. Phys. Lett.* **1999**, *313*, 795–804. [CrossRef]
19. Guo, C.; Wang, C. A theoretical study on the hydrogen storage properties of planar (AlN)n clusters (n = 3–5). *Struct. Chem.* **2017**, *28*, 1717–1722. [CrossRef]
20. Solimannejad, M.; Kamalinahad, S.; Shakerzadeh, E. Selective detection of toxic cyanogen gas in the presence of O2, and H2O molecules using a AlN nanocluster. *Phys. Lett. A* **2016**, *380*, 2854–2860. [CrossRef]
21. Takahashi, N.; Matsumoto, Y.; Nakamura, T. Investigations of structure and morphology of the AlN nano-pillar crystal films prepared by halide chemical vapor deposition under atmospheric pressure. *J. Phys. Chem. Solids.* **2006**, *67*, 665–668. [CrossRef]
22. Makarov, Y.N.; Avdeev, O.V.; Barash, I.S.; Bazarevskiy, D.S.; Chemekova, T.Y.; Mokhov, E.N.; Nagalyuk, S.S.; Roenkov, A.D.; Segal, A.S.; Vodakov, Y.A.; et al. Experimental and theoretical analysis of sublimation growth of AlN bulk crystals. *J. Cryst. Growth.* **2008**, *310*, 881–886. [CrossRef]
23. Yin, L.W.; Bando, Y.; Zhu, Y.C.; Li, M.S.; Li, Y.B.; Golberg, D. Growth and field emission of hierarchical single-crystalline Wurtzite AlN nanoarchitectures. *Adv. Mater.* **2010**, *17*, 110–114. [CrossRef]
24. Sumathi, R.R. Bulk AlN single crystal growth on foreign substrate and preparation of free-standing native seeds. *Cryst. Eng. Comm.* **2013**, *15*, 2232–2240. [CrossRef]

25. Ma, X.; Zhao, Y.; Zhao, X.; Gao, T.; Chen, H.; Liu, X. Influence mechanisms of Cu or Fe on the microstructures and tensile properties at 350 °C of network AlN$_p$ reinforced Al composites. *J. Alloy. Compd.* **2018**, *740*, 452–460. [CrossRef]
26. Perdew, J.P.; Burke, K.; Ernzerhof, M. Generalized gradient approximation made simple. *Phys. Rev. Lett.* **1996**, *77*, 3865. [CrossRef]
27. Hohenberg, P.; Kohn, W. Inhomogeneous electron gas. *Phys. Rev.* **1964**, *136*, B864. [CrossRef]
28. Kohn, W.; Sham, L.J. Self-consistent equations including exchange and correlation effects. *Phys. Rev.* **1965**, *140*, A1133. [CrossRef]
29. Kresse, G.; Joubert, D. From ultrasoft pseudopotentials to the projector augmented-wave method. *Phys. Rev. B* **1999**, *59*, 1758. [CrossRef]
30. Kresse, G.; Furthmuller, J. Efficient iterative schemes for ab initio total-energy calculations using a plane-wave basis set. *Phys. Rev. B* **1996**, *54*, 11169. [CrossRef]
31. Hartmann, C.; Wollweber, J.; Dittmar, A.; Irmscher, K.; Kwasniewski, A.; Langhans, F.; Neugut, T.; Bickermann, M. Preparation of bulk AlN seeds by spontaneous nucleation of freestanding crystals. *Jap. J. Appl. Phys.* **2013**, *52*, 08JA06. [CrossRef]
32. Marsman, M.; Paier, J.; Stroppa, A.; Kresse, G. Hybrid functionals applied to extended systems. *J. Phys-Condens. Mat.* **2008**, *20*, 064201. [CrossRef]
33. Ouyang, T.; Qian, Z.; Ahuja, R.; Liu, X. First-principles investigation of CO adsorption on pristine, C-doped and N-vacancy defected hexagonal AlN nanosheets. *Appl. Surf. Sci.* **2018**, *439*, 196–201. [CrossRef]
34. Ouyang, T.; Qian, Z.; Hao, X.; Ahuja, R.; Liu, X. Effect of defects on adsorption characteristics of AlN monolayer towards SO2 and NO2: Ab initio exposure. *Appl. Surf. Sci.* **2018**, *462*, 615–622. [CrossRef]

© 2019 by the authors. Licensee MDPI, Basel, Switzerland. This article is an open access article distributed under the terms and conditions of the Creative Commons Attribution (CC BY) license (http://creativecommons.org/licenses/by/4.0/).

Article

A Study of the Shock Sensitivity of Energetic Single Crystals by Large-Scale Ab Initio Molecular Dynamics Simulations

Lei Zhang [1,2,*], Yi Yu [1] and Meizhen Xiang [2]

1 CAEP Software Centre for High Performance Numerical Simulation, Beijing 100088, China
2 Laboratory of Computational Physics, Institute of Applied Physics and Computational Mathematics, Beijing 100088, China
* Correspondence: zhang_lei@iapcm.ac.cn

Received: 6 August 2019; Accepted: 27 August 2019; Published: 3 September 2019

Abstract: Understanding the reaction initiation of energetic single crystals under external stimuli is a long-term challenge in the field of high energy density materials. Herewith, we developed an *ab initio* molecular dynamics method based on the multiscale shock technique (MSST) and reported the reaction initiation mechanism by performing large-scale simulations for the sensitive explosive benzotrifuroxan (BTF), insensitive explosive triaminotrinitrobenzene (TATB), four polymorphs of hexanitrohexaazaisowurtzitane (CL-20) pristine crystals and five novel CL-20 cocrystals. A theoretical indicator, $t_{initiation}$, the delay of decomposition reaction under shock, was proposed to characterize the shock sensitivity of energetic single crystal, which was proved to be reliable and satisfactorily consistent with experiments. We found that it was the coupling of heat and pressure that drove the shock reaction, wherein the vibrational spectra, the specific heat capacity, as well as the strength of the trigger bonds were the determinants of the shock sensitivity. The intermolecular hydrogen bonds were found to effectively buffer the system from heating, thereby delaying the decomposition reaction and reducing the shock sensitivity of the energetic single crystal. Theoretical rules for synthesizing novel energetic materials with low shock sensitivity were given. Our work is expected to provide a useful reference for the understanding, certifying and adjusting of the shock sensitivity of novel energetic materials.

Keywords: BTF; TATB; CL-20; cocrystal; energetic materials; shock sensitivity; large-scale ab initio molecular dynamics simulations

1. Introduction

Energetic materials (EMs) such as explosives, oxidizers and propellants are of significant importance in aerospace, oil-well drilling and other military and civilian applications. In this field, understanding the sensitivity of single crystals under shock or impact has long been a challenge. In engineering, shock sensitivity can be identified by the shock initiation threshold pressure, P_{90}, which is obtained from the gap test and produces a detonation 50% of the time. P_{90} results are generally reproducible and reliable. However, because of the high complexity of the test, the measurements have been performed for only limited types of energetic single crystals [1,2] and the test is not easily applicable to newly synthesized EMs. On the other hand, the drop-weight impact test, which characterizes the impact sensitivity of EMs by height $h_{50\%}$, is easy to implement. Therefore, there are abundant values of $h_{50\%}$ in the literature compared to P_{90} values. However, the $h_{50\%}$ value is generally not reproducible as the results significantly vary depending on the conditions under which the tests are performed. For example, for benzotrifuroxan (BTF), the reported $h_{50\%}$ values vary from 21 [3] to 50 [4] cm; for hexanitrohexaazaisowurtzitane/trinitrotoluene (CL-20/TNT) cocrystal, the values vary from 30 [4] to 99 [5] cm. Therefore, the $h_{50\%}$ values derived from the same experiments are comparable, while those from different equipment can only be used for a quantitative comparison of the mechanical sensitivity among various energetic single crystals.

With the rapid increase of computational capability and the development of material modeling methods, large-scale atomistic simulation becomes a powerful tool for understanding the physical processes of materials under extreme conditions [6,7]. However, in recent decades, there has been a strong tendency in the literature to elucidate the sensitivity of energetic single crystals by the electron density properties in a separate molecule of EM, such as electrostatic potential, molecular electronegativities, partial atomic charges, molecular weights, vibrational states, oxygen balance of the molecules, detonation gas concentrations and heats of detonation [1]. These quantities ignore the deformation of the chemical bonds, the motion of the molecules, the inner/inter molecular chemical reactions and the symmetrical structure of the crystals, and thereby cannot comprehensively characterize the reaction initiation of an energetic single crystal under shock.

To this end, we developed an *ab initio* molecular dynamics method [8] and extended its computational capability by improving the code's parallel calculation efficiency. We performed shock wave simulation tests on 11 types of energetic single crystals, with each simulated model composed of ~1000 atoms. On the basis of the calculations, we proposed a theoretical indicator to characterize the shock wave sensitivity of energetic single crystals, which is expected to be useful for the evaluation and adjustment of the shock sensitivity of novel EMs. We also revealed the shock reaction initiation mechanism, found factors that can inhibit the shock sensitivity of EMs and provided theoretical rules for synthesizing novel EMs with low shock sensitivity.

2. Methodology

2.1. Multiscale Simulation Method of Shock Wave Tests

Figure 1a schematically shows the simulated dynamical shock process in a single crystal. At the beginning when the shock wave reaches the single crystal, the simulation region starts to undergo lattice and molecular deformation. With the increase of simulation time, the simulation region gradually leaves the wave front relative to the unshocked material. The molecules in the simulated region are then compressed to react, eventually reaching the Chapman–Jouguet (CJ) state when the shock wave propagates steadily. We implemented the multiscale shock technique (MSST) [9,10] in the High Accuracy atomistic Simulation for Energetic Materials (HASEM) package [8], so as to capture the atomic motions and the chemical reactions of inner/inter molecules on the density functional theory (DFT) accuracy level.

2.1.1. Continuum Theory Description of the Shock Wave Structure

The shock wave propagation was modelled using the one-dimensional Euler equations for compressible flow [9,10], which were represented by the conservation of mass, momentum, and energy, respectively, in the regions before and after the shock wave interface.

$$u = v_{shock}(1 - \frac{\rho_0}{\rho}), \tag{1}$$

$$p - p_0 = v_{shock}^2 \rho_0 (1 - \frac{\rho_0}{\rho}), \tag{2}$$

$$e - e_0 = p_0(\frac{1}{\rho_0} - \frac{1}{\rho}) + \frac{v_{shock}^2}{2}(1 - \frac{\rho_0}{\rho})^2, \tag{3}$$

where v_{shock} is the shock wave speed, u is the particle velocity, ρ is the material density, e is the energy per unit mass and p is the negative component of the stress tensor along the shock propagation direction. The quantities with 0 subscript describe the states of the unreacted material.

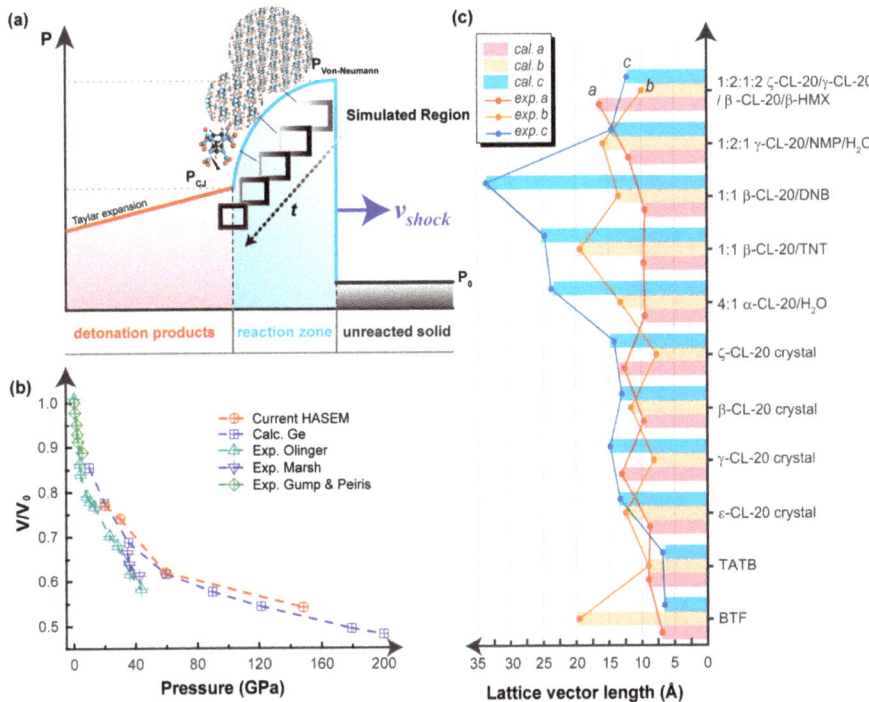

Figure 1. Dynamics simulation method of shock wave tests. (**a**) Schematics of the simulation method. (**b**) Validation of the code by comparing the Hugoniot curve of a single crystal model of octogen (HMX) single crystal to the reported experimental and calculational results in the literature. (**c**) Accuracy verification of the code by comparing the lattice lengths of the studied eleven energetic single crystals to the experimental values obtained by X-ray crystallography.

2.1.2. Molecular Dynamics Description of the Atomic Motions

In the simulated region, the atomic motions were simulated using the molecular dynamics (MD) method. The Lagrangian per unit mass is

$$L = T_e\left(\left\{\vec{r}_i\right\}\right) - V_e\left(\left\{\vec{r}_i\right\}\right) + \frac{1}{2}Q\dot{v}^2 + \frac{1}{2}\frac{v_{shock}^2}{v_0^2}(v_0 - v)^2 + p_0(v_0 - v), \tag{4}$$

where $v = 1/\rho$ is the specific volume; T_e and V_e are kinetic and potential energies, respectively, and their sum equals to e; Q is a parameter related to the mass of the simulated cell. When the volume of the simulated cell was fixed, that is, when $\dot{v} = 0$, the Lagrangian expression is equivalent to the continuum Hugoniot relation of Equation (3).

The atomic position and velocity at each time step were obtained from the control equation of the simulated cell volume:

$$Q\ddot{v} = \frac{\partial T}{\partial v} - \frac{\partial V}{\partial v} - p_0 - \frac{v_{shock}^2}{v_0^2}(v_0 - v), \tag{5}$$

which degenerates to the Rayleigh line of Equation (2) when the cell volume changes uniformly [9,10]. Thus, the system is restrained to fit the shock Hugoniot and the Rayleigh line of the material by changing the volume of the simulated cell.

2.1.3. Density Functional Theory Description of the Electronic Structure

The atomic force of each time step was updated according to the DFT calculations of the electronic structure using HASEM software. The generalized gradient approximation was used for the exchange-correlation functional in the Perdew–Burke–Ernzerhof form. Norm-conserving pseudopotentials specialized for EM crystals were used to replace the core electrons. The valence electrons, described by linear combinations of numerical pseudoatomic orbitals, were calculated on a three-dimensional real-space grid. The reliability of the DFT calculations to describe the structures, energetics, dynamics, mechanical properties, detonation performance and sensitivity of EM crystals has been extensively confirmed in previous work [8,11–15].

In order to improve the computational capability of the dynamics simulation method, we reconstructed the HASEM software based on the J parallel adaptive structured mesh applications infrastructure (JASMIN), which has successfully accelerated many parallel programs for large scale simulations of complex applications on parallel computers [16]. Through this, the calculation efficiency of HASEM software was improved by one order of magnitude. Simulations of large-scale systems containing ~1000 atoms can thereby be achieved by using extended central processing units (CPUs) on supercomputers.

2.1.4. Verification and Validation of the Dynamics Simulation Method

We constructed a single crystal model of octogen (HMX) and performed shock wave tests using the newly developed dynamics simulation method. A series of shock waves, with a speed smaller than 5 km/s, were applied on the HMX model. As shown in Figure 1b, the obtained Hugoniot curve satisfactorily agreed with the experiments [17–19] and other calculations [20], thereby confirming the reliability of the current method.

2.2. Simulation Models of Eleven EM Single Crystals

CL-20 has been proven to show excellent performance since it was first synthesized by the Naval Air Warfare Center China Lake 30 years ago [21]; however, it has not been widely used until now because of the sensitivity problems of its ε, β, γ and ζ polymorphs [21]. Cocrystallization, which mixes several components on a molecular level, has been considered a promising technique to obtain advanced EMs with good detonation performance and low sensitivity to accidental initiation. Under that circumstance, five novel cocrystals—CL-20/H_2O, CL-20/TNT, CL-20/1,3-dinitrobenzene (CL-20/DNB), CL-20/N-methyl-2-pyrrolidone/H_2O (CL-20/NMP/H_2O) and CL-20/HMX, have been recently synthesized.

Herewith, we studied the four CL-20 polymorphs and the five novel CL-20 cocrystals, as well as the sensitive explosive BTF and the insensitive explosive triaminotrinitrobenzene (TATB). For the 11 EMs studied, we optimized the crystal geometries using the conjugate gradient method on a DFT level, with the initial inputs taken from the lattice parameters and atomic coordinates from single-crystal X-ray diffraction analysis [5,22–29]. The structures were considered as optimized when the stress components were less than 0.01 GPa and the residual forces were less than 0.03 eV/Å. As shown in Figure 1c, the calculated lattice lengths showed satisfactory agreement [30] with the experimental measurements (σ = 0.18 Å; R^2 = 0.9992), thereby confirming the reliability of the current method.

Subsequently, we built large-scale supercells of the eleven EM crystals for the shock simulations. As shown in Table 1, the supercells generally contained more than 1000 atoms, with the lattice length in the range of 15~30 Å. There were 24~64 molecules in each supercell, containing 72~192 trigger chemical bonds. The number of the chemical bonds included here was an order of magnitude larger than the traditional *ab initio* MD simulations, and thereby better reflects the randomness and probability characteristics of the chemical reaction kinetics.

Table 1. Lattice lengths (Å) of the supercells used for shock simulations. The type of the trigger bonds for each crystal, as well as the number of the composed atoms, molecules and trigger bonds in each supercell, are also given. EMs = energetic materials, BTF = benzotrifuroxan, TATB = insensitive explosive triaminotrinitrobenzene, CL-20 = hexanitrohexaazaisowurtzitane, TNT = trinitrotoluene, HMX = single crystal model of octogen.

EMs	a	b	c	Trigger Type	Number of		
					Atoms	Molecules	Triggers
Sensitive							
BTF	20.86	19.89	19.63	N–O	648	36	108
Insensitive							
TATB	18.18	27.31	19.44	C–N	576	24	72
Pure CL-20							
ε-polymorph	17.83	25.26	26.70	N–N	1152	32	192
γ-polymorph	26.11	16.75	29.66	N–N	1152	32	192
β-polymorph	19.47	23.03	26.44	N–N	1152	32	192
ζ-polymorph	26.74	16.22	29.37	N–N	1152	32	192
CL-20 cocrystal							
4:1 γ-CL-20/H_2O	19.15	27.02	23.36	N–N	1176	40	192
1:1 β-CL-20/TNT	19.33	19.65	25.14	N–N	912	32	96
1:1 β-CL-20/DNB	18.94	13.48	33.57	N–N	832	32	96
1:2:1 γ-CL-20/NMP/H_2O	23.50	15.82	28.88	N–N	1136	64	96
1:2:1:2 ζ-CL-20/γ-CL-20/β-CL-20/β-HMX	16.56	19.81	24.06	N–N	800	24	96

2.3. Control Parameter of the Shock Simulation Tests

All the 11 systems were shocked with the same speed V_{shock} = 9 km/s at a direction perpendicular to the lattice vector. We used this shock wave speed based on a previous classical MD study, in which the breaking of the CL-20 trigger bonds was apparent at this shock speed [31]. The time step for the *ab initio* MD simulation was set to be 0.1 fs. As shown in Table 1, under this shock condition, 10 of the systems (excluding the insensitive explosive TATB) were initiated within 10,000 steps, that is, 1000 fs.

We defined the chemical bonds as broken when they were stretched to a cutoff percentage relative to each equilibrium state. The cutoff criterion was 20% on average, but it varied from 10% to 40% for different types of bonds. The criterion for each type of bond was determined by the statistics of the reaction products of the TATB explosive when the number of product molecules best agreed with the number of the stable clusters with a life span more than hundreds of time steps during the kinetics simulation.

3. Results and Discussion

3.1. Shock Dynamics of the 11 EM Crystals

The 11 crystals studied were rapidly compressed under shock, as shown in Figure 2. The temperature and pressure of the systems increased as a function of time during the shock process. The molecules were packed more densely in space and the chemical bonds plastically deformed to break, leading to the decomposition of material. According to our simulation, the N–NO_2 bond, N–O bond and C–NO_2 bond were the most active chemical bonds to deform and break in the CL-20 molecule, BTF molecule and TATB molecule, respectively, under shock. We therefore denoted these bonds as the "trigger bonds".

Take the shocked ε-CL-20 crystal as an example. The N–NO_2 bonds with the exo-spatial orientation with respect to the five-membered imidazolidine ring were the trigger bonds. During the shock process, the molecular conformations at different times are shown in Figure 3a. Both the increase of the trigger bond length and the decrease of the trigger bond strength went in an exponential manner, as shown in Figure 3b,c. When the stretching of the chemical bond reached the cutoff percentage relative to the equilibrium state, we defined it as broken. Therefore, the first breaking of the trigger bond of the shocked ε-CL-20 crystal occurred at t_3 = 145.8 fs.

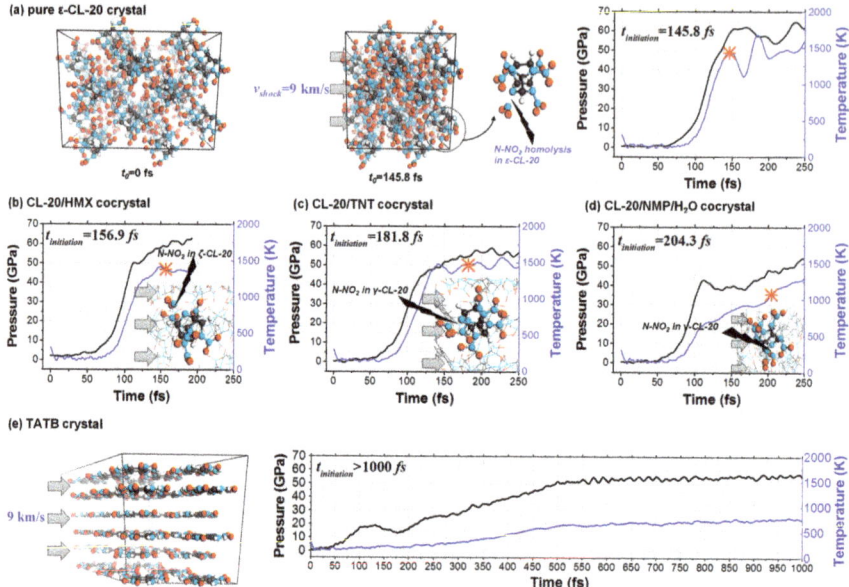

Figure 2. Shock dynamics of the (**a**) ε-CL-20 crystal, (**b**) CL-20/HMX cocrystal, (**c**) CL-20/TNT cocrystal, (**d**) CL-20/NMP/H_2O cocrystal and (**e**) TATB crystal. In each panel, the temperature and pressure are shown as a function of time. The molecular conformations at the beginning of each decomposition reaction are also plotted.

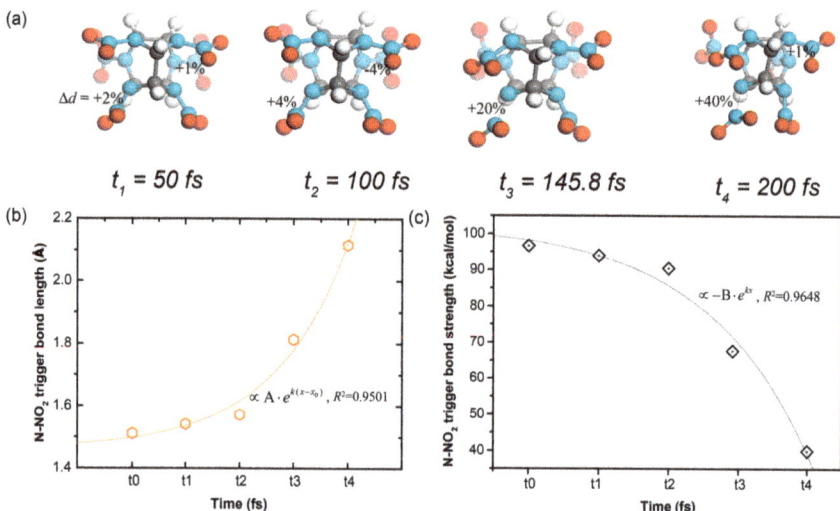

Figure 3. (**a**) Conformation vs time plots for a molecule in the shocked ε-CL-20 crystal, along with the corresponding (**b**) length and (**c**) strength of the trigger bond N–NO_2 in each snapshot of (**a**).

The chemical bonds of the eleven crystals studied included N–N, H–C, H–N, H–O, C–C, C–N, C–O, N–O and O–O bonds. For all types of chemical bonds in each crystal, we counted their number as a function of time during the shock process. As shown in Figure 4, the covalent bonds' breaking and recombination of shocked material was a dynamical process. For example, the trigger bonds N–NO_2

in CL-20/NMP/H$_2$O started to break at 124.8 fs, but they recombined to the initial state at 189.3 fs. We thereby defined the initiation of the chemical reaction by the time $t_{initiation}$, from when the breaking of the chemical bonds was always more than their recombination and the number of the trigger bonds decreased continuously. Therefore, the decomposition reaction began at $t_{initiation}$ = 103.6 fs for BTF and at 204.3 fs for CL-20/NMP/H$_2$O.

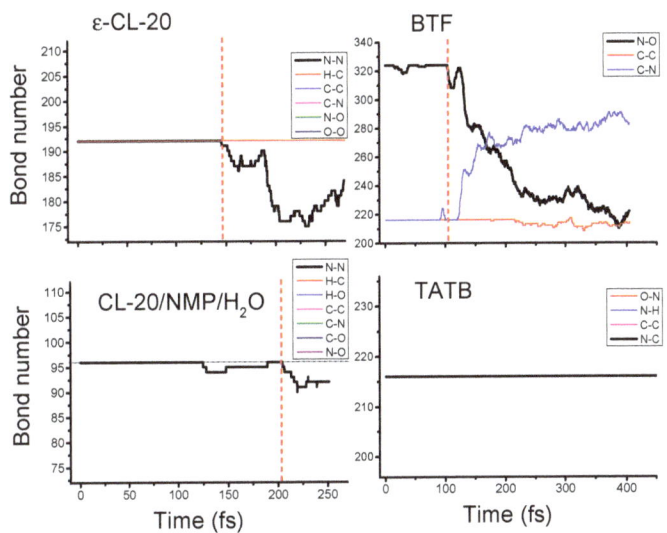

Figure 4. Bond number vs time plots for the shocked ε-CL-20 crystal, BTF crystal, CL-20/NMP/H$_2$O cocrystal and TATB crystal. The chemical bonds counted included N–N, H–C, H–N, H–O, C–C, C–N, C–O, N–O and O–O bonds. The sign of the initiation of the reaction under shock was a continuous reduction in the number of trigger bonds.

Generally speaking, for the energetic crystals containing CL-20 molecules, both the temperature and the pressure increased slowly at the beginning of the shock process, as shown in Figure 2. Then, at ~100 fs, the temperature and the pressure started to drastically increase to higher than 1000 K and higher than 50 GPa. Next, at ~150 fs, the systems reached a gently varied stage, during which the trigger chemical bonds started to break and the decomposition reactions of material began. For the shocked insensitive explosive TATB, both temperature and pressure varied uniformly, and no chemical reactions occurred before 1000 fs, while for the shocked sensitive explosive BTF, the chemical reaction already started at 103.6 fs.

3.2. Theoretical Indicator of Shock Sensitivity: $t_{initiation}$

Because of the lack of experimental shock sensitivity for most of the EMs studied, we proposed using $t_{initiation}$ as a theoretical indicator to characterize the ease of the shock reaction initiation. Because shock sensitivity has been proven to have a satisfactory correlation with the impact sensitivity [1,2], we used the experimental value $h_{50\%}$ as a reference to compare with $t_{initiation}$, as shown in Table 2. We note comparing $h_{50\%}$ values from the same experiments can well reflect the relative sensitivity of different compounds, while comparing those from different experiments was only qualitatively reasonable because of the influence of different experimental conditions used.

Table 2. The experimental $h_{50\%}$ (cm) values and the initiation time of the shock reaction $t_{initiation}$ (fs) of the 11 EMs studied. The strength of the trigger bond $S_{trigger}$ (kcal/mol), the temperature $T_{initiation}$ (K) and the temperature rising rate TRR (K/fs) for each crystal are also given for the study of the mechanism of the shock reaction initiation.

EMs	$h_{50\%}$							$S_{trigger}$	$t_{initiation}$	$T_{initiation}$	TRR
	Expt 1 [4]	Expt 2 [27]	Expt 3 [28]	Expt 4 [5]	Expt 5 [3]	Expt 6 [32]	Expt 7 [26]	Current Calculation			
Sensitive											
BTF	50			21				42	103.6	2246	21.7
Insensitive											
TATB	>320							125	>1000.0	754	0.8
Pure CL-20											
ε-polymorph		14	29	47		12–21	13	112	145.8	1431	9.8
γ-polymorph								112	138.0	1600	11.6
β-polymorph					14			111	139.3	1149	8.2
ζ-polymorph								110	116.9	1048	9.0
CL-20 cocrystal											
CL-20/H$_2$O		16						113	174.4	1484	8.5
CL-20/TNT				99			30	112	181.8	1464	8.1
CL-20/DNB							55	111	174.6	1400	8.0
CL-20/NMP/H$_2$O		112						115	204.3	1074	5.3
CL-20/HMX			55					112	156.9	1377	8.8

As shown in Table 2, there was a satisfactory agreement between the $t_{initiation}$ and the $h_{50\%}$ values derived from the same experiment. For example, in experiment 1, BTF had the highest sensitivity with $h_{50\%}$ = 50 cm, and TATB had the lowest sensitivity with $h_{50\%}$ > 320 cm. Correspondingly, BTF had the shortest delay of shock reaction at $t_{initiation}$ = 103.6 fs, while TATB had the longest delay with $t_{initiation}$ > 1000.0 fs. In experiment 2, the sensitivity order characterized by $h_{50\%}$ was ε-CL-20 > CL-20/H$_2$O > CL-20/NMP/H$_2$O. Correspondingly, the sensitivity order quantified by $t_{initiation}$ was exactly the same, which was 145.8 fs for ε-CL-20, 174.4 fs for CL-20/H$_2$O and 204.3 fs for CL-20/NMP/H$_2$O. In experiments 3 and 4, the $h_{50\%}$ for ε-CL-20 was less than those for CL-20/HMX and CL-20/TNT. Consistent with this, $t_{initiation}$ = 145.8 fs for ε-CL-20 was also smaller than $t_{initiation}$ = 156.9 fs for CL-20/HMX and $t_{initiation}$ = 181.8 fs for CL-20/HMX.

According to all the above comparisons between the calculated $t_{initiation}$ and the measured $h_{50\%}$ values, $t_{initiation}$ is a reproduceable and reliable indicator to calibrate the shock sensitivity.

3.3. Mechanism of Shock Reaction Initiation

The shock can be simplified into a perfect impulse f(t), which has an infinitely small duration. Its Fourier transform $F(\omega) = \int_{-\infty}^{+\infty} f(t)e^{-j\omega t}dt = F_0$ implies that the shock causes a constant amplitude response in the entire frequency domain. Therefore, the more characteristic peaks in the vibrational spectra of an energetic crystal, the more modes can be excited and the more heat can be generated under the same shock condition. We plotted the vibrational spectra for the ε, β, γ and ζ polymorphs of CL-20 crystals in Figure 5. The number of characteristic peaks of the vibrational spectra of ζ-CL-20 was 25, which was the least among the four polymorphs. Correspondingly, the generated temperature $T_{initiation}$ = 1048 K was also the lowest. On the other hand, that peak number was the most for γ-CL-20 (29) and consistent with this, the temperature of $T_{initiation}$ = 1600 K was the highest. For the other two polymorphs, both the peak number and the temperature fall in the middle.

In order to study the mechanism of the initiation of shock reaction, we calculated for the 11 crystals the strength of the trigger bond $S_{trigger}$ and the temperature rising rate (TRR) under shock, as shown in Table 2. In this, the bond strength was quantified by the integrated value of the crystal orbital Hamilton population (COHP) at band energy, and the temperature rising rate was calculated by dividing the temperature (when t = $t_{initiation}$) by $t_{initiation}$.

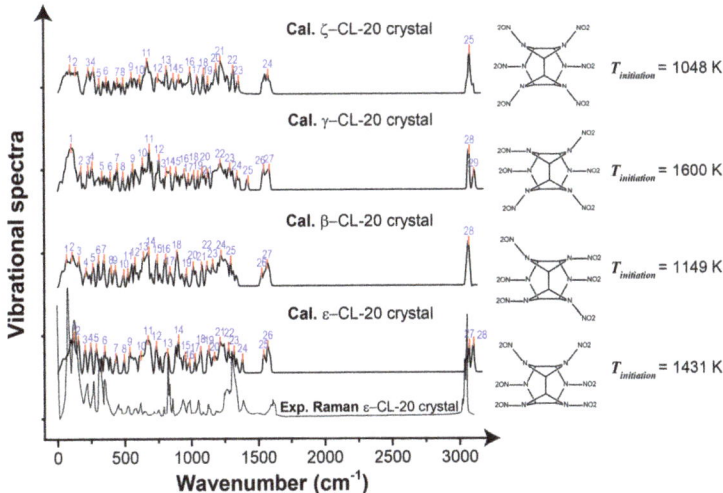

Figure 5. Vibrational spectra for ε, β, γ and ζ polymorphs of CL-20 crystals. The characteristic peaks of the vibrational spectra are marked by vertical red lines and are indexed by blue texts. For each of the four polymorphs, the molecular conformation and the temperature ($T_{initiation}$) the crystal starts to decay under shock are also shown.

As shown in Table 2, the trigger bond strength of the sensitive explosive BTF was $S_{trigger}$ = 42 kcal/mol, while that of the insensitive explosive TATB was three times higher. As well as this, the temperature rising rate of shocked BTF was TRR = 21.7 K/fs, while that of TATB was 29 times smaller. For the other EMs containing CL-20 molecules, both the trigger bond strength and the TRR fell in the range between BTF and TATB. In addition, we found that the $t_{initiation}$–TRR correlation showed a satisfactory power function $y = 804 \times x^{-0.76}$, with the coefficient of determination R^2 = 0.995, as shown in Figure 6. Therefore, the ease of the shock reaction initiation was apparently determined by the trigger bond strength and the temperature rising rate under shock. As a simplification, the specific heat capacity of a compound was the amount of heat needed per unit mass in order to raise the temperature by ΔT. Therefore, a compound with a larger specific heat capacity generally has a smaller TRR. We thereby propose that stronger covalent bonds and higher specific heat capacity are beneficial for delaying the time of shock initiation $t_{initiation}$, that is, reducing the shock sensitivity.

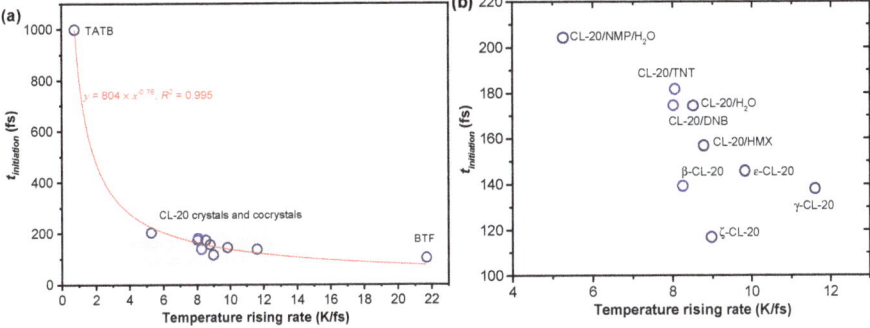

Figure 6. (a) Correlation between the shock sensitivity and the temperature rising rate under shock; (b) is an enlarged plot in a focused range, as marked by gray in plot (a).

Among the five CL-20 cocrystals, CL-20/NMP/H$_2$O had the lowest TRR = 5.3 K/fs, while the other EMs had their TRR values in the range of 8.0–8.8 K/fs, as shown in Figure 6b. At the same time, the trigger bond strength of CL-20/NMP/H$_2$O was also the highest, which was $S_{trigger}$ = 115 kcal/mol, as shown in Table 2. Therefore, CL-20/NMP/H$_2$O was able to obtain the lowest shock sensitivity among the five cocrystals.

From the explanation above, the initiation of the shock reaction of energetic single crystals is shown to be a process driven by the coupling of heat and pressure. The heat is derived from the mechanical work of the shock compression and is transferred into the vibration of the lattice, the molecules and the chemical bonds of the shocked material. Denser characteristic peaks of vibrational spectrum correspond to a larger amount of heat generated by the shock. Driven by the heat, the temperature of the system quickly increases and the stretch vibrational modes of the chemical bonds are activated. While vibrating, the chemical bonds also endure plastic deformation under the shock compression. When the deformation of the trigger bond is beyond the critical level, the shock reaction begins.

3.4. Shock Sensitivity Buffer: Intermolecular Hydrogen Bond

On the basis of the calculated $t_{initiation}$, we were able to predict the relative sensitivity of all the 11 EMs studied, which was shown to be BTF > ζ-CL-20 > γ-CL-20 > β-CL-20 > ε-CL-20 > CL-20/HMX > CL-20/H$_2$O > CL-20/DNB > CL-20/TNT > CL-20/NMP/H$_2$O > TATB. The predicted order shows a close relationship between the shock sensitivity and the hydrogen bonding amount. For example, BTF contains no hydrogen and it owns the highest sensitivity, whereas TATB contains the most hydrogen and it has the lowest sensitivity.

In Figure 7 we show quantitatively the relationship between the shock sensitivity and the hydrogen bonding amount for the EMs containing CL-20 molecules, wherein the hydrogen bonding amount is represented by the occupied percentage in the Hirshfeld surface of the CL-20 molecules. The correlation is a satisfactory exponential function, in which $t_{initiation} \propto 1/\left(1+e^{-k(x-x_0)}\right)$, with the coefficient of determination R^2 = 0.9998. The correlation implies that the more hydrogen bonding occurs, the lower the shock sensitivity.

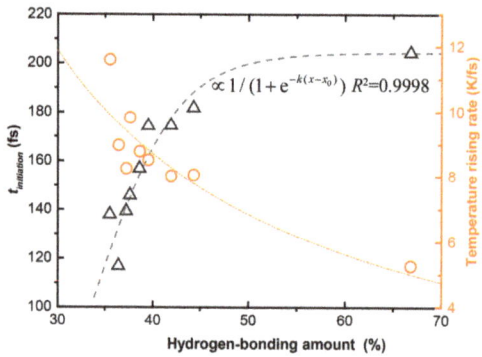

Figure 7. Shock sensitivity vs amount of hydrogen bonding for the CL-20 composed EMs studied.

The intermolecular hydrogen bond A:H–D (with ":" representing the electron lone pair, A for acceptor and D for donor) integrates the H–D polar-covalent bond, the A:H nonbond, and the A–D repulsive coupling interaction. Under shock, the hydrogen bonds show their elasticity—the covalent bond segment contracts and the nonbond elongates [33,34]. The special elasticity allows hydrogen bonds to vibrate in a continuous frequency region (<200 cm^{-1}) so that the crystal can absorb more energy from the shock before reaching a temperature that is too high. This is analogous to the function of hydrogen bonds on improving the specific heat capacity of liquid H$_2$O [35]. In order to confirm our hypothesis, we show the relationship between the TRR and the hydrogen bonding amount in

Figure 7, which is roughly a power function. This relationship suggests that the hydrogen bonding has a buffering effect on the heating of the system under shock, thereby delaying the initiation time of the chemical reaction $t_{initiation}$. This is the fundamental reason why cocrystallization with low-sensitive EM components can effectively reduce the sensitivity of CL-20 crystals.

4. Conclusions

To conclude, we have developed an *ab initio* molecular dynamics method based on the multiscale shock technique and performed shock wave simulation tests for the sensitive explosive BTF, insensitive explosive TATB, four polymorphs of CL-20 crystals and five novel CL-20 cocrystals, with each model containing ~1000 atoms. The main conclusion includes:

(1) We proposed a theoretical indicator $t_{initiation}$ to characterize the shock sensitivity of an energetic single crystal, which has been proven to be reliable and satisfactorily consistent with experiments.
(2) The shock reaction initiation was found to be a process driven by heat and pressure coupling and the vibrational spectra, the specific heat capacity, as well as the strength of the trigger bonds being the determinants of the shock sensitivity of energetic single crystals.
(3) Intermolecular hydrogen bonds were found to effectively buffer the system from heating, thereby delaying the trigger bonds from breaking and ultimately reducing the shock sensitivity of the energetic crystal.
(4) To synthesize advanced energetic materials with low shock sensitivity, small characteristic peak density of the crystal vibrational spectra, high specific heat capacity, strong trigger chemical bonds and high hydrogen bond amounts were theoretically recommended.

Our work is expected to provide a theoretical reference for the understanding, certifying and adjusting of the mechanical sensitivity of the single crystals of novel energetic materials.

Author Contributions: L.Z. designed and wrote the original manuscript. Y.Y. and M.X. developed and verified the *ab initio* molecular dynamics code.

Funding: This research was funded the National Natural Science Foundation of China [grant number 11604017].

Acknowledgments: L.Z. thanks Hui Huang from the China Academy of Engineering Physics for fruitful discussion.

Conflicts of Interest: The authors declare no conflict of interest.

References and Note

1. Brill, T.B.; James, K.J. Thermal decomposition of energetic materials. 61. Perfidy in the amino-2,4,6-trinitrobenzene series of explosives. *J. Phys. Chem.* **1993**, *97*, 8752. [CrossRef]
2. Gray, P. *Chemistry and Physics of Energetic Materials*; Springer: Dordrecht, The Netherlands, 1990.
3. Zhang, H.; Guo, C.; Wang, X.; Xu, J.; He, X.; Liu, Y.; Liu, X.; Huang, H.; Sun, J. Five Energetic Cocrystals of BTF by Intermolecular Hydrogen Bond and pi-Stacking Interactions. *Cryst. Growth Des.* **2013**, *13*, 679. [CrossRef]
4. Shukla, M.K.; Boddu, V.M.; Steevens, J.A.; Damavarapu, R.; Leszczynski, J. *Energetic Materials: From Cradle to Grave*; Springer International Publishing: Cham, Switzerland, 2017; Volume 25.
5. Weinan, E. *Principles of Multiscale Modeling*; Cambridge University Press: Cambridge, UK, 2011.
6. Sosso, G.C.; Miceli, G.; Caravati, S.; Giberti, F.; Behler, J.; Bernasconi, M. Fast Crystallization of the Phase Change Compound GeTe by Large-Scale Molecular Dynamics Simulations. *J. Phys. Chem. Lett.* **2013**, *4*, 4241. [CrossRef] [PubMed]
7. Kuhne, T.D.; Pascal, T.A.; Kaxiras, E.; Jung, Y. New Insights into the Structure of the Vapor/Water Interface from Large-Scale First-Principles Simulations. *J. Phys. Chem. Lett.* **2011**, *2*, 105. [CrossRef] [PubMed]
8. Zhang, L.; Jiang, S.L.; Yu, Y.; Long, Y.; Zhao, H.Y.; Peng, L.J.; Chen, J. Phase Transition in Octahydro-1,3,5,7-tetranitro-1,3,5,7-tetrazocine (HMX) under Static Compression: An Application of the First-Principles Method Specialized for CHNO Solid Explosives. *J. Phys. Chem. B* **2016**, *120*, 11510–11522. [CrossRef] [PubMed]

9. Reed, E.J.; Fried, L.E.; Joannopoulos, J.D. A Method for Tractable Dynamical Studies of Single and Double Shock Compression. *Phys. Rev. Lett.* **2003**, *90*, 235503. [CrossRef] [PubMed]
10. Reed, E.J.; Fried, L.E.; Henshaw, W.D.; Tarver, C.M. Analysis of simulation technique for steady shock waves in materials with analytical equations of state. *Phys. Rev. E* **2006**, *74*, 056706. [CrossRef] [PubMed]
11. Zhang, L.; Wu, J.Z.; Jiang, S.L.; Yu, Y.; Chen, J. From intermolecular interactions to structures and properties of a novel cocrystal explosive: A first-principles study. *Phys. Chem. Chem. Phys.* **2016**, *18*, 26960–26969. [CrossRef] [PubMed]
12. Jiang, C.; Zhang, L.; Sun, C.; Zhang, C.; Yang, C.; Chen, J.; Hu, B. Response to Comment on "Synthesis and characterization of the pentazolate anion cyclo-N_5^- in $(N_5)_6(H_3O)_3(NH_4)_4Cl$.". *Science* **2018**, *359*, eaas8953. [CrossRef]
13. Zong, H.H.; Zhang, L.; Zhang, W.B.; Jiang, S.L.; Yu, Y.; Chen, J. Structural, mechanical properties and vibrational spectra of LLM-105 under high pressures from a first-principles study. *J. Mol. Model.* **2017**, *23*, 275. [CrossRef]
14. Zhang, L.; Jiang, S.L.; Yu, Y.; Chen, J. Revealing Solid Properties of High-energy-density Molecular Cocrystals from the Cooperation of Hydrogen Bonding and Molecular Polarizability. *Sci. Rep.* **2019**, *9*, 1257. [CrossRef] [PubMed]
15. Zhang, L.; Yao, C.; Yu, Y.; Jiang, S.L.; Sun, C.Q.; Chen, J. Stabilization of the Dual-Aromatic cyclo-N_5^- Anion by Acidic Entrapment. *J. Phys. Chem. Lett.* **2019**, *10*, 2378. [CrossRef] [PubMed]
16. Mo, Z.; Zhang, A.; Cao, X.; Liu, Q.; Xu, X.; An, H.; Pei, W.; Zhu, S. JASMIN: A parallel software infrastructure for scientific computing. *Front. Comput. Sci. China* **2010**, *4*, 480. [CrossRef]
17. Olinger, B.; Roof, B.; Cady, H. The linear and volume compression of β-HMX and RDX. In *Proceedings of the Symposium (Intern.) on High Dynamic Pressures*; Commissariat a l'Energie Atomique: Paris, France, 1978.
18. Marsh, S.P. *LASL Shock Hugoniot Data*; University of California Press: Berkeley/Los Angeles, CA, USA, 1980.
19. Gump, J.C.; Peiris, S.M. Isothermal equations of state of β-octahydro-1,3,5,7-tetranitro-1,3,5,7-tetrazocine at high temperatures. *J. App. Phys.* **2005**, *97*, 053513. [CrossRef]
20. Ge, N.N.; Wei, Y.K.; Song, Z.F.; Chen, X.R.; Ji, G.F.; Zhao, F.; Wei, D.Q. Anisotropic Responses and Initial Decomposition of Condensed-Phase β-HMX under Shock Loadings via Molecular Dynamics Simulations in Conjunction with Multiscale Shock Technique. *J. Phys. Chem. B* **2014**, *118*, 8691. [CrossRef] [PubMed]
21. Klapötke, T.M. *Chemistry of High-Energy Materials*, 3rd ed.; De Gruyter: Berlin, Germany, 2015.
22. Cady, H.H.; Larson, A.C.; Cromer, D.T. The Crystal Structure of Benzotrifuroxan (hexanitrosobenzene). *Acta Cryst.* **1966**, *20*, 336. [CrossRef]
23. Bolotina, N.B.; Hardie, M.J.; Speer, R.L., Jr.; Pinkerton, A.A. Energetic materials: Variable-temperature crystal structures of γ- and ε-HNIW polymorphs. *J. Appl. Crystallogr.* **2004**, *37*, 808. [CrossRef]
24. Nielsen, A.T.; Chafin, A.P.; Christian, S.L.; Moore, D.W.; Nadler, M.P.; Nissan, R.A.; Vanderah, D.J.; Gilardi, R.D.; George, C.F.; Flippen-Anderson, J.L. Synthesis of polyazapolycyclic caged polynitramines. *Tetrahedron* **1998**, *54*, 11793. [CrossRef]
25. Millar, D.I.A.; Maynard-Casely, H.E.; Kleppe, A.K.; Marshall, W.G.; Pulham, C.R.; Cumming, A.S. Putting the squeeze on energetic materials-structural characterisation of a high-pressure phase of CL-20. *CrystEngComm* **2010**, *12*, 2524. [CrossRef]
26. Wang, Y.; Yang, Z.; Li, H.; Zhou, X.; Zhang, Q.; Wang, J.; Liu, Y. A Novel Cocrystal Explosive of HNIW with Good Comprehensive Properties. *Propellants Explos. Pyrotech.* **2014**, *39*, 590. [CrossRef]
27. Yang, Z.; Zeng, Q.; Zhou, X.; Zhang, Q.; Nie, F.; Huang, H.; Li, H. Cocrystal explosive hydrate of a powerful explosive, HNIW, with enhanced safety. *RSC Adv.* **2014**, *4*, 65121. [CrossRef]
28. Bolton, O.; Simke, L.R.; Pagoria, P.F.; Matzger, A.J. High Power Explosive with Good Sensitivity: A 2:1 Cocrystal of CL-20:HMX. *Cryst. Growth Des.* **2012**, *12*, 4311. [CrossRef]
29. Cady, H.H.; Larson, A.C. The crystal structure of 1,3,5-triamino-2,4,6-trinitrobenzene. *Acta Cryst.* **1965**, *18*, 485. [CrossRef]
30. ζ-CL-20 (measured at 3.3 GPa) is not accounted for fitting.
31. Xue, X.; Wen, Y.; Zhang, C. Early Decay Mechanism of Shocked ε-CL-20: A Molecular Dynamics Simulation Study. *J. Phys. Chem. C* **2016**, *120*, 21169. [CrossRef]
32. Simpson, R.L.; Urtiew, P.A.; Ornellas, D.L.; Moody, G.L.; Scribner, K.J.; Hoffman, D.M. CL-20 performance exceeds that of HMX and its sensitivity is moderate. *Propellants Explos. Pyrotech.* **1997**, *22*, 249. [CrossRef]

33. Huang, Y.; Zhang, X.; Ma, Z.; Zhou, Y.; Zheng, W.; Zhou, J.; Sun, C.Q. Hydrogen-bond relaxation dynamics: Resolving mysteries of water ice. *Coord. Chem. Rev.* **2015**, *285*, 109. [CrossRef]
34. Sun, C.Q. *Relaxation of the Chemical Bond: Skin Chemisorption Size Matter ZTP Mechanics H2O*; Springer: Berlin/Heidelberg, Germany, 2014; Volume 108.
35. Sun, C.Q.; Zhang, X.; Fu, X.; Zheng, W.; Kuo, J.l.; Zhou, Y.; Shen, Z.; Zhou, J. Density and Phonon-Stiffness Anomalies of Water and Ice in the Full Temperature Range. *J. Phys. Chem. Lett.* **2013**, *4*, 3238. [CrossRef]

© 2019 by the authors. Licensee MDPI, Basel, Switzerland. This article is an open access article distributed under the terms and conditions of the Creative Commons Attribution (CC BY) license (http://creativecommons.org/licenses/by/4.0/).

MDPI
St. Alban-Anlage 66
4052 Basel
Switzerland
Tel. +41 61 683 77 34
Fax +41 61 302 89 18
www.mdpi.com

Nanomaterials Editorial Office
E-mail: nanomaterials@mdpi.com
www.mdpi.com/journal/nanomaterials

www.ingramcontent.com/pod-product-compliance
Lightning Source LLC
LaVergne TN
LVHW070719100526
838202LV00013B/1131